Imagery in Scientific Thought

Imagery in Scientific Thought
Creating 20th-Century Physics

Arthur I. Miller

The MIT Press
Cambridge, Massachusetts
London, England

First MIT Press edition, 1986

Printed and bound in the United States of America

Library of Congress Cataloging-in-Publication Data

Miller, Arthur I.
  Imagery in scientific thought.

  Bibliography: p.
  Includes index.
  1. Science—Methodology. 2. Science—History.
3. Creative thinking. 4. Imagery (Psychology)
I. Title.
Q175.M628 1986   502.8   85-23212
ISBN 0-262-63104-0 (paperback)

For

Lori and Scott

Contents

Preface

Throughout my research in the history of science I have been struck by the interest of many key scientists in the origins of scientific concepts and the process of creative thinking, particularly its intuitive dimension. These scientists saw apparently disparate subjects as being related because the depth of their research had led them to consider the process of thinking itself. This book explores the connection of creative scientific thinking with the origins of scientific concepts and the ways in which this connection may provide a better understanding of scientific progress. Thus my concern here is with individuals, with the detailed structure of scientific change, and not with its macrostructure.

I have chosen to study Niels Bohr, Ludwig Boltzmann, Albert Einstein, Werner Heisenberg, and Henri Poincaré. Through their work, the period 1900–1950 was one in which our customary notions of space, time, causality, and substance were transformed as never before. These philosopher-scientists were chiefly responsible for setting the intellectual milieu of the twentieth century.

I have developed the history, philosophy, psychology, and science contained herein with the goal of reaching the widest possible audience. Every effort has been made to render this book self-contained. Parts I and II contain historical case studies of Bohr, Boltzmann, Einstein, Heisenberg, and Poincaré from which emerge the philosophical-scientific currents of their times. The analyses strive to elucidate their styles of thinking, particularly their modes of mental imagery. These results are input for the cognitive psychological analyses in Part III that explore creative scientific thinking. Thus, in Part III, the history of science is used as a laboratory for cognitive psychology.

My recent book, *Albert Einstein's Special Theory of Relativity: Emergence (1905) and Early Interpretation (1905–1911)*, probed the problem of creativity for a particular case, with emphasis on scientific and philosophical aspects. I concluded that book by suggesting the need for further investigation into the imagery in Albert Einstein's thought experiments. This is among the topics developed here. Like that book, this one concludes with problems for further work. That is the manner in which my essays often end, and that is the way it should be when the history of science is defined broadly enough to be considered part of the history of ideas.

It is my hope that this book will serve as a catalyst for increasing interaction between cognitive psychologists and historians of science, so that we may amplify each other's intellectual strengths for a multidisciplinary approach to a fascinating problem in the history of ideas: creative scientific thinking.

ARTHUR I. MILLER

Acknowledgments

For suggestions on early versions of Chapter 1 I thank, in particular, Gerald Holton, Robert E. Innis, and Sheldon Krimsky.

Chapter 3 benefitted from Gerald Holton's perceptive comments.

For Chapter 4 I acknowledge conversations particularly with Felix Bloch, Stephen G. Brush, Abner Shimony, and Victor Weisskopf. For Chapter 5 I thank Michael Wertheimer, Lise Wertheimer Wallach, Rand Evans, S. E. Asch, Gerald Holton, Joseph Phelan, and Mrs. John Hornbostel (formerly Mrs. Max Wertheimer) for informative conversations and assistance. I gratefully acknowledge Rudolf Arnheim's insightful comments on an early version. I am particularly indebted to Mr. Valentine Wertheimer for guiding me through the Wertheimer papers on deposit at the New York Public Library.

It is a pleasure to acknowledge comments on Chapter 6 from Robert E. Innis, Jon Madian, Donald A. Norman, Herbert A. Simon, and particularly Martha Farah and Stephen M. Kosslyn for their insightful guidance through the whys and wherefores of cognitive science.

I was fortunate to have the comments of Susan Bloch and Kurt Fischer on very early drafts of Chapter 7. Through the criticisms of Howard E. Gruber, based on his deep knowledge of Jean Piaget's genetic epistemology, I was able to bring Chapter 7 into its current form.

For Gerald Holton's encouragement to extend my research into cognitive psychology I am deeply grateful. Our interactions over the years have been an inspiration to me.

This book was written in Harvard University's Department of Physics, whose hospitality I gratefully acknowledge.

For permission to quote from their archives, I thank the Jewish National and University Library, Hebrew University of Jerusalem, the Estates of Henri Poincaré and Max Wertheimer, and the Center for History of Physics of the American Institute of Physics.

For the travel and research funds that were essential to me over the period in which I gathered materials for this book, I acknowledge grants from the Section for History and Philosophy of Science of the National Science Foundation, the Centre National de la Recherche Scientifique, the American Philosophical Society, the University Professor Fund of the University of Lowell, and a fellowship (1979–1980) from the John Simon Guggenheim Memorial Foundation.

For editorial assistance I thank Mrs. Geraldine Stevens.

Author's Notes to the Reader

So as to avoid unnecessary nestings of footnotes that contain no information other than a page number, I use abbreviations of the sort "Poincaré (1903)," which means the paper listed in the Bibliography (pp. 315–338) under Poincaré and dated 1903. In Chapters 4, 6, and 7 the abbreviation AHQP appears. Quotations from *AHQP* (*Archive for History of Quantum Physics*) are taken from the interviews of Werner Heisenberg by Thomas S. Kuhn. AHQP materials are on deposit at the American Institute of Physics in New York City, the American Philosophical Society in Philadelphia, the University of California, Berkeley, and at the Niels Bohr Institute in Copenhagen.

Imagery in Scientific Thought

Introduction

Laws of thought have evolved according to the same laws of evolution as the optical apparatus of the eye, the acoustic machinery of the ear and the pumping device of the heart . . . . We must not aspire to derive nature from our concepts, but must adapt the latter to the former.

L. Boltzmann (1904)

Mr. Russell will tell me no doubt that it is not a question of psychology, but of logic and epistemology; and I shall be led to answer that there is no logic and epistemology independent of psychology.

H. Poincaré (1909)

Scientific thought is a development of pre-scientific thought.

A. Einstein (1934a)

Indeed, we find ourselves here on the very path taken by Einstein of adapting our modes of perception borrowed from the sensations to the gradually deepening knowledge of the laws of Nature. The hindrances met on this path originate above all in the fact that, so to say, every word in the language refers to our ordinary perception.

N. Bohr (1928)

According to our customary intuition [we attributed to electrons the] same sort of reality as the objects of our daily world . . . . In the course of time this representation has proved to be false [because the] electron and the atom possess not any degree of direct physical reality as the objects of daily experience.

W. Heisenberg (1926b)

W HAT ARE THE ORIGINS of scientific concepts? How are scientific concepts transformed as science progresses? What is the role of mental imagery in scientific research? How do scientists invent or discover theories? Because these problems go right to the heart of the age-old inquiry of how we construct knowledge through interacting with the world we live in, they have long occupied scientists and philosophers and, more recently, historians of science and cognitive psychologists.

Here I propose a fresh approach. First the relation between creative scientific thinking and the construction of scientific concepts from prescientific knowledge is explored through historical case studies in nineteenth- and twentieth-century mathematics and physics. Then the scenarios that emerge from these analyses are examined by means of contemporary cognitive psychology so that both the role of mental imagery in the research of twentieth-century science and the dynamics of creative scientific thinking may be assessed.

Scientists whose work set the foundations of twentieth-century science have emphasized the importance that considerations of the origins of scientific concepts have had in their research. The epigraphs to this introduction indicate that this was a guiding theme for

Ludwig Boltzmann, Henri Poincaré, Albert Einstein, Niels Bohr, and Werner Heisenberg. How it came about and how it affected their scientific work is developed here.

The problem of whether scientists invent or discover theories concerns the extent of the influence of empirical data on their thinking. Einstein eloquently stated this problem in a letter of 6 January 1948 to his old friend and confidant, Michele Besso (Einstein, 1972):

> Mach's weakness, as I see it, lies in the fact that he believed more or less strongly, that science consists merely of putting experimental results in order; that is, he did not recognize the free constructive element in the creation of a concept. He thought that somehow theories arise by means of *discovery* [durch Entdeckung] and not by means of *invention* [nicht durch Erfindung]. (italics in original)

By invention Einstein meant the mind's ability to leap across what he took to be the essential abyss between perceptions and data on the one side and the creation of concepts and axioms on the other. I shall be less abrupt than Einstein by defining discovery as "putting experimental results in order" according to already existing models or mental images. This is the distinction between invention and discovery that I use in this book. Although Einstein sometimes interchanged the terms invention and discovery, he deemed invention to be the route to creative scientific thinking.

BACKGROUND

The suggestion that cognitive psychology might shed further light on the history of science has been made most notably by Jacques Hadamard (1954), Gerald Holton (1973), Thomas S. Kuhn (1962), Peter Medawar (1969), Jean Piaget (1970a), and Max Wertheimer (1959). In this book the application is actually made.

An early example of modern investigations by historians of science into the psychology of scientific discovery is Gerald Holton's article, "On Trying to Understand Scientific Genius" (1973g), in which he explores Einstein's style of thinking. Holton's assessment of most studies of scientific creativity by psychologists still holds true (see also Holton, 1978):

> What is meant by genius in science? What are its characteristics? Can one understand it, or is that a contradiction in terms? I am not speaking merely of "creative" people, nor of men of "high attainment." I am aware of the large amount of literature on creativity, and of some fine studies of men of genius in the arts or in political affairs. But I do not

find them very helpful for understanding the life or the work of a Fermi or an Einstein, and even less for discerning how his personality and his scientific achievements interact.

For example, studies by Mahoney (1976) and Mansfield (1981) comprise results of various intelligence tests administered to scientists and students, and then of interviews and surveys concerning race, marital status, politics, education, frequency of publication, refereed reports on their work, and discussions of how and whether these results fit into scenarios proposed by Robert K. Merton, Imre Lakatos, Thomas S. Kuhn, and Sir Karl Popper. The most informative investigation based on surveys of scientists remains Roe's (1952).[1] A model for psychobiography is Manuel's (1968) work on Newton.[2] In Arieti's otherwise interesting book (1976), the discussion of scientific creativity is brief and mostly second-hand. For example, for Einstein he depends on Wertheimer's book *Productive Thinking* (1959), in which the scenarios turn out to have been reconstructed according to the Gestalt psychological theory of thinking; before I brought to light the relevant archival documents in 1974, this facet of Wertheimer's analysis was unknown (as discussed in Chapter 5). Beveridge's survey of methods of scientific research is aptly entitled *The Art of Scientific Investigation*. Although useful, the historical narrative is mostly anecdotal, there is no discussion of current theories of psychology, and no new conclusions are drawn.[3] There are serious and well-intentioned efforts to form a new discipline called the psychology of science (see Tweney, 1981), whose goal is to "investigate the cognitive mechanisms that underlie scientific thinking." In my opinion a new name is not necessary for history of science properly defined as the history of ideas. The 1945 book by the mathematician Jacques Hadamard is an informative survey of creative scientific thinking that focuses on mathematicians, particularly on Hadamard's teacher, Poincaré. Among the psychologists today who have entered seriously into the examination of scientific thinking by immersing themselves in a science is Howard E. Gruber, whose book on Darwin (1974) is a landmark study.

METHODS OF ANALYSIS

Part I is a comparative study of the origin and development of philosophical views of science that affected the direction of research in the twentieth century. Chapters 1 and 2 explore the extent of the reciprocal interactions between science and philosophy in the thinking of

Boltzmann, Einstein, and Poincaré. Besides setting straight the differences and similarities in the philosophical and scientific views of Poincaré and Einstein, Part I delineates the central role played by the style of visual thinking that was characteristic of currents in German philosophy dating back at least to Kant. From Part I we learn that Poincaré's philosophy of science affected his scientific research with little inverse reaction, for Boltzmann, that the converse was the case; and how Einstein realized the necessity for mutual interaction between philosophy and science in order to formulate a consistent physics.

The theme of visual thinking is developed further in Part II through case studies of major developments in twentieth-century physics. Chapter 3 presents a scenario of Einstein's invention of the special theory of relativity that is consistent with available archival data and both primary and secondary sources. This scenario places Einstein within the currents of philosophy, science, and electrical engineering in 1905 and shows how he drew from these disciplines a new view of physical theory. It is essential to bear in mind that Einstein's first paper on electrodynamics was initially appreciated largely for the wrong reasons, if at all. In other words, in 1905 there was no scientific revolution. Einstein worked in the Patent Office in Bern, Switzerland until 1909, when he received his first academic appointment on the basis of his research on the quantum theory of solids—not on relativity.

From Chapter 4 emerges the transformation of mental imagery required by research into a realm beyond visualization, the world of the atom. Chapter 4 traces the rise and fall of Bohr's atomic theory from 1913 to 1925, Heisenberg's invention of the quantum mechanics in 1925, the struggles of Bohr and Heisenberg during 1926 and 1927 to understand what constitutes physical reality in submicroscopic physics, and then Heisenberg's further dazzling research in the period 1926–1943, that resulted in the exchange force, the beginnings of modern nuclear physics, and quantum electrodynamics. This research of Heisenberg is conjectured to have been among the antecedents of modern-day elementary particle physics. Whereas Bohr ultimately arrived at a hybrid form of positivism—the so-called Copenhagen interpretation—Heisenberg's predilection for mathematics led him to a Platonic idealism. All these startling developments between 1913 and 1943 are related to transformations in, and abstractions of, mental imagery. Bohr and Heisenberg offer examples of how a scientist's philosophical view could be determined by his scientific research, with inverse interaction. They learned well from Einstein. Another result of Part III is that the mode of mental imagery of Bohr, Boltzmann, Einstein, and Heisenberg is culturally-based.

In these analyses the scientists' use of language is taken seriously, for it is right to assume that a Bohr, an Einstein, or a Heisenberg would choose his words carefully when writing a seminal paper. In fact, the shifts in the meanings of certain terminologies in their scientific papers and correspondence signal transformations of physical reality. We shall see that Bohr, taking his cue from Einstein, emphasized the importance of examining the language in which scientific theories are formulated. And besides Bohr's influence on Heisenberg, we recall that Heisenberg's father was an authority on Greek philology and held a position at the University of Munich.

Part III uses results of the previous case studies as data for the Gestalt theory of thinking, cognitive science, and Jean Piaget's genetic epistemology. Thus in Part III the history of science is used as a laboratory for cognitive psychological theories of thinking. The psychology required for these analyses is presented and the histories are based on the detailed scenarios from Parts I and II. Chapter 5 analyzes an oft-cited psychological analysis by Einstein's colleague at the University of Berlin, Max Wertheimer, of Einstein's thinking that supposedly led to the relativity of simultaneity. Wertheimer was one of the founders of Gestalt psychology. Documents I found among Wertheimer's papers, in connection with my exegesis of his work on Einstein, reveal that his reconstruction of Einstein's thinking followed Gestalt principles as he had meant it to. In addition to discussing the possible pitfalls in psychological analysis of a historical case study, Chapter 5 proposes other episodes in Einstein's thinking to be investigated by Gestalt psychology; these are taken up in Chapter 7.

I have found that neither Gestalt psychology nor genetic epistemology treats mental imagery in a manner consistent with its importance in creative scientific thinking. Thus, as a prelude to Chapter 7, Chapter 6 explores the transformations of mental imagery in creative scientific thinking during the late nineteenth and twentieth centuries. Historical scenarios of the thinking of Poincaré, Einstein, Bohr, and Heisenberg are analyzed, using cognitive science, which is a relatively new branch of psychology that integrates philosophy, linguistics, cognitive psychology, and computer science.

Chapter 7 contains reconstructions of Einstein's thinking that led to the special theory of relativity and the thinking of the small group of scientists that resulted in the formulation of atomic theory in the period 1913–1927. These reconstructions are written according to the guidelines of the Gestalt psychological theory of thinking and with genetic epistemology. I chose this method, the inverse of the one in Chapter 6, because I have found it to be the best way to test these two psychological theories. Thus Chapter 7 strives to maintain historical accuracy while conforming to the guidelines of each psychological theory.

Among the comments I have received in the decade-long journey of Chapter 7, there is one that I should like to deal with straightaway—namely, the criticism that since Piaget's theory concerns how children construct knowledge, it is inapplicable to the thinking of a Bohr or an Einstein. Although this criticism misses the point of genetic epistemology, it is well taken. The ultimate goal of Piaget's research of sixty-odd years was to explain the growth of knowledge through combining his psychological research with concepts from biology, his original field of study. In 1970 he wrote that since "we are not very well informed about the psychology of Neanderthal man . . . we shall do as biologists and turn to ontogenesis . . . . There are children all around us. It is with children that we have the best chance of studying the development of logical knowledge, mathematical knowledge, physical knowledge, and so forth." Thus he formulated a "genetic epistemology" that "attempts to explain knowledge and in particular scientific knowledge" by searching for those evolving structures in scientific thought that exhibit parallelisms in structures in formative psychological processes. But in order to employ genetic epistemology for this purpose, I have found that certain extensions and redefinitions of some of its basic notions are required. These hypotheses, proposed in Chapter 7, turn out to be fruitful because they yield new insights into the way in which scientific research is done.

The concluding remarks and outlook discusses results of the analyses in Part III and, particularly, the close coupling between the dynamics of creative scientific thinking and transformations of mental imagery. A view of scientific invention and discovery is then offered that differs from those of Norwood Hanson, Kuhn, and Herbert A. Simon.

NOTES

1. An early statistical survey of scientists' thinking is contained in a book by the founder of eugenics, Francis Galton (1883). Among

the criticisms of Galton's work, Hadamard's is pertinent to the theme of this book. Although only about 10 percent of Galton's respondents to his "breakfast-table survey" considered themselves to be nonimagists, most of them were successful academics and scientists, so Galton concluded that the "visualizing facility" was inferior to the "higher intellectual operations." Galton was surprised at this result because he considered himself to be a visual thinker. Hadamard (1954) differs with Galton's conclusion. In particular, Hadamard notes that in the face of Galton's statistical study, Galton did not give precise percentages of those who thought in images rather than words; this may have been due to Galton's realization that he needed a larger sample to distinguish between creative thinking and the imageless "tense thought" that sometimes accompanies routine problem solving and which is also the predominant mode of thought in the average person.

2. Part III does not contain psychobiographical studies of the genre of Frank E. Manuel's *Portrait of Isaac Newton*. Consequently, I omit that dimension of human thinking in which resides personal anxieties, quirks, and vicissitudes that have often contributed to or been in part the catalysts to great works of art and literature. These aspects of personality lie in the realm of psychoanalysis, and I shall address no conjectures in that direction. This omission restricts only the scope of the investigation, for Freudian psychoanalysis does not pretend to explore the construction of knowledge.

3. For example, Beveridge (1950) writes, "Most mathematicians are the speculative type . . . . Scientists may be divided broadly into two types according to their method of thinking. At one extreme is the speculative worker [who tries] to arrive at the solution by use of imagination and intuition . . . . The other extreme is the systematic worker who progresses slowly by carefully reasoned stages and who collects most of the data before arriving at the solution. Research commonly progresses in spurts . . . ."

PART I

Studies
in Comparative
Epistemologies

CHAPTER 1

Poincaré
and Einstein

The logical point of view alone appears to interest [Hilbert]. Being given a sequence of propositions, he finds that all follow logically from the first. With the
foundation of this first proposition, with its psychological origin, he does not
concern himself.

H. Poincaré (1903)

The whole of science is nothing more than a refinement of everyday thinking. It
is for this reason that the critical thinking of the physicist cannot possibly be restricted to the examination of concepts of his own specific field. He cannot proceed without considering critically a much more difficult problem, the problem
of analyzing the nature of everyday thinking.

A. Einstein (1936)

Henri Poincaré (1854–1912) is shown here circa 1900. Straightaway from his Ph.D. dissertation in 1878 Poincaré opened up new areas of mathematics, critiqued, extended and made fundamental contributions to all areas of the physical sciences, all of this in addition to spinning a new view of the philosophy of science. Poincaré was elected to l'Académie des Sciences in 1887 and to its Presidency in 1906, and to l'Académie Française in 1908 and to its Directorship in 1912. He was the only member of l'Académie des Sciences whose research admitted him to all its sections— geometry, mechanics, astronomy, physics, geography and navigation. Poincaré published over thirty books and five hundred papers. Gaston Darboux (1913) rightly eulogized him as "à la fois mathématicien hors de pair, physicien pénétrant et profond philosophe." (Courtesy of AIP Niels Bohr Library.)

Albert Einstein (1879–1955) as a Patent Clerk in Bern during the period 1902–1909 in which he introduced the notion of light quanta, solved the problem of Brownian motion, invented the special theory of relativity, realized the equivalence between mass and energy, initiated studies on the quantum theory of solids, and began to generalize the special relativity theory. In 1909 he became an Associate Professor at the University of Zurich on the basis of his work on the quantum theory of solids, not relativity. From 1911 to 1912 Einstein was a Professor at the German University, Prague, then he returned to his alma mater the Zurich Polytechnic Institute, for a year, from there he assumed a Professorship at the University of Berlin and the Directorship of the Kaiser Wilhelm Institute for Physics during the period 1914–1933, and then he emigrated to the United States where he spent the remainder of his life at The Institute for Advanced Study, Princeton. (Courtesy The Hebrew University of Jerusalem and the AIP Niels Bohr Library.)

Henri Poincaré and Albert Einstein are exemplars of the fundamental investigator referred to in the grand manner as philosopher-scientist. Their philosophical and scientific thoughts were linked and their interest in fundamental issues led them to probe the process of thinking itself. The quotations in the epigraphs to this chapter and to the Introduction bespeak their commitment to this connection as a guiding theme in their scientific research. This Ariadne's thread is best thrown into perspective through the realization that their individual philosophies of science are composed of an epistemology of the origin of knowledge and an epistemology of scientific theories. In fact, the case of Poincaré and Einstein indicates that it is necessary—and, in my opinion, revealing of the structure of scientific theories—to separate the broad discipline of epistemology into two parts: (1) the construction of prescientific knowledge, that is, the origins of knowledge, which may be referred to as theory of knowledge; and (2) the relations of the scientist's knowledge of the world of perceptions to the structure of a scientific theory and the study of what a scientific theory is, hereinafter to be referred to as epistemology or scientific epistemology.[1] Both aspects of epistemology include an analysis of concept formation.

This distinction enables us to approach problems such as why it was that even though in 1905 Poincaré and Einstein were in possession of the same empirical data, they dealt with them so differently. This distinction also is revealing of the dynamics inherent in the link between philosophy and science, which will be explored further in succeeding chapters. Poincaré worked in the tradition of prerelativistic philosopher-scientists whose philosophical views influenced their science, but with little inverse reaction. Einstein, conversely, found that his scientific research required a balanced interplay between his developing philosophical view and his science. Thus, the special relativity theory of 1905 became a virtual watershed for twentieth-century philosophy of science. Taking their cues from Einstein, the mutual tempering of philosophy and science would turn out to be of critical importance in the research of Niels Bohr and Werner Heisenberg. How this occurred and how Einstein's philosophy of science affected his view of the quantum theory is explored in Chapters 4, 6, and 7. In this chapter, the philosophies of science of Poincaré and Einstein are analyzed.

There have been many intrinsically interesting historical and philosophical studies of Poincaré, though their scope is limited. Some of them treat his philosophy, or his work on physical theory, or his work on mathematics; others treat Poincaré's philosophy along with one of the other two topics. Particularly good examples are the essays of J.J.A. Mooij (1966) on mathematics and philosophy, of Gerald Holton (1973b) on Poincaré's philosophy and his attitude toward Lorentz and Einstein, and the survey of Anne-François Schmid (1978), that emphasizes Poincaré's mathematics and philosophy while reaching conclusions on his science that agree with those in my (1973) essay. Certain historians of science even deny that Poincaré's scientific research in electron theory reflects his philosophy (e.g., Goldberg, 1970). There is a plethora of philosophical analyses of Poincaré's view of geometry and space (e.g., Grünbaum (1973), Sklar (1974), Guidymyn (1978)), that move toward comparing Poincaré's philosophical stance to the views of Eddington, Einstein, Reichenbach, and Whitehead, among others. However, by not treating Poincaré's researches as a whole, these analyses lack the depth needed to come to grips with the problems discussed here—namely, the origins of Poincaré's philosophy of science, the presuppositions to his views on geometry, classical mechanics, and physical theory, and of the relation among Poincaré's research in these subjects.

On the pedagogical side, this chapter strives to fill the regretful lack in current texts on the philosophy of science of any historical development of Poincaré's philosophical view and its relation to his scientific research. I began to develop this relation in my (1973, 1980, 1981b) publications.

POINCARÉ'S THEORY OF KNOWLEDGE

Ernst Mach defined the proper domain for science as the study of data of the senses and the laboratory. He developed the view that the goal of science was merely to describe data economically: "Science itself therefore should be regarded as a minimal problem consisting of the completest possible presentation of facts with the least expenditure of thought" (1883). Among reasons for Mach's positive, or positivistic, view of science was a desire to combat the defeatist sentiment among scientists that grew out of the criticism that science could not respond to such truly pressing and meaningful problems as the nature of force, or velocity, or the human brain. Mach's positivism declared these problems illegitimate and thus of no concern to the serious scientist. The initial resistance to positivism was eventually overcome in most quarters by Mach's withering criticisms of the foundations of mechanics in his classic *Science of Mechanics*.

Although in his 1894 book on mechanics Heinrich Hertz expressed his intellectual debt to Mach, he was one of the new wave of philosopher-scientists who used positivism as a springboard to philosophical views that emphasized the role of mathematics and the primacy of the imagination.

Another advocate of a different sort of positivism was France's greatest living mathematician, Henri Poincaré, who also was counted among the first rank of philosophers and scientists. Poincaré's view was positivist to the extent that it emphasized sense perceptions and empirical data, although he went on to weave it around organizing principles that he assumed to be innate.

Like Mach and others, Poincaré considered the process by which knowledge is constructed to be an active one. But unlike Mach, Poincaré assumed that the mind contained two synthetic a priori intuitions that acted to organize perceptions into knowledge: the principle of mathematical induction—"this rule, inaccessible to analytical proof and to experiment, is the exact type of the *a priori* synthetic intuition" (1894)[2]; and the intuition of continuous groups, which "exists in our mind prior to all experience" (1898a). These two synthetic a priori intuitions lead us to discover that "nothing is objective that is not transmissible, and consequently . . . the relations between the sensations can alone have an objective value . . . . In sum, the sole objective reality consists in the relations of things whence results the universal harmony" (1902b). Poincaré went on to conclude that since we construct our knowledge of the world of sensations in an active manner, then "a reality completely independent of the mind which conceives it, sees or feels it, is an impossibility. A world as exterior as that, even if it existed, would for us be forever inaccessible." In an even stronger tone, Poincaré said at the International Conference of Philosophy in 1900 that the "question of the *réalité du monde extérieur* would be best placed in section I (Metaphysics)" (1900a).

The first step in constructing the world about us (i.e., the real external world for Poincaré) is to discover that objects exist external to ourselves. These objects are "not fleeting and fugitive appearances because they are not only groups of sensations, but groups cemented by a constant bond. It is this bond, and this bond alone, which is the object in itself, and this bond is a relation" (1905b). Further assimilation of sensations leads us to realize that there are objects whose changes of position (i.e., changes external to our own body) can be compensated for by "correlative" movements of our body (1902a). Poincaré referred to these objects as "solid bodies" and to their changes of position as "displacements." The relative displacements of solid bodies are stored in the mind where they are surveyed by the two synthetic a priori intuitions.

To summarize thus far: Poincaré's theory of knowledge is neo-Kantian since it is based on two synthetic a priori intuitions that serve to organize into knowledge the potpourri of assimilated sensations. Only the relations among sensations are objectively real; thus, only the positions of objects relative to other objects are real. Effects are caused by contiguous actions. In order to continue our discussion of how, according to Poincaré, we establish the existence of the real external world, I turn next to Poincaré's view of the origins of geometry.

POINCARÉ ON THE ORIGINS OF GEOMETRY

The axioms of geometry, reasoned Poincaré, are neither synthetic nor analytic a priori judgments; nor do they reach us directly via our sensations (1902a). His argument against the first statement, which contains Kant's point of view, is that if the axioms of Euclidean geometry were imposed on our minds, we could not conceive of an alternative geometry. Yet consistent non-Euclidean geometries had been formulated, presumably by mathematical reasoning alone. His argument against the second statement was that if it were true, then the axioms of Euclidean geometry would be open to constant revision, and this has not been the case. Thus the problems Poincaré faced were (1) how to maintain the portion of Kant's theory of knowledge that he deemed to be of lasting value (i.e., the organizing principles), and (2) how to account for both non-Euclidean geometries and the fact that only Euclidean geometry (i.e., an approximate version) applied to the real external world. He found the resolution in the notion of continuous groups that had been set forth in rigorous mathematical form by the Norwegian mathematician Marius Sophus Lie. In 1887 Poincaré showed that in two dimensions any geometry could be generated from a Lie group (see note 4, below). Thus, taking the notion of continuous groups to be a synthetic a priori intuition, Poincaré could argue that all geometries pre-exist potentially in our minds, and that non-Euclidean geometries could be discovered by thought alone. But, continued Poincaré, why does Euclidean geometry appear to be the most natural one? What "was the reasoning of Euclid" (1902a) that led him to set down the axioms of his geometry and not another? In replying to these questions, Poincaré proceeded as follows: By 1891 he had concluded that the definitions and proofs of Euclidean geometry contain implicit axioms. As an example he gave the definition "of the equality of two figures. Two figures are equal when they can be superposed" (1891). However, continued Poincaré, to "superpose them, one of them must be displaced until it coincides with the other. But how must it be displaced? If we asked this question, no doubt we should be told that it ought to be done without deforming it, and as an invariable solid is displaced." The hidden axiom is the existence of invariable solids, and thus this definition is a "vicious circle" that "defines nothing." Furthermore, this definition "has no meaning to a being living in a world in which there are only fluids." Thus, neither is the "possibility of the motion of an invariable figure . . . a self-evident truth." Then how do we make the mental leap to solid bodies "absolutely invariable"? According to Poincaré, the process of constructing a geometry from the sensations assimilated from the real external world would begin thus: *"The laws of these phenomena* [i.e., the displacements of solid bodies] *are the object of geometry"* (italics in original). The displacements of solid bodies are surveyed by the synthetic a priori intuition of the continuous groups, and the mind discovers that they are a good approximation only to the continuous group from which is generated the three-dimensional Euclidean geometry. However, the axioms of Euclidean geometry are exact and not open to revision; thus it is at this juncture, according to Poincaré, that the mind creates the "concept" of the ideal solid. These solids constitute the content or matter of the particular continuous group from which the Euclidean geometry is generated. The principle of mathematical induction permits the finite displacements that led to the rough Euclidean geometry to be decomposed into a limitless number of infinitesimal displacements. This continuum of displacements is the grist for the mill of the proper Lie group with which to construct three-dimensional axiomatic Euclidean geometry.[3]

Before the sensations from which the mind constructs the concept of the solid object were assimilated, this particular continuous group possessed only form and not content. It was unknowingly along these lines that Euclid conceived his geometry. Of the many passages in which Poincaré summarized his view of geometry, the one that best expresses his neo-Kantian position is (1898a):

> Geometry is not an experimental science; experience forms merely the occasion for our reflecting upon the geometrical ideas which pre-exist in us. But the occasion is necessary; if it did not exist we should not reflect; and if our experiences were different, doubtless our reflexions would also be different. Space is not a form of our sensibility; it is an instrument which serves us not to represent things to ourselves, but to reason upon things.
>
> What we call geometry is nothing but the study of formal properties of a certain continuous group; so that we may say, space is a group. The notion of this continuous group exists in our mind prior to all experience; but the assertion is no less true of the notion of many other continuous groups; for example, that which corresponds to the geometry of Lobatchévski. There are, accordingly, several geometries possible, and it remains to be seen how a choice is made between them. Among the continuous mathematical groups which our mind can construct, we choose that which deviates least from that rough group, analogous to the physical continuum, which experience has brought to our knowledge as the group of displacements.
>
> Our choice is therefore not imposed by experience. It is simply guided by experience. But it remains free; we choose this geometry rather than that geometry, not because it is more *true*, but because it is the more *convenient* (italics in original).

Thus, of the many different geometries that can be generated from the continuous group—and hence that "pre-exist in us"—experience in the world of sensations guides us to Euclidean geometry as the most convenient choice. The existence of infinite varieties of *forms* of continuous groups renders possible the mathematical formulation of an infinity of geometries. Until the organism actively probes its environment in order to feed perceptions to the mind, among them displacements, these groups have only form and no content. By abstracting from the solid bodies of the real external world, models can be constructed which visualize non-Euclidean geometries. Hermann von Helmholtz (e.g., 1876) and Poincaré himself described models to illustrate non-Euclidean geometries (e.g., 1902a).[4] For Poincaré the mathematical statement of the compatibility of several geometries is that "the existence of one group is not incompatible with that of another" (1887). However, only the continuous group that is the generator of Euclidean geometry is convenient for understanding the world we live in; for only that group "deviates least from the rough group" that characterizes the relative displacements of solid bodies. To Poincaré, the existence of the solid body is of paramount importance for the selection of the most convenient geometry from all those that potentially pre-exist in our mind (1902a). *"If, then, there were no solid bodies in nature there would be no geometry"* (italics in original).

In the passage quoted from his essay of 1898, Poincaré referred to the Euclidean geometry as the most *"convenient"* geometry. In 1887 Poincaré had referred to the axioms of Euclidean geometry as "conventions." In both these essays he emphasized that "our choice among the many geometries that pre-exist in our minds is . . . not imposed by experience. It is simply guided by experience. But it remains *free* . . . ." Consequently we are led to conclude that the conventions of geometry, which are statements that do not refer to the world of sensations, are free creations of the human mind. The section "Poincaré's Epistemology" considers how free these creations are in the light of Poincaré's view of the origins of knowledge.

POINCARÉ ON THE NATURE OF SPACE

After discussing the origins of geometry, Poincaré turned to the notion of space. During the initial period of groping about the world we live in, we develop the notion of representative space that has "a triple form—tactile, visual and motor" (1902a). This space is not isotropic, homogeneous, or infinite, and furthermore it can have as many dimensions as we have muscles.

Once we have created the invariable ideal solid body and have organized our knowledge by use of the Euclidean group of continuous displacements, then the following exact assertions follow concerning the properties of the space of these bodies, that is, geometrical space. Since displacements in geometrical space can be described by a continuous group, geometrical space must be homogeneous—that is, all points in space are equivalent. The homogeneity of space is equivalent to what Poincaré referred to as the "relativity of space" (1902a) and as the "principle of the relativity of position." Since this principle does not discuss phenomena in the world of perceptions, then it is unfalsifiable, and Poincaré referred to it as a "convention." In representative space there is a rough version of this principle, which Poincaré called the "principle of the relativity of space" (1908a); the space this principle assumes is a three-dimensional space in which the displacements of solid bodies obey a rough version of Euclidean geometry. Geometrical space is homogeneous, and so it is infinite in extent. Since a finite displacement or a general change of orientation can be accomplished in a number of operations repeated indefinitely during which the ideal body is left unchanged, then geometrical space is also isotropic.[5] Since the possibility of repeating an operation over and over is what leads us to our concept of space, Poincaré concluded, "repetition presupposes time; this is enough to tell that time is logically anterior to space" (1905b).

Thus, in Poincaré's view, we arrive at the concept of geometrical space after having ascertained which continuous group best describes the world of perceptions. Geometrical space is Euclidean since the objects in it and their displacements are consistent with three-dimensional Euclidean geometry. Since geometrical space is three-dimensional, so is representative space, although the latter can have as many dimensions as we have muscles. Since for Poincaré there is no relation between the objects that Euclidean geometry discusses and the solid objects of the real external world, then Euclidean geometry cannot be tested experimentally; neither can the physical nature of space be tested experimentally. Eugenio Beltrami, Arthur Cayley, and Felix Klein demonstrated that any statement in a non-Euclidean geometry with constant curvature could be translated into one in Euclidean geometry. These researches were, in addition to Poincaré's results, reported in his 1887 paper, the basis for his assertion that an experimental result could be interpreted according to any geometry (cf. Guidymin (1977)).[6] However, our movements about the physical world convince us that Euclidean geometry is the most convenient geometry to use in physical theory, and is therefore the only meaningful choice. Other geometries have intrinsic interest for the mathematician who is concerned only with form, or with speculating on the inhabitants of a world with no solid bodies as we know them; such creatures would arrive at a non-Euclidean geometry as the most convenient geometry (1898a):

> To ask whether the geometry of Euclid is true and that of Lobatchévski is false, is as absurd as to ask whether the metric system is true and that of the yard, foot, and inch, is false. Transported to another world we might undoubtedly have a different geometry, not because our geometry would have ceased to be true, but because it would have become less convenient than another.

Thus, if empirical data disagreed with the axioms of Euclidean geometry—for example, if apart from errors introduced by the use of solid bodies, it were found that there were not 180° in a triangle—then the physical theory that underlies the analysis rather than the Euclidean geometry would have to be altered. Hence, Poincaré was led to make his famous assertion (1902a):

> To sum up, whichever way we look at it, it is impossible to discover in geometric empiricism a rational meaning. Experiments only teach us the relations of bodies to one another. They cannot give us the relations of bodies and space, nor the mutual relations of the different parts of space.

Poincaré emphasized these points by asserting, for the benefit of philosophers such as Bertrand Russell (1897) who defended a priori concepts of distance, that space is "neither Euclidean nor non-Euclidean and the words straight line, equality of two figures, form, and distance are meaningless" without designating a convention for measurement (1899).

Since "it is impossible to discover in geometric empiricism a rational meaning," and since "experiments only teach us the relations of bodies to one another," Poincaré concluded reasonably that geometrical "space is in reality amorphous and it is only the things in it that give it a form . . . . It is impossible to picture empty space . . . whoever speaks of absolute space uses a word devoid of meaning" (1908a). The notions of empty space and of absolute space are alien to our thought because of the active way in which we construct the notion of physical space and then abstract to the notion of geometrical space. Our movements about the physical world impress on us the belief that only relative displacements have meaning. Thus we can state the "principle of the relativity of space" (1908a).

Poincaré described the relativity of space in a particularly graphic way. If the dimensions of the world and everything in it were to change overnight, we would not be aware of that occurrence because the dimensions of our measuring instruments would also change. Thus absolute size or absolute distance can have no meaning. One can speak meaningfully only of the relation between two points, and then only by stating how distances are measured. He emphasized that measuring instruments were at first—for us when we were children and for primitive man—our own bodies, and then other solid bodies external to ourselves, followed perhaps by light rays. In 1912 Poincaré summarized his views on space as follows (1912c):

> In reality, space is therefore amorphous, a flaccid form, without rigidity, which is adaptable to everything, it has no properties of its own. To geometrize is to study the properties of our instruments, that is, of solid bodies.

But, Poincaré argued, that means that geometry is subject to constant revision because our instruments are imperfect. However, to define an ideal instrument by means of geometry would lead to a vicious circle. Thus, he concluded, as he had before, that the axioms of geometry must be considered to be conventions that are not open to falsification.

It is a mistake to represent, as do certain philosophers of science, that in Poincaré's view of the foundations of geometry, one geometry is as good as another and Euclidean geometry is only the one we find most convenient. Interpreting Poincaré's position thus, usually in order to contrast it with Einstein's general relativity theory, does not come to grips with Poincaré's position as he developed it from 1887 to his death in 1912. Poincaré's view of the foundations of geometry is based on a theory of knowledge and not on a scientific epistemology; consequently his position transcends holding that the statements of non-Euclidean geometries can be translated into Euclidean geometry. To Poincaré three-dimensional Euclidean geometry eclipsed all possible geometries because of its necessity for the survival of organisms in our world.

Poincaré analyzed how organisms construct the principle of the relativity of space thus: I can only defend myself against a menacing object, A, if I know where it is relative to my body (1908a):

> . . . I know that, in order to reach the object A, I have only to extend my right arm in a certain way; even though I refrain from doing it, I represent to myself the muscular and other analogous sensations which

accompany that extension, and that representation is associated with that of the object A . . . this complex system of associations [is] so to speak . . . our whole geometry.

He referred to such associations in Darwinian overtones as "conquests," and some of these conquests have become innate:

But these associations are not for the most part conquests made by the individual, since we see traces of them in the newly-born infant; they are conquests made by the race. The more necessary these conquests were, the more quickly they must have been brought about by natural selection.

Thus, according to Poincaré, only those species survived that made the necessary conquests, that experimented actively on their environment (cf. Nye, 1970). For example, someone from a four-dimensional world would not survive a sudden transportation to our world. Such beings would not be able to "defend themselves against the thousand dangers by which they would be assailed."

These passages appear to open the way to what can perhaps best be called an evolutionary apriorism: No matter what theory of knowledge one espouses, we are not born with a tabula rasa; rather, there exist in our mind traces of the conquests made by our ancestors from time immemorial.

To discuss whether Euclidean geometry is testable empirically, Poincaré developed an epistemological analysis of classical mechanics, for the possibility of testing Euclidean geometry cannot be analyzed without discussion of the state of a body in relation to the states and dispositions of neighboring bodies.

POINCARÉ'S EPISTEMOLOGY

A BACKGROUND NOTE

Since his death, most of Poincaré's letters and manuscripts have been missing. In 1976 I found the main collection in Paris. Among the documents in Paris is an Introductory note that he contributed to a new (1881) edition of Leibnitz's *Monadology* that is of interest to the philosophy he subsequently developed. This essay does not appear in Lebon's (1912) catalog of Poincaré's works. Poincaré (1881) compared Descartes's and Leibnitz's notions of the conservation laws of momentum and energy. For example, whereas Descartes had assumed incorrectly that in a closed system the magnitude of the momentum remains unchanged, Leibnitz concluded that what is conserved are the components of the momentum and also the kinetic energy. For this to be the case, Leibnitz invoked the law of action and reaction as a definition because, wrote Poincaré (1881), "there is a certain harmony in mechanical phenomena that affects the different parts of a system . . . . That is why, as Leibnitz wrote in paragraph 80 of the *Monadology,* if Descartes had known the true laws of mechanics, he would have hit upon the system of pre-established harmony."[7,8]

Most likely it was this passage that led A. Calinon to seek out Poincaré. Calinon worked on the fringes of French philosophical-scientific circles. He lived near Poincaré's home city, Nancy. From their correspondence we know that Calinon and Poincaré met on 8 August 1886 in the Nancy area, and the next day Calinon sent Poincaré a copy of his 1885 book entitled, *Étude Critique sur la Mécanique,* in which Calinon wrote that "there is a rigorous rational mechanics, a genuine geometry of motion that shall subsist even if the universe ceases to exist or exists otherwise." Armed with an idealist view, Calinon sought the true laws of nature. Calinon was among the first in French physics to subject the laws of mechanics to the sort of critical investigation in which dynamics was separated from kinematics, and kinematics was in turn closely linked to geometry in a geometrical kinematics—that is, a "genuine geometry of motion." As Calinon wrote to Poincaré on 14 August 1886, in France the ideas of Jean Marie Constant Duhamel were accepted, while his own were resisted. For example, Duhamel defined mass as force divided by acceleration, without much further ado (Jammer, 1957). Calinon was happy that Poincaré agreed with his notions of time, mass, and force (evidently Poincaré had read Calinon's book carefully in the intervening five days since their meeting and had commented on it in some detail), because there are "teachers, even distinguished ones, who obstinately resist them; there are some minds that refuse absolutely to revise acquired ideas, who refuse to subject to a philosophical investigation some notions that they have always considered evident." It would take hard-hitting positivistic critiques such as Mach's *Science of Mechanics* followed by such books as Poincaré's *Science and Hypothesis* to begin to shake what Einstein would recall as the "dogmatic faith" (1946) in assured foundations of Newtonian mechanics. Whereas Poincaré's subsequent analyses of mechanics diverged from Calinon's and Mach's, the effect of Calinon on Poincaré can be noticed in several places in *Science and Hypothesis*—for example, Poincaré's insistence that to define force as the cause of motion is insufficient and merely metaphysical, how in defining the equality of two forces Newton's third law enters as a definition, and that we are driven to suppose that the proper definition of mass is the constant coefficient in Newton's second law (he would write otherwise when discussing an electromagnetic foundation for mechanics). Most important, though, is Calinon's distinction between time and simultaneity. I have not seen this point made as forcefully elsewhere. Calinon took the definition of time to be a quantitative problem, whereas simultaneity was a qualitative problem since simultaneity (1885) "is independent of the position of the observer and of the choice of axes." The proper definition of time, he wrote to Poincaré on 14 August 1886, is the one that permits "simplification of the formulae" of mechanics. Poincaré incorporated these results into his essay, "Measurement of Time" (1898b). There he cited a passage from Calinon's book of 1897 on the vicious circle involved in defining velocity in terms of time measurements. Similar passages can be found in Calinon's 1885 book. Poincaré went on to recommend Calinon's 1897 book, along with works of the better-known Jules Andrade, who was among the (by then) acceptable wave of Frenchmen including Poincaré who were analyzing the foundations of science.[9]

The publication of Heinrich Hertz's book *Principles of Mechanics* (1894) roused Poincaré to begin publishing his view of the foundations of classical mechanics (1897).

THE STRUCTURE OF A SCIENTIFIC THEORY

In his Author's Preface to *Science and Hypothesis,* Poincaré differentiated between the standings of statements in arithmetic, geometry, classical mechanics, and the physical sciences. "A mathematical entity [e.g., a theorem] can exist provided there is no contradiction implied in its definition" (1902a). Obviously the same criterion cannot be applied to a "material point." Basic to mathematical reasoning is the principle of mathematical induction, which is the "power of the mind" to "pass from the finite to the infinite." This principle is an essential ingredient of the mind's invention of new symbols, such as incommensurable numbers, that lead to the invention of the mathematical continuum (1902a).

The axioms of Euclidean geometry, though created from our study of the displacements of solid objects, predict no consequences in the real external world. Poincaré took this sort of convention to be both unfalsifiable and undiscardable. Conversely, the principles of mechanics make predictions about phenomena in the real external

world. In the course of analyzing the empirical testability of geometries, Poincaré emphasized the necessity of extending into classical
mechanics the principle of the relativity of position from geometry;
in 1899 he referred to the corresponding statement in classical mechanics as the "law of relativity," and then in 1900 he called it the
"principle of relative motion" (1900d):

> The movement of any system whatever ought to obey the same laws,
> whether it is referred to fixed axes or to the movable axes which are
> implied in uniform motion in a straight line.

This principle is a convention that contains Newton's principle of
relativity (Newton's Corollary V) and Newton's second law as well.
Poincaré asserted in no uncertain terms the necessity for maintaining
this principle at all costs: "The commonest experience confirms it";
its contradiction "is singularly repugnant to our mind." Poincaré's
first reason refers to the firm experimental foundation for the principle (e.g., Newtonian celestial mechanics); his second points up the
reflection of his theory of knowledge into his epistemology. Poincaré
proposed two arguments in support of the principle of relative motion. First, it is possible to consider formally that Euclidean geometry can be derived from the laws of mechanics. Thus, if the principle
of relative motion were not a convention, we might have to recon-
sider the status of the axioms of Euclidean geometry—a geometry
necessary for survival in the real external world. Second, our movements in the real external world have led us to believe that only the
relative motion of ponderable matter is meaningful.

A further discussion of Poincaré's views on classical mechanics and
the physical sciences requires studying his epistemology itself. The
reason is because analysis in classical mechanics, and especially in the
physical sciences, unlike geometrical analysis, requires that data be
obtained by complex instruments, and thus the role of mathematical
physics becomes important. Just as Poincaré's theory of knowledge
was rooted in sensationism, so is his epistemology rooted in empiricism: "Experiment is the sole source of truth" (1902a). From experiment can be obtained what Poincaré referred to as a useful crude
fact—a fact that occurs many times and is related to other facts.

Fig. 1.1 illustrates Poincaré's epistemology, using his analysis of
Newton's mechanics as an example. From this figure it is clear that
Poincaré considered Newton's three principles to be fundamentally
different from those of Euclidean geometry. Newton's principles are
unfalsifiable, but they can be discarded if no longer "useful to predict

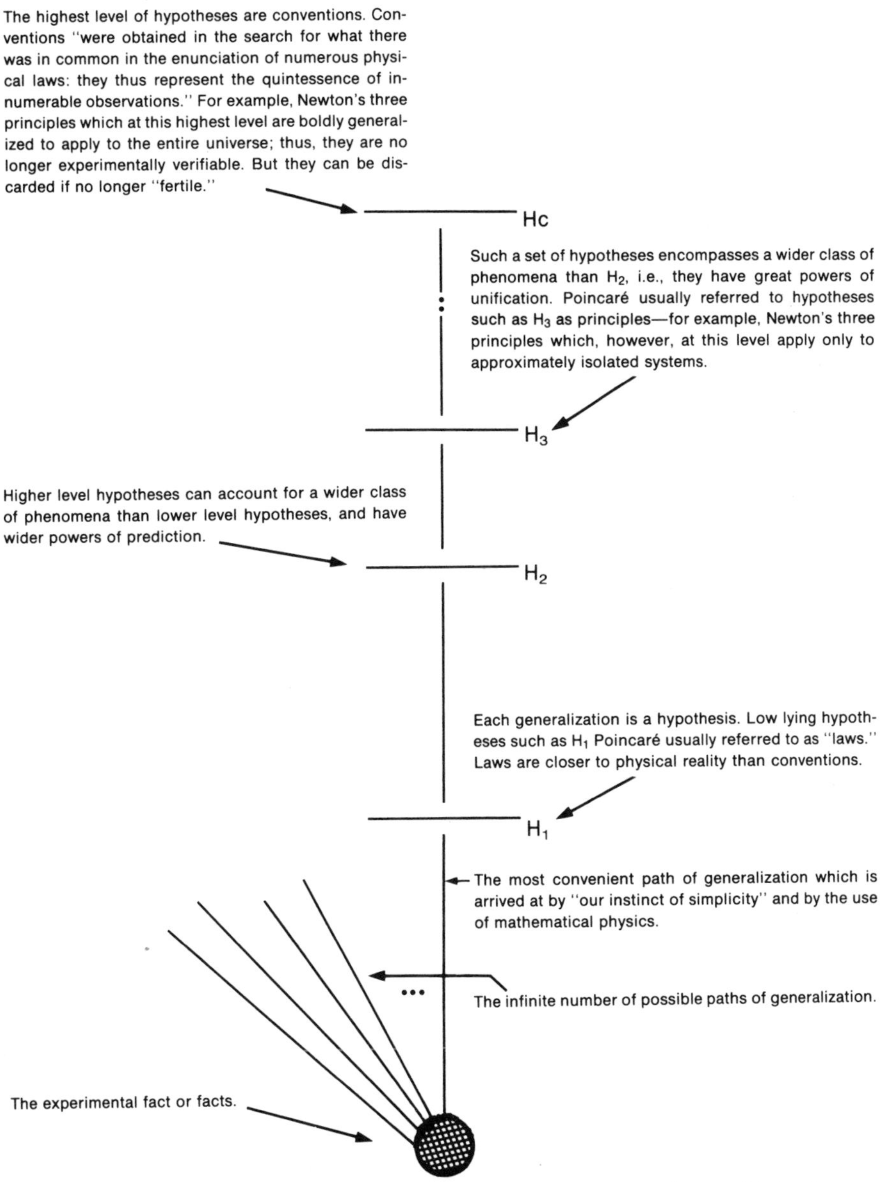

The highest level of hypotheses are conventions. Conventions "were obtained in the search for what there was in common in the enunciation of numerous physical laws: they thus represent the quintessence of innumerable observations." For example, Newton's three principles which at this highest level are boldly generalized to apply to the entire universe; thus, they are no longer experimentally verifiable. But they can be discarded if no longer "fertile."

Hc

Such a set of hypotheses encompasses a wider class of phenomena than H_2, i.e., they have great powers of unification. Poincaré usually referred to hypotheses such as H_3 as principles—for example, Newton's three principles which, however, at this level apply only to approximately isolated systems.

H_3

Higher level hypotheses can account for a wider class of phenomena than lower level hypotheses, and have wider powers of prediction.

H_2

Each generalization is a hypothesis. Low lying hypotheses such as H_1 Poincaré usually referred to as "laws." Laws are closer to physical reality than conventions.

H_1

The most convenient path of generalization which is arrived at by "our instinct of simplicity" and by the use of mathematical physics.

• • •   The infinite number of possible paths of generalization.

The experimental fact or facts.

Fig. 1.1.   A schematic to illustrate Poincaré's epistemology. This figure attempts also to show why he entitled his 1902 book, *Science and Hypothesis*. Although Poincaré was not always consistent with his nomenclature, he was clear as to which statements are conventions, for example, the law of relativity is a convention. All quotations are from Poincaré's (1902a).

new phenomena" and thus "experiment without directly contradict-
ing [these principles] will nevertheless have condemned [them]."
Furthermore, the conventions of classical mechanics are not synthetic a priori intuitions.[10] Thus we must interpret with care Poincaré's statement in the Author's Preface to *Science and Hypothesis*
that Newton's principles "share the conventional character of the
geometrical postulates." The asymmetry of the relation between empirical data and conventions is worth emphasizing because it clashes
with Sir Karl Popper's criterion for the falsification of theories. According to Poincaré, classical mechanics can never be empirically
falsified but only altered if no longer fruitful.

Although Poincaré differentiated between the status of conventions in geometry and classical mechanics, he denied "tracing
artificial frontiers between the sciences . . . , separating by a barrier
geometry so called from the study of solid bodies" (1902a). Poincaré's reason reflects his biologically based theory of knowledge,
namely, that just as the experiments that led to the adoption of
Euclidean geometry are "pre-eminently physiological experiments
which refer, not to space which is the object that geometry must
study, but to our body—that is to say, to the instrument which we
use for that study"—so "the experiments" that "prove to us" that
the "conventions of mechanics" are "convenient certainly refer to
the same objects or to analogous objects."

POINCARÉ'S NOTIONS OF INDUCTION AND OF SCIENTIFIC INVENTION

An examination of Poincaré's theory of knowledge—and my conjecture that there lies the basis of his epistemology—suggests to me that
Poincaré did not expect to be interpreted literally when he wrote that
the choice of the path of generalization from an experimental fact is
"free," and that we are led to the most convenient path by "our
instinct for simplicity" and by mathematical physics (1902a). Consider first Poincaré's notion of the origins of geometry, which can
be discussed using only his theory of knowledge. Considerations
of epistemology enter when the empirical testability of the geometry
is in question. According to Poincaré, the knowledge of a rough
geometry is necessary for the survival of any form of life. The only
presuppositions to theorizing at this level are the two synthetic a
priori intuitions. Thus the choice of the path of generalization is as
free as it could ever be; the only external constraint is survival.
Poincaré speculated, however, that this complete freedom very
likely existed only for primitive forms of life (see the section "Poin-

caré on the Nature of Space"); through the millenia of evolution of the species the mind has become imbued with a predilection for a three-dimensional Euclidean geometry—that is, an evolutionary a priori. Once the path of generalization is chosen, mathematical reasoning takes over because the most convenient geometry pre-exists in our mind. Since the three-dimensional Euclidean geometry was discovered before all other scientific knowledge, then there was not an infinite number of paths that generalization could take from experimental data to, say, the principles of classical mechanics. This is because the form *and* content of the most convenient mathematical physics are also fixed by the choice of geometry; for example, the differential equations of classical mechanics are covariant under the Galilean group. Hence in the sciences the path of generalization is limited to one—the one that allows a mathematical physics formulated in a three-dimensional Euclidean space.[11] The progress along this path is linear and is by mathematical reasoning alone. The path is broken, however, at those points at which a hypothesis is invented. In his essay, "Mathematical Invention" (1908b), Poincaré emphasized that there was no logic of discovery in the narrow sense of the term logic, because it is "practically impossible" to state in "precise language" the rules that guide us to invention: "They must be felt rather than formulated." Inventions, continued Poincaré, are made by people who possess a "special aesthetic sensibility." In other essays he referred to the intuition leading to invention as an "indefinable something," or a "vague instinct" (1903). For these reasons, hypotheses, according to Poincaré's theory of knowledge and epistemology, are constructed toward understanding the world of perceptions—for example, in this manner we invent Euclidean geometry and then Newton's laws. The gap around each hypothesis in Fig. 1.1 indicates that the act of invention cannot be quantified (see Chapter 6 for further discussion of Poincaré's view of invention in mathematics and science).

In summary, Poincaré's inductivism is of a special sort. Fig. 1.1 demonstrates that although there are gaps in the bridge that connects facts from the world of sensations with their scientific descriptions, nevertheless a well-defined direction does exist. It is just this aspect of Poincaré's epistemology that is shared with inductivism in the Baconian sense.

Poincaré considered that the *"meaning of physical theories"* (1902a) is that they express in the most convenient manner the "true relations between . . . real objects" and the "invariant laws which are relations between crude facts" (1905b). These relations constitute the truth content of a theory.[12] Theories may come and go, but the relations between objects appear in successive theories. That was true for forces acting at a distance—but Poincaré considered that hypothesis to be merely a convenient foundation for theories that relate many apparently diverse phenomena (1902a). He believed differently about the notion of an ether.[13]

POINCARÉ ON THE REALITY OF THE ETHER

Through the years Poincaré offered several justifications for an ether. It is the ether, he reasoned, that supports light in transit (1902a); the ether is the seat of the compensating mechanisms that in Lorentz's electromagnetic theory preserve Newton's principle of action and reaction (ca. 1900 Poincaré considered this principle a convention); "if we did not believe in the ether, which serves as a connection between phenomena, then the differential equations of mechanics would have to be replaced by difference equations";[14] and Poincaré's concern over the possibility of discovering the effect on optical phenomena of the earth's motion led him to write that "it would be necessary to have an ether in order that these so-called absolute movements should not be their displacements with respect to empty space, but with respect to something concrete" (1902a). Thus, instead of motion relative to absolute space, one could speak of motion relative to the ether.[15] To Poincaré relativism was crucial, as would be expected from his theory of knowledge.

A more fundamental justification for an ether appeared first in 1889, when his theory of knowledge was in its nascent form (1889):

> Whether the ether exists or not matters little—let us leave that to the metaphysicians; what is essential for us is that everything happens as if it existed, and that this hypothesis is found to be suitable for the explanation of phenomena. After all, have we any other reason for believing in the existence of material objects? That, too, is only a convenient hypothesis; only, it will never cease to be so, while some day, no doubt, the ether will be thrown aside as useless.

One must, however, not jump to hasty conclusions at this point, for Poincaré continued:

> But at the present moment the laws of optics, and the equations which translate them into the language of analysis, hold good—at least as a first approximation. It will therefore be always useful to study a theory which brings these equations into connection.

Clearly the phrase, "the ether will be thrown aside as useless," is more than partly negated by the "but" that begins the latter passage. The ether is the connector in the equations of optics that describe phenomena transmitted at a finite speed; hence such phenomena must be supported in transit. Poincaré in the passage quoted above posed the question: Why do we believe in the existence of solid objects? His reply: It is a convenient hypothesis, since everything happens as if such objects do exist.

In 1889 Poincaré judged the ether a convenient hypothesis. Over the years he thought deeply about the formation of concepts and of what constitutes physical reality. By 1902 he had incorporated the concept of an ether into his theory of knowledge, and (in 1905b) he rewrote the passage of 1889 quoted above as:

> It may be said, for instance, that the ether is no less real than any external body; to say this body exists is to say there is between the color of this body, its taste, its smell, an intimate bond, solid and persistent; to say the ether exists is to say there is a natural kinship between all optical phenomena, and neither of the propositions has less value than the other.

Since to Poincaré the sole objective reality is the relation among sensations, and consequently among facts and equations, the ether is not simply a convention but is real. The concept of an ether was common to the successful electromagnetic theories of circa 1900. Poincaré's ether accorded with his belief that theory should "express true relations" among phenomena and that, while the theories themselves may be transitory, the true relations survive. Moreover, all the theories that include an ether discuss only contiguous action and not forces acting at a distance; since these theories postulate that phenomena occur because of the intervention of an external agent, they thus reflect closely prescientific concepts. In addition, Lorentz's electromagnetic theory, based on an ether, was able to explain a wealth of empirical phenomena such as the Zeeman effect, in 1897 Lorentz's electron was discovered, and in 1900 Lorentz had speculated on reducing gravitational interactions to those of electromagnetism. These were the chief developments that led Wilhelm Wien to propose research toward an "electromagnetic basis for mechanics" (1900)—that is, an electromagnetic world-picture, as an alternative to the apparently sterile inverse research effort toward a mechanical world-picture. In the electromagnetic world-picture, mechanics and then all of physical theory would be derivable from Lorentz's electromagnetic theory. A far-reaching implication of this research program was that the electron's mass originates in its own electromagnetic field quantities as a self-induction effect. Consequently, the electron's mass should be a velocity-dependent quantity. Consistent with emphasis on empirical data, Wien called for verification of this consequence. The challenge was taken up by Walter Kaufmann (1901a) at Göttingen, who used electrons emitted from radium salts at velocities exceeding .9c, to demonstrate experimentally in 1901 the nonconstancy of the electron's mass (1901b). Max Abraham's 1902 theory of a rigid spherical electron and Lorentz's 1904 theory of a deformable electron accounted for Kaufmann's data (1902, 1903). Poincaré opted for Lorentz's theory, despite the epistemological problems that it posed for him.

THE REALM OF THE PHYSICAL SCIENCES

In (1904a) Poincaré praised Lorentz's new theory of the electron because it was able to explain the results of all positive and negative optical experiments accurate to any order in (v/c), where v could be the measurable velocity of matter relative to the laboratory (such as in Fizeau's 1851 measurement of the velocity of light propagating through moving water—a positive experiment), or v could be the velocity of matter moving through the ether to be determined (such as in the Michelson and Morley experiment of (1887)—a negative, or null, experiment).[16] According to Lorentz's theory the failure of the ether-drift experiments could be explained in terms of the dynamical interactions between the electrons that constitute matter with the ether. These interactions caused moving rods to contract, and mass to be a function of velocity, and thus explained why the velocity of light was always measured to be c in inertial reference systems.

However, many of the successes of Lorentz's theory of the electron conflicted with Poincaré's view of science. For example, Lorentz's theory violated Newton's third principle;[17] it violated Lavoisier's principle (conservation of mass); and Newton's second principle could not be deduced exactly.

In (1904a) Poincaré distinguished between the status of Newton's principles in classical mechanics, the axioms of Euclidean geometry, and certain statements in the physical sciences. He wrote that "If we pass to mechanics [from geometry] we still see great principles whose origin is analogous" but whose "radius of action" is smaller. Poincaré continued: "In physics [i.e., the physical sciences], finally, the role of the principles is still more diminished." This is because data in the physical sciences are necessarily gathered by complex instruments and so the systems under study (e.g., electrons) are no longer directly accessible to the scientist's own perceptions.

However, one principle whose status was of great concern to Poincaré was a principle of relative motion in the physical sciences, which he named the "principle of relativity":

> The principle of relativity, according to which the laws of physical phenomena must be the same for a stationary observer as for an observer carried along in a uniform motion of translation; so that we have not and can not have any means of discerning whether or not we are carried along in such a motion.

The principle of relativity is an extension into the physical sciences of the principle of relative motion from classical mechanics, and is a generalized form of Lorentz's 1904 theorem of corresponding states (see the section, "Electromagnetic Theory," in Chapter 3). Much as he had emphasized the importance of maintaining the principle of relative motion (its contradiction is "repugnant to our mind") in 1900, now, in 1904, Poincaré said of the principle of relativity that "it is irresistibly imposed upon our good sense." Then, as he had done for the principle of relative motion, he went on to buttress this assertion with a reason based on his epistemology: that "experiment is the sole source of truth" (1902a), "[A]nd it is not merely a principle which it is a question of saving, it is the indubitable results of the experiments of Michelson" (1904a).

The importance to Poincaré of maintaining both a principle of relativity and an ether may at first glance seem to be inconsistent, since the principle of relativity implies that the ether could never be observed directly. In addition, Poincaré had asserted many times that quantities that could not be measured were metaphysical and, hence, had no place in a scientific theory. For example, on the definition of force, Poincaré had written (1902a):

> When we say force is the cause of motion, we are talking metaphysics . . . . For a definition to be of any use it must tell us how to measure force.

Yet, taking account of Poincaré's theory of knowledge, according to which effects occur because of dynamical causes, we may conjecture that he held to an ether because its presence implied contiguous action, with the result that phenomena such as length contraction would have a dynamical explanation. In fact, in 1908 Poincaré asserted, "Beyond the electrons and the ether there is nothing" (1908a).

Nor is the Lorentz contraction directly measurable; however, its postulation permits explanation of several experimental results within a dynamical ether-based theory (Michelson-Morley, Rayleigh and Brace, and Trouton-Noble).[18] Another instance of Poincaré's relaxing his sharp operational definition of physical quantities is his symmetrical version of Lorentz's transformation of 1904 that relates an ether-fixed reference system, with whose observers one could apparently not communicate, to a reference system having no physical content. One cannot do physics this way.[19] So in practice the ether-based system was cavalierly taken to be the laboratory system with the tacit assumption that the ratio of the velocity of the earth relative to the ether, to the velocity of light in vacuum, was very much smaller than unity. This tacit assumption indicates that the notion of "at rest," relative to an inertial reference system, was not well defined in the dynamical ether-based electrodynamics of moving bodies.

Poincaré pursued his discussion of Lorentz's theory of the electron in two classic papers with the same title: "On the Dynamics of the Electron" (1905a, 1906). There Poincaré not only corrected certain technical errors in Lorentz's paper (1904b) but, more importantly, proved that of all possible electromagnetic descriptions of the electron, only Lorentz's was consistent with the principle of relativity (see Miller 1973, 1981b for details). Yet, owing to the disagreement of Lorentz's prediction for the electron's transverse mass with Kaufmann's 1905 data, Poincaré was unable to consider the principle of relativity as a convention, but only as a high-level hypothesis, i.e., a principle. Even after 1908, when Alfred Bucherer's data appeared to have vindicated Lorentz's theory of the electron (which, incidentally, most physicists now called the Lorentz-Einstein theory), Poincaré's epistemology still prevented his raising the principle of relativity to a convention. The reason is that the Lorentz-covariant *field* theory of gravitation, which Poincaré had formulated in his paper of 1906, predicted a value for the advance of the perihelion of the planet Mercury that did not agree with the measured value.[20]

In Poincaré's theory of gravitation the gravitational force propagates through the ether with the same velocity as do electromagnetic disturbances. In his view, this could not happen by chance but, rather, resulted because the ether is the connector of things. Furthermore, such a theory of gravitation was surely more satisfactory to his view of science than was Newton's theory, which discussed forces acting at a distance.

ON POINCARÉ'S POST-1905 THOUGHTS ON GEOMETRY, CLASSICAL MECHANICS, AND THE PHYSICAL SCIENCES

In *Science and Method* Poincaré wrote that acceptance of the Lorentz contraction hypothesis brought into the realm of physical reality a version of Poincaré's parable of a world whose dimensions changed overnight but which remained geometrically similar to itself. The only difference was that in the Lorentz contraction, the dimensions of a moving object are not geometrically similar to the absolutely resting object. However, Poincaré continued, the change in dimensions cannot be observed with light rays because such a comparison of lengths just repeats the Michelson-Morley experiment. Congruence measurements were also insufficient since all objects change in shape. But, owing to the contraction, might not Euclidean geometry, on further analysis, not turn out to be the most convenient geometry? Poincaré conceived of this new development as exhibiting "in how large a sense we must understand the relativity of space." In order to underscore, and also to preserve, his view of the necessity for a three-dimensional Euclidean geometry to describe the world of perceptions, Poincaré took recourse in his biologically based analysis of the origins of geometry that I have described in the earlier section, "Poincaré on the Nature of Space."

In his lecture of 4 May 1912 at the University of London, entitled "Space and Time," Poincaré discussed a principle of physical relativity that is verifiable because it refers to the classical mechanics of approximately isolated systems (1912c). Since, however, classical mechanics analyzes the motion of "natural solids," it uses a geometry that "must escape revision." In order for this to be true, the principle of physical relativity is seen as an approximate version of a principle of relativity in classical mechanics, which is a convention. Owing to the "recent progress in physics," Poincaré continued, speaking of research toward an electromagnetic world-picture, the principle of relativity from mechanics "has had to be abandoned; it is replaced by the principle of relativity according to Lorentz." Thus the differential equations of the new mechanics transform under the "group of Lorentz" rather than the Galilean group. Perhaps in order to underscore his commitment to an electromagnetic world-picture, Poincaré entitled this section of his paper (1912a), "Gravitation is of Electromagnetic Origin." There Poincaré discussed his theory of gravitation and its disagreement with the advance of Mercury's perihelion.

Nevertheless, Poincaré believed that Lorentz's principle of relativity would achieve the status of a convention, and he began to reconcile this prospect with his theory of knowledge and epistemology along the following lines. The solids that the Lorentz group discusses are not the ideal solids of Euclidean geometry—"we must admit that all bodies become deformed" (1912c). Furthermore, the geometry associated with the Lorentz group is a four-dimensional geometry and "everything happens as if time were a fourth dimension of space . . . ." The space and time coordinates have to be considered as "two parts of the same whole" because the Lorentz-invariant distance measure is $x^2 + y^2 + z^2 - c^2 t^2$. In fact, Poincaré, in "On the Dynamics of the Electron" (1906), had used a four-dimensional vector space as an aid toward constructing Lorentz-invariant quantities; moreover, in 1906 Poincaré had also proven that the generators for infinitesimal Lorentz transformations along the $x, y,$ and z directions are the generators for a Lie group for infinitesimal rotations in four dimensions. But, since to Poincaré Lie groups constituted one of our synthetic a priori intuitions, then this result may well have made him pause to consider whether life forms in our world could construct the pseudo-Euclidean four-dimensional geometry from sensations. Of course, he argued otherwise (1912c): "Today some physicists want to adopt a new convention" for simultaneity, and to assert that "everything happens as if time were a fourth dimension of space . . . ."

Poincaré's opposition was consistent with his theory of knowledge:

> It is not that they are constrained to do so; they consider this new convention more convenient; that is all. And those who are not of this opinion can legitimately retain the old one in order not to disturb their old habits. I believe, just between us, that this is what they shall do for a long time to come.

This was the way that Poincaré, two months before his death on 17 July 1912, assessed the developments in physical theory that ever since 1908 had compelled him to emphasize his view of the unity of the foundations of geometry, classical mechanics, and the physical sciences. In the face of these new developments Poincaré stood steadfast and held to his position that in the world of perceptions there was an absoluteness of simultaneity, and that three-dimensional Euclidean geometry must remain a convention—for Euclidean geometry is necessary to the survival of the species. Indeed, what better way was there for Poincaré to advertise his epistemological-scientific position in 1912, one year after most physicists recognized the difference between Einstein's special relativity and Lorentz's theory of the electron, than to refer to the "principle of relativity according to Lorentz"?

EINSTEIN'S THEORY OF KNOWLEDGE

Einstein's analysis of the notions of time and simultaneity in the 1905 relativity paper is an early indicator of his interest in the origin and formation of prescientific knowledge and its influence on the formulation of scientific knowledge. This is also one of the themes in Poincaré's *Science and Hypothesis,* which Einstein had read between 1902 and 1904 as part of the curriculum of the informal study group called the "Olympia Academy," whose charter members were Conrad Habicht, Maurice Solovine, and Einstein. Solovine recalled the "profound impression on us" made by Poincaré's book (Einstein, 1956). Among the other philosophical works they discussed were Mach's *Science of Mechanics* (to which Einstein had already been introduced by Michele Besso in 1897) and David Hume's *Treatise on Human Nature* in connection with which Solovine recalled discussing Hume's "singularly sagacious critique of the notions of substance and causality." From Carl Seelig's biography of Einstein we know that he read Kant's *Critique of Pure Reason* at age sixteen, and then at the Kanton Schule in Aarau during 1895–1896.[21]

Einstein's earliest known published essay which discusses philosophy is his (1916b) obituary of Mach, in which he emphasized that research scientists and serious students should be interested in the "theory of knowledge [and] the aims and methods of science" (quoted in Holton, 1973). Einstein went on to caution that a "theory of knowledge" in which "concepts" are quantities "useful for ordering things" may "easily assume so great an authority over us that we forget their terrestrial origin and accept them as unalterable facts."

The Gestalt psychologist Max Wertheimer recalled the "hours and hours" of intense discussions on creative thinking he had with Einstein, focusing on the thinking that led to Einstein's 1905 special relativity theory (1959). (But Wertheimer's reconstruction of these reminscences must be treated with care; see Chapter 5.) Additional evidence for Einstein's early and continuing interest in the origins of knowledge is Jean Piaget's recollection that in 1928 Einstein suggested to him "that it would be of interest to study the origins in children of notions of time and in particular of notions of simultaneity" (1970a). This supports Solovine's recollection of the manner in which Einstein analyzed philosophical works (Einstein, 1956): "Einstein employed with predilection the genetic method in examining fundamental notions. He used it in order to clarify what he could observe in children."

Further evidence for the importance of philosophical analysis to Einstein's thought is that he often began his essays and monographs on science or philosophy of science not with an exposition of physics or epistemology, but with an analysis of how prescientific knowledge is constructed. For example, he began his essentially technical monograph, *The Meaning of Relativity* (1921b) thus:

> The theory of relativity is intimately connected with the theory of space and time. I shall therefore begin with a brief investigation of the origin of our ideas of space and time. . . . How are our customary ideas of space and time related to the character of our experiences?

He prefaced his reply by criticizing, as he had in his 1916 obituary of Mach, those philosophers who have "had a harmful effect upon scientific thinking in removing certain fundamental concepts from the domain of empiricism, where they are under our control, to the intangible heights of the *a priori* . . . . This is particularly true of our concepts of time and space." Einstein credited Poincaré with the realization that we must "pay strict attention to the relation of experience to our concepts. It seems to me that Poincaré clearly recognized the truth in the account he gave in his book, *La Science et l'Hypothèse.*"

Einstein continued with Poincaré's description of the difference between a change in position and a change in state, which leads to the discovery of the existence of solid bodies. The laws that govern the relative displacements of these bodies are the laws of a rough geometry, whereas the axioms of exact Euclidean geometry are "free creations of the human mind" (1921a). The "concept of space" (1921b) relative to a solid body comes about through extending the solid body. But soon the solid body that served as reference was discarded and there developed an "abstract notion of space which certainly cannot be defended." Thus Einstein's notion of the origin of geometry and space is parallel to Poincaré's, differing in the rejection of Poincaré's a priori organizing principles.

Compared with the *Meaning of Relativity,* the essentially philosophical intent of "Geometry and Experience" gave Einstein the opportunity to probe deeper into his view of the origin and testability of geometry and then to compare his view with the view of the "acute and profound thinker H. Poincaré." Einstein (1921a) referred to the solid body from the world of perceptions as the "practically-rigid body" and to the geometry associated with it as the "practical geometry." By rejecting the "relation between the body of axiomatic Euclidean geometry and the practically-rigid body of reality," Poincaré arrived at his notion that of all possible axiomatic geometries, Euclidean geometry is the most convenient and "is to be retained as the simplest." It is easy to understand Poincaré's stance since, owing to effects of temperature, "external forces, etc., . . . real solid bodies in nature are not rigid." Einstein "characterized Poincaré's standpoint" on Euclidean geometry thus: The purely axiomatic geometry (G) predicates nothing about the world of perceptions. But the sum of (G) with a set of physical laws (P), that is, (G) + (P), "is subject to experimental verification [since] all these laws are conventions." In order to preserve (G), Poincaré would change (P) so that (G) + (P) described the experimental facts: Envisaged in this way, axiomatic geometry and the part of natural law which has been given a conventional status appear as epistemologically equivalent." Einstein's version does not square with Poincaré's own view that the conventions of axiomatic geometry are *not* epistemologically equivalent to those of physics and that the laws of mechanics, for example, cannot be empirically disconfirmed. But Einstein's terse assessment of Poincaré's view of geometry hits the mark: "*sub specie aeterni* Poincaré in my opinion is right."

In 1905 and, most spectacularly, when he proposed the generalized theory of relativity (in 1915), Einstein found it necessary to modify Poincaré's epistemological position regarding geometry. In the special relativity theory the geometry (G) is the practical Euclidean geometry, and the physics (P) is the laws of mechanics and electromagnetism suitably reinterpreted; in this way the geometry (G) becomes testable empirically. For example, in a key section of the 1905 relativity paper, Einstein linked the "methods of Euclidean geometry" to the "rigid measuring rods" that were the basis of his in-principle operational definitions for position relative to an inertial reference system.

In 1923 Einstein proposed a criterion for the admissibility of a concept whose germ can be discerned in the 1905 special relativity paper. He wrote (1923):

> [concepts] are only admissible to the extent that observable facts can be
> assigned to them without ambiguity (stipulation that concepts and

distinctions should have meaning). This postulate, pertaining to epistemology, proves to be of fundamental importance.

For example, if we consider "the rigid measuring body as an object which can be experienced," then the " 'system of coordinates' concept" and the "concept of the motion of matter relative thereto" both satisfy the "stipulation of meaning." Similarly, by associating the rigid body with the practically rigid body, as he had done in 1905, Euclidean geometry "has been adapted to the requirements of the physics of the 'stipulation of meaning'":

> The question of whether Euclidean geometry is valid becomes physically significant; its validity is assumed in classical physics and also later in the special theory of relativity.

"Geometry thus completed," wrote Einstein, "is evidently a natural science; we may in fact regard it as the most ancient branch of physics [1921b]. I attach special importance to the view of geometry which I have just set forth, because without it I should have been unable to formulate the theory of relativity" (1923).

But neither the rigid measuring rods nor the clocks of special relativity "find their exact correspondence" in the world of perceptions. The "logically more correct way" to have proceeded in 1905, recalled Einstein (1923), was to have made assumptions on the constitution of matter using the "whole of the laws" to which the stipulation of meaning could be applied; rather, it was necessary to sacrifice the stipulation of meaning in its exact form by introducing into the 1905 special relativity theory the rigid body as an "irreducible" element, and hence fulfilling the "unambiguous relation to the world of experience . . . in an imperfect form."[22]

Thus, consistent with his philosophical view, Poincaré in 1908 resorted to biologically based arguments to support the privileged position of Euclidean geometry, despite realizing that the Lorentz contraction might invalidate that claim. However, by 1909 Einstein was independently driving the consequences of the contraction of moving bodies toward generalizing special relativity through analysis of a uniformly rotating disk.[23] The generalization, however, would require an even more drastic break with the world of perceptions than was required by special relativity.

CONCEPTS, GEOMETRY, AND PHYSICS

Describing Max Planck's thinking in an almost autobiographical vein, Einstein wrote that while the "world of phenomena uniquely determines the theoretical system," there is "no logical bridge between them" (1918b).

Particularly in essays written in the last twenty-five years of his life, Einstein developed this theme further. In a 1934 essay bearing a title heavy with scientific overtones—"The Problem of Space, Ether and the Field in Physics"—he opened with: "Scientific thought is a development of pre-scientific thought." And in a 1936 essay entitled "Physics and Reality," he wrote:[24]

> The whole of science is nothing more than a refinement of everyday thinking. It is for this reason that the critical thinking of the physicist cannot possibly be restricted to the examination of the concepts of his own specific field. He cannot proceed without considering critically a much more difficult problem, the problem of analyzing the nature of everyday thinking.

Einstein had employed this tack fruitfully in thinking toward the 1905 special theory of relativity. The product of three decades of distilling this technique is in "Physics and Reality," in which he explored the origin and formation of concepts in greater detail than he had previously. There is an abyss, he wrote, between "concepts" and "experiences," and their connection cannot be articulated. A good concept is one that enables us to organize perceptions into "knowledge." In Einstein's assessment of whether a concept is useful, "success is alone the determining factor." Although concepts play a role analogous to Kant's categories, they are not unalterably fixed. This had been the case, for example, when the light axiom from the special theory of relativity was excluded from the generalized relativity theory.

Einstein elaborated further on this analysis in what he referred to as his "obituary," the "Autobiographical Notes" (1946). As we have come to expect in his writings on the foundations of science, the "Autobiographical Notes" promptly addresses the question: "What, precisely, is 'thinking'?" Einstein's descriptions of the "setting of the 'real external world'" in "Physics and Reality" and in the "Autobiographical Notes" are based on the connection between concepts and another important ingredient in his creative scientific thinking that he had begun to explore in the 1930s, visual thinking (Einstein, 1946):

> What, precisely, is "thinking"? When, at the reception of sense-impressions, memory-pictures emerge, this is not yet "thinking." And when such pictures form series, each member of which calls forth

another, this too is not yet "thinking." When, however, a certain picture turns up in many such series, then—precisely through such return—it becomes an ordering element for such series, in that it connects series which in themselves are unconnected. Such an element becomes an instrument, a concept. I think that the transition from free association or "dreaming" to thinking is characterized by the more or less dominating rôle which the "concept" plays in it. It is by no means necessary that a concept must be connected with a sensorily cognizable and reproducible sign (word); but when this is the case thinking becomes by means of that fact communicable.

Thus our first impressions of a world external to ourselves are the "sense impressions" from which "memory-pictures emerge," and certain memory pictures form series. However, for Einstein, this did not yet constitute thinking. Thinking begins when one memory picture occurs a great many times in several different series. The picture serves as an "ordering element" for the different series, an element that could also relate heretofore unconnected series. Einstein referred to this ordering element as a "concept." Thinking is "operations with concepts, i.e., the creation and use of definite functional relations between concepts and the coordination of sense experiences to these concepts."

Although Einstein took the connection between perceptions or data and concepts or axioms to be necessarily vague, he was quite specific on their roles in the structure of geometry and in a scientific theory.

EINSTEIN'S EPISTEMOLOGY AND THE STRUCTURE OF A SCIENTIFIC THEORY

In a letter written in 1952 to his old friend from the days of the Olympia Academy in Bern, Maurice Solovine, Einstein made it clear that his scientific research reflected his theory of knowledge. Fig. 1.2 is Einstein's own schematic depiction of his epistemology.[25] He explained that the axiomatic structure (A) of a theory is built psychologically on the experiences (E) of the world of perceptions. Inductive logic cannot lead from the (E) to the (A). The (E) need not be restricted to experimental data, nor to perceptions; rather, the (E) may include the data of Gedanken experiments. Pure reason (i.e., mathematics) connects (A) to theorems (S). But pure reason can grasp neither the world of perceptions nor the ultimate physical reality because there is no procedure that can be reduced to the rules of logic to connect the (A) to the (E). Physical reality can be grasped

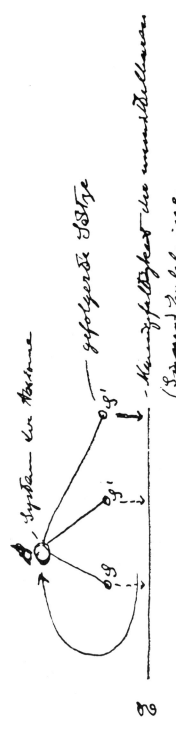

Fig. 1.2.   In a letter of 7 May 1952 to his old friend Maurice Solovine, Einstein wrote: "I probably expressed myself badly [in previous attempts at explaining to Solovine his philosophy of science]. I view such matters schematically thus . . .". The figure is Einstein's own sketch from this letter, where "*System der Axiome*" = system of axioms, "*gefolgerte Sätze*" = deduced laws, and "*Mannigfaltigkeit der unmittelbaren (Sinnes) Erlebnisse*" = totality of (sense) experiences. (Quoted with permission of the Jewish National and University Library, Hebrew University of Jerusalem)

not by pure reason (as Kant had asserted), but by "pure thought" (1933). Less certain, continued Einstein, is the connection between the (S) and (E). If at least one correspondence cannot be made between the (A) and (S) and the (E), then the scientific theory is only a mathematical exercise. Einstein referred to the demarcation between concepts or axioms and perceptions or data as the "metaphysical 'original sin'" (1949); and his defense of it was its usefulness. For whereas the problem of the relation between perceptions and mental images or concepts may well be interesting physiologically (e.g., How do neural firings lead to images?) or philosophically (e.g., philosophy of mind or metaphysics), it is of no concern to the working scientist—at least not to Einstein, who also displayed a good nose for philosophical problems.

Thus, for example, in special relativity there are among the (E) data from the ether-drift experiments, electromagnetic induction, stellar aberration, and Fizeau's experiment. As Einstein recalled (1946), the set of "known facts" was too restricted to discover the "true laws." So he added to them the data of the Gedanken experimenter as well as symmetry considerations of an essentially aesthetic sort. Then he leaped across the abyss between these (E) to invent the (A), which comprises the two principles of special relativity. The experimental predictions (S) included the "laws according to which electrons must move" as well as the (then unappreciated) exact solution for stellar aberration.

Comparison of Figs. 1.1 and 1.2 throws into relief the differences in emphasis on empirical data and on perceptions by Poincaré and Einstein. Figure 1.1 depicts the development of classical mechanics, where H_1, H_2, . . . also could be sets of hypotheses which constitute a series of theories that are increasingly wide in scope. So, for example, H_1 could be Galileo's early theory of the one-dimensional motion of free fall. Then, H_2 could be Galileo's realization that two-dimensional motion can be considered as if it were composed of two one-dimensional motions; and so on, up to Newton's mechanics.

In Einstein's view, however, science is composed of a hierarchy of theories where each one of them can be depicted schematically, as he himself did in Fig. 1.2. This sequence of theories converges to the unified description of nature that captures the true physical reality that is "in an external world independent of the perceiving subject" (1931). Poincaré considered such an ontological view mere metaphysics because of his emphasis on sense data. To be sure, historical analysis has revealed that scientific discoveries were never made by the scientist focusing only on the data at hand (i.e., by strict induction). Yet, close inspection of available empirical data had been immensely successful. But, in 1905 it turned out that only by breaking the link between perceptions and concepts could Einstein go beyond perceptions to realize that simultaneity is a relative quantity. This realization resulted from taking seriously the effects due to the very large but finite velocity of light instead of, as Poincaré had done, neglecting relevant time delays in electromagnetic signal propagation as being beyond our perceptions (e.g., 1898b). Similarly, even more basic and much less intuitive is the second axiom of special relativity, which states that the velocity of light is a definite velocity c no matter what the relative motion is between the observer and the light. Such statements turned out to be only the tip of the iceberg, indicating that further developments in physical theory could require additional restrictions or extensions of intuitions that are, after all, based on objects that we have actually perceived. (This theme is pursued in Chapter 6.) This realization became ever more meaningful to Einstein through his work on the structure of light, and again around 1913 when he first began to immerse himself in complex mathematics. For, at that time, his scientific research led him to raise the "question of the physical meaning (principally, the measurability) of the coordinates x_1, x_2, x_3, x_4," and then, on further investigation, to confer physical reality instead on the metric tensor $g_{\mu\nu}$, which characterizes the gravitational field (1913).

Einstein's modification of Poincaré's stance on geometry required his breaking with a heretofore successful formula for scientific research; namely, to build a theory step-by-step by posing hypotheses to explain hard-won experimental data. The hypotheses were by degrees closely linked with the data as in Figure 1.1. Einstein referred to this sort of theory as a "constructive theory" (1919), and placed a high premium on it: "When we say we understand a group of natural phenomena, we mean that we have found a constructive theory that embraces them." A constructive theory explains phenomena; for example, Lorentz's theory of the electron is such a theory. Einstein, however, would realize by the end of 1904 that the "known facts" (1946) were too restricted and contemporary physics and philosophy were inadequate to solve the problems that by consensus defined the frontier of physics—the formulation of an electromagnetic world-picture. Einstein has recollected that the way out of this "despair" was to turn to "theories of principle [whose] merit . . . is their logical perfection, and the security of their foundation (within their domain of applicability) (1946)." Thermodynamics and special relativity, based as they are on axioms, are theories of principle. This sort of theory accounts for, but does not explain, phenomena because it makes no assumptions on the constitution of matter. In other words, Einstein set out to focus immediately on the upper part of the pyramidal structure of a scientific theory.

Such an approach could not have been successful without a first attempt at a constructive theory. As is the case with theories of principle, many of the postulates of a constructive theory (e.g., Lorentz's theory of the electron) could be deduced from a more compact theory of principle (e.g., special relativity theory). Thus, in the special relativity theory, Lorentz's transformations and the hypothesis of contraction are deduced from the theory's two axioms.

Since Einstein's shift in the meaning of physical reality is rooted in a combination of concepts, visual thinking and axiomatics, it is apropos to explore the background to his unique manner of mixing these notions. This is accomplished by placing Einstein into the currents of the philosophy of 1905, where the notion of a "concept" was much discussed, as were visual thinking and axiomatization.

VISUAL THINKING, CONCEPTS, AND AXIOMATICS
VISUAL THINKING AND CONCEPTS

The matrix of science, philosophy, and technology in which Einstein was educated and worked placed a high premium on visual thinking, a mode of thought that he preferred for creative scientific thinking.[26] While he was a student at the Zurich Polytechnic Institute (1896–1900), we may assume that he read Volume I of Ludwig Boltzmann's *Lectures on Mechanics* (1897a), where Boltzmann's "characterization of the method" to set the foundations of mechanics emphasized using "mental pictures." Boltzmann went on to suggest that "unclarities in the principles of mechanics [derive from] not starting at once with hypothetical mental pictures but trying to link up with experience from the outset." Elsewhere (in 1897) Boltzmann delved into an analysis of perceptions in order to set the notion of useful mental pictures as best as he could. He defined useful mental pictures to be common to various perceptual complexes or groups of phenomena, thereby providing an understanding of the complexes (1897b). Boltzmann referred to useful mental pictures as "concepts."

Solovine recalled that the Olympic Academy members dutifully read some "papers and lectures of Helmholtz." Thus Einstein may well have read von Helmholtz's 1894 essay, "The Origin and Correct

Interpretation of Our Sense Impressions," in which he discussed the important role of mental pictures and nonverbal thinking in the creative process, and defined the notion of a concept analogous to Boltzmann's:

> Thus the memory images of pure sense impressions can also be used as elements in combinations of ideas, where it is not necessary or even possible to describe those impressions in words and thus to grasp them conceptually . . . .
>
> Indeed, the idea of a three-dimensional figure has no content other than the ideas of the series of visual images which can be obtained from it, including those which can be produced by cross-sectional cuts.
>
> In this sense, we may rightly claim that the idea of the stereometric form of a material object plays the role of a concept formed on the basis of the combination of an extended series of sensuous intuition images. It is a concept, however, which, unlike a geometric construct, is not necessarily expressible in a verbal definition. It is held together or united only by the clear idea of the laws in accordance with which its perspective images follow one another.

The similarity is unmistakable between Boltzmann's and von Helmholtz's definition of a concept and the descriptions in Einstein's writings of 1916, 1936, and 1946.[27] Einstein's debt to philosopher-scientists of the late nineteenth and early twentieth centuries was best described by Philipp Frank, who wrote that from their works, Einstein "learned how one builds up the mathematical framework and then with its help constructs the edifice of physics" (1947).

But Einstein's combination of visual thinking with concepts went beyond sense perceptions so that he could invent, for example, the relativity of simultaneity. In special relativity the measuring rods and inertial reference systems of the thought experimenter play the role of concepts in the meaning of Boltzmann and von Helmholtz.

AXIOMATICS

By the end of the nineteenth century, Mach's criticisms in his 1883 *Science of Mechanics* of Newtonian absolute space and time as metaphysical obscurities had been heeded by most major physicists. Mach's empiricist emphasis, however, did not deter philosopher-scientists such as Hertz and Poincaré from exploring the Kantian notion of a priori organizing principles. More than Mach, these men placed a strong premium on the primacy of the imagination and the deep meaning of mathematics. Before 1905 Einstein had read at least the introduction to Hertz's 1894 *Principles of Mechanics* (Sauter, 1965).

As Wien put it in 1900, Hertz's program was diametrically opposite to the goal of an electromagnetic world-picture. Yet, Wien
hoped that in due course the new research effort would measure up
to the logical structure of Hertz's mechanics. Abraham wrote similarly of Hertz in the *Annalen* paper of 1903, in which he presented his
theory of the rigid sphere electron. Wien's suggestion for an electromagnetic world-picture appeared in the 1900 Lorentz *Festschrift,* as
well as in the *Annalen* of 1901. We can conjecture that before 1905
Einstein had read the *Annalen* version of Wien's paper. The trend
toward axiomatics was in the air in 1905, very much along the lines
of Fig. 1.1. The ultimate theoretical basis for the electromagnetic
world-picture was to be achieved by degrees, after all extant data
were analyzed. Yet, owing to his research on fluctuation phenomena, Einstein knew by 1905 that the electromagnetic world-picture
could not succeed.[22]

That the trend toward axiomatization was mentioned at key places
in the scientific literature permits us to begin to fill in Einstein's 1946
recollection of a turning point in his thinking toward special relativity. We recall that by the end of 1904 he had convinced himself that
electromagnetic theory and mechanics could not serve as the basis for
all of physics. He recalled (1946) that he "despaired of discovering
the true laws by means of constructive efforts based on known
facts." Then, rather tersely in his "Autobiographical Notes," he
wrote that in the midst of the "despair" of 1904, he decided to try his
hand at a purely axiomatic theory that could offer a "universal formal principle, . . . the example I saw before me was thermodynamics." A problem ripe for historical analysis is how Einstein realized
that the principles of thermodynamics were exemplars for the double-edged sword that he would use in his Gordian resolution of
problems confronting the physics of 1905—that is, his realization
that the laws of thermodynamics could be wielded as restrictive
principles that also make no assumptions on the constitution of matter. In the "Autobiographical Notes" he mentioned only their
usefulness as restrictive principles. A possible reply to this problem
is: Does not everyone know these two properties of the laws of
thermodynamics? Although this reply may be the case, it fails to
assess the problem in historical context. For implicitly knowing the
double-edged power of thermodynamics is quite different from applying it to the physics of 1905. I add that the straight-forward reply
is also uninteresting to the historian who seeks to place Einstein in
the matrix of science in 1905. A conjecture that may be closer to the

mark is that in Abraham's 1904 *Annalen* paper, "On the Theory of Radiation and of Radiation Pressure" (1904b), Einstein saw the laws of thermodynamics applied specifically because they make no assumptions on the constitution of matter. Abraham resorted to this strategy because he refused to permit his theory of the electron to be falsified by the data of Rayleigh and Brace. After acknowledging this disagreement between data and theory, Abraham went on to emphasize the need for further research on the optics of moving bodies. His route was to investigate the thermodynamics of radiation in order to analyze a quantity indigenous to every theory of the electron—namely, the Poynting vector—all the while working within a theory that made no assumptions on the constitution of matter.

A total of 70-odd pages after Abraham's paper in the *Annalen* (1904b) is Einstein's (1904) paper, in which he calculated that black-body radiation could exhibit observable fluctuations. Then, as I conjectured earlier (with support, in part, from Einstein's own testimony in 1907), Einstein went on in unpublished calculations to conclude that classical electromagnetism failed in spatial regions as small as the electron. Thus, Abraham's elegant applications of the thermodynamics of radiation may well have struck a responsive chord in someone who "despaired of discovering the true laws [through] constructive efforts based on known facts."

We can depict Einstein's approach to an axiomatic formulation of the special relativity theory as a hybrid version of the views of Boltzmann, who emphasized mental pictures but was not daring enough in the raising of concepts to axiomatic status owing to his anti-Kantian stance; of Hertz's brilliant use of axioms as organizing principles, but within a scheme that was inapplicable to real mechanical phenomena; of Poincaré's far-reaching neo-Kantian organizing principles, but with their overemphasis on perception and empirical data; of Mach, who, with Poincaré, presented to Einstein paths not to be followed, except with care (i.e., not to overemphasize perceptions); of Wien's suggestion of axiomatics as a goal; and of Abraham's 1904 paper that suggested an approach both to fundamentals and to not permitting data to decide the issue. This last lesson was soon useful to Einstein when the first published citation of his 1905 electrodynamics paper disconfirmed empirically what was assumed to be its principal prediction—that is, empirical data of Kaufmann (1905, 1906) disagreed with the prediction of the Lorentz-Einstein theory of the electron for the electron's transverse mass. (This episode is explored in detail in Miller, 1981b.)

PRINCIPLES OF RELATIVITY

Einstein's statement of the 1905 principle of relativity is:

> The laws by which the states of physical systems undergo changes are
> independent of whether these changes of state are referred to one or the
> other of two coordinate systems moving relatively to each other in
> translational motion.

The second axiom of the special relativity theory is:

> Any ray of light moves in the "resting" coordinate system with the
> definite velocity c, which is independent of whether the ray was
> emitted by a resting or by a moving body.

Einstein's principle of relativity is a version of Newton's principle of
relativity enlarged to embrace all branches of physical theory.
Twelve years later Einstein wrote (1917a) that "we should retain the
principle of relativity, which appeals so convincingly to the intellect
because it is so natural and simple." (He referred here to the apparent
inconsistency between the enlarged Newtonian principle of relativity
and the second axiom of the relativity theory.)[28] The similarity is
striking between this comment and those of Poincaré (1902a) on the
necessity of maintaining a principle of relative motion (its contradiction "is singularly repugnant to our mind") and a principle of relativity ("it is irresistibly imposed upon our good sense"). This is reasonable because both their theories of knowledge are based on an
active approach to the construction of knowledge from assimilated
sensations. This notion is reflected in their epistemologies by the
principles of relativity.

Although the contents of the principles of relativity of Poincaré
and Einstein differ, their statements and their intents are parallel: The
manner in which knowledge is constructed imposes on us the notion
that only relative motion is meaningful. Anchored to the world of
sensations, Poincaré could put forth his principle of relativity only
provisionally, since in his opinion there was insufficient experimental support in 1904 to raise his principle of relativity to a convention.

For Einstein, however, the principle of relativity was only "suggested by the world of experience" (1936). Einstein's principle of
relativity is a restrictive principle that asserts a priori the impossibility of detecting motion relative to the ether; no reasons are required.

It is of interest to elaborate on this point. Consistent with his
theory of knowledge and epistemology, Poincaré's principle of relativity is a constructive principle based on dynamical explanations

for the empirical data of the ether-drift experiments. These explanations had their origins in the interaction with the ether of the bound electrons which were assumed to be the fundamental constituents of matter. In contrast, Einstein's approach, using rigid bodies as irreducible elements, provided a Gordian solution to the leading problems of contemporaneous physics (e.g., the failure of ether-drift experiments) by setting up a physics in which such problems have no place. Thus Einstein's two axioms of relativity theory do not explain the failure of the ether-drift experiments, or equivalently why the measured velocity of light always turns out to be *c*. Rather, by axiom the ether-drift experiments must fail, and by axiom the space of every inertial system is homogeneous and isotropic for the propagation of light.

Besides Poincaré's predilection for dynamical explanations at every stage of scientific theorizing, his view of the foundations of geometry would have prevented him from asserting axiomatically that the space of laboratory inertial systems is homogeneous and isotropic for the propagation of light because these are the properties of mathematical space. Poincaré rejected any connection between the objects of Euclidean geometry and those in the real external world, and thus he was led to reject any connection between representative space and mathematical space. This is the distinction on which, wrote Einstein (1921a), Poincaré was correct, *sub specie aeternitatus.* Yet Poincaré's (1898b) analysis of methods by which astronomers and electrodynamicists treat the velocity of light can be interpreted to betray his diffidence on the unbridgeable gap between mathematical and physical space. For the space of the astronomer (i.e., the space of classical mechanics) is mathematical space, whereas that of the electrodynamicist is representative space which, owing to repeated failures of ether-drift experiments, was turning out to have the properties of mathematical space.

Einstein believed otherwise. He associated the concept of the practically rigid body with the rigid body of axiomatic geometry because it satisfied the stipulation of meaning; furthermore, to a good approximation, representative space is isotropic and homogeneous. These realizations permitted Einstein to consider a practical Euclidean geometry that could be merged with physical laws, and then to proceed with an axiomatic approach to the propagation of light.

The two principles of relativity are, however, similar in their mathematical content. Einstein interpreted his principle of relativity to be also a "valuable heuristic aid in the search for general laws of nature" (1917a). A law of nature must be Lorentz covariant. An examination of Poincaré's mathematical techniques in 1906 for deriving a gravitational force law reveals that he, too, used Lorentz covariance as a guideline.

Thus my line of inquiry suggests that Einstein's theory of knowledge was an ingredient essential to his formulation of a special relativity theory based on two principles, or axioms.

One of the consequences of the two axioms of the special relativity theory is the relativity of simultaneity that appears to be at variance with prescientific knowledge. However, Einstein realized that simultaneity was a relative quantity before proposing the light axiom (see Chapter 3). This realization, Einstein recalled in his "Autobiographical Notes," required freeing oneself from "the practical experience of everyday life," where the high value of the velocity of light tricks one into believing in the absoluteness of simultaneity.

Conversely, Poincaré, anchored to the world of sensations, wrote in his (1898b) essay, "The Measurement of Time," that the definition of simultaneity is a matter of convention, and the "most convenient" definition is the one that is consistent with our sensations, which leads to an absoluteness of simultaneity. Poincaré supported this conclusion with an almost Einsteinian analysis of the nature of simultaneity and time. For instance, Poincaré located the vicious circles inherent in defining time in terms of cause and effect, and in the analysis of a one-way electromagnetic signal-exchange experiment to synchronize two widely separated clocks, for example, those in Paris and Berlin. Poincaré adroitly noted that to be rigorous would require the delay time in transmission to be taken into account, although in practice it is disregarded because our sensations cannot distinguish between the signal emitted from Paris and the one received in Berlin (see Miller, 1981b and 1983, for further discussion). Poincaré summarized his 1898 essay thus (1902a):

> Not only have we no direct intuition of the equality of two periods, but we have not even direct intuition of the simultaneity of two events occurring in two different places. I have explained this in an article entitled "Mesure du Temps."

So far as we know, Einstein never alluded to this passage, or to Poincaré's essay of 1898. In Einstein's special relativity theory the problems of the measurement of time and of simultaneity are both quantitative.[29]

ACTION-AT-A-DISTANCE, CONCEPTS, ETHER, AND RELATIVITY THEORY

Einstein's 1905 declaration that the ether is "superfluous" appears to contradict my conjecture that his scientific researches reflect his theory of knowledge; but Einstein spoke only about Lorentz's ether.

By 1911 the views of Lorentz and Einstein had been clearly defined, and the term ether had taken on at least two different connotations. Lorentz and Poincaré persisted in discussing Lorentz's ether. Other scientists such as Paul Langevin (1911) and Emil Wiechert (1911) proposed ether-based theories of relativity. Their ether provided dynamical causes for relativistic effects; for example, in the so-called clock paradox, the ether was to be the dynamic cause of the breaking of the symmetry between the inertial systems involved (see Miller, 1981b). In Einstein's relativity paper of 1905, however, effects such as time dilation and length contraction were presented as having kinematical causes.

In a letter dated 17 June 1916, Einstein wrote to Lorentz from Berlin:

> I agree with you that the general relativity theory admits of an ether hypothesis as does the special relativity theory. But this new ether theory would not violate the principle of relativity. The reason is that the state $g_{\mu\nu}$ = Aether is not that of a rigid body in an independent state of motion, but a state of motion which is a function of position determined through the material phenomena.

This is the first indication I have found that Einstein could quantify an ether by means of the higher-level general relativity theory.

To the best of my knowledge the (1918a) essay, "Dialogue on Objections against the Relativity Theory," constituted the first appearance in print of Einstein's quantification of an ether through the gravitational potentials, as well as his concern that the special relativity theory contained no medium to mediate disturbances. In the article a theme discussed by two interlocutors—one a relativist and the other a learned critic of relativity—was the interpretation of the so-called clock paradox. The particular case they analyzed was the linear outward and return trajectories at a constant velocity of a clock, U^2, while another clock, U^1, remained at rest. The problem is how the symmetry is broken between the inertial reference system that contains U^2 and the reference system of U^1. By proper application of the relativistic space and time transformations, the relativist (speaking for Einstein) demonstrates that there is no paradox. The "critic," however, is dissatisfied with the account offered by the special theory of relativity. The relativist agrees, and gives a detailed resolution of the clock paradox based on the principle of equivalence. Since the general relativity theory is closer to what Einstein took to be physical reality than is the lower-level special relativity theory, we can assume that Einstein, too, preferred a dynamic cause for the effect of time dilation.

Einstein concluded this essay by emphasizing that "the special relativity theory has disavowed the former meaning of the ether" (i.e., Lorentz's). Then, consistent with his letter of 1916 to Lorentz, Einstein wrote that it was the gravitational potentials of the general relativity theory that characterized the "physical qualities of empty space" (i.e., space devoid of matter).

In his 1920 lecture at Leiden, "Relativity and the Ether," Einstein discussed the need for an ether in physical theories. This ether is not Lorentz's but one that serves as a medium to fill space and transmit disturbances because, as Einstein wrote (1920):

> Non-physical thought knows nothing of forces acting at a distance. When we try to subject our experiences of bodies by a complete causal scheme, there seems at first sight to be no reciprocal interaction except what is produced by means of immediate contact, e.g., the transmission of motion by impact, pressure or pull, heating or inducing combustion by means of a flame, etc.

Newton, perhaps out of necessity, continued Einstein, did not take account of the origins of our knowledge of space and force. In 1927 Einstein had written that Newton formulated his theory of gravitation as action-at-a-distance because in his time it was the only way he could have arrived at a consistent gravitational theory (1927):

> The introduction of forces acting directly and instantaneously at a distance into the representation of the effects of gravity is not in keeping with the character of most of the processes familiar to us from everyday life. Newton meets this objection by pointing to the fact that his law of reciprocal gravitation is not supposed to be a final explanation but a rule derived by induction from experience.

The first sentence is clearly interesting from the point of view of Einstein's theory of knowledge, and the second bears on his epistemology. Just as Newton believed his theory of gravitation was yet to be perfected, so did Einstein in 1905 hold the same of special relativity. And just as little was Newton's theory derived by "induction from experience," so was the case for special relativity.

Expressing the connections between his theory of knowledge, his epistemology, and his physics, Einstein then wrote that although "people got into the habit of treating Newton's law of force as an irreducible axiom, . . . the ether hypothesis was bound always to play a part, even if it was mostly a latent one at first in the thinking of physicists." In view of Einstein's assertion that "non-physical thought knows nothing of forces acting at a distance" (1920), we can conjecture that what he meant by labeling the ether hypothesis as "latent in the thinking of physicists" could only have been that the physical theories that best describe nature are those based on contiguous action, which is consistent with prescientific concepts.[30]

As a postscript to dismissing Lorentz's ether on the grounds that it leads to asymmetries that are "intolerable to the theorist," Einstein wrote (1920):

> The most obvious line to adopt in the face of this situation seemed to be the following: —There is no such thing as the ether.

After all, he continued, electromagnetic radiation could be considered an independent reality, not attached to any substratum. Even in Lorentz's theory it has energy and momentum like ponderable matter. Electromagnetic radiation could even be conceived to be propagated through space like a hail of shot or light quanta.

Einstein went on to describe an ether suitable for use in the special theory of relativity. It is a ghostly medium and hence similar to Lorentz's; however, Einstein's ether possesses no state of motion—it simply supports energy in transit: "It is an extended physical object to which the concept of motion cannot be applied." This is a highly unsatisfactory concept; indeed, it is metaphysical in the worst sense of the term. Einstein thought so, too:

> From the point of view of the special theory of relativity the ether hypothesis does certainly seem an empty one at first sight.

However, as Einstein had explained at the start of this lecture, theories containing no medium to transmit disturbances are unsatisfactory. He continued by explaining why even prerelativistic mechanics requires something to fill space. I quote in extenso because in just a few sentences Einstein encapsulates the approaches of Newton, Mach, Lorentz, and Poincaré to the vexing problem of the absoluteness of rotatory motion (1920):

On the other hand, there is an important argument in favor of the hypothesis of the ether. To deny the existence of the ether means, in the last analysis, denying all physical properties to empty space. But such a view is inconsistent with the fundamental facts of mechanics. The mechanical behavior of a corporeal system floating freely in empty space depends not only on the relative positions (intervals) and velocities of its masses, but also on its state of rotation, which cannot be regarded physically speaking as a property belonging to the system as such. In order to be able to regard the rotation of a system at least formally as something real, Newton regarded space as objective. Since he regards his absolute space as a real thing, rotation with respect to an absolute space is also something real to him. Newton could equally well have called his absolute space 'the ether'; the only thing that matters is that in addition to observable objects another imperceptible entity has to be regarded as real, in order for it to be possible to regard acceleration, or rotation, as something real.

Thus, besides supporting light in transit, a key role of the ether in Lorentz's theory was to be the ultimate reference system to which all motion was referred.

Einstein then discussed Mach's attempt to "avoid the necessity of postulating an imperceptible real entity by substituting in mechanics a mean velocity with respect to the totality of masses in the world for acceleration with respect to absolute space" (see Miller, 1981b). However, Einstein continued, this assertion "presupposes action at a distance," which "the modern physicist does not consider himself entitled to assume . . . this view brings him back to the ether, which has to act as the medium of inertial interaction." Thus in Einstein's view the modern physicist cannot, with the same right as had Newton, formulate theories based on action-at-a-distance.

In conclusion Einstein wrote (1920):

> In accordance with the general theory of relativity space without an ether is inconceivable. For in such a space there would not only be no propagation of light, but no possibility of the existence of scales and clocks, and therefore no spatiotemporal distances in the physical sense.

The state of the ether of general relativity theory is causally conditioned by the theory's field equations. Hence, in the higher-level generalized relativity theory, space-time is not empty, and "forces acting at a distance" are explicitly excluded since "non-physical thought" knows nothing of them.[31]

CONCLUSION

Although Einstein's ether in the general relativity theory was of a different sort than the ether of Lorentz and Poincaré, it served the same purpose: to fill space and hence to set theories in the framework of contiguous action. Furthermore, as has been shown, this type of theory was important to Poincaré and Einstein because the directions of their scientific researches were influenced by their respective theories of knowledge.

Important differences and similarities characterize the theories of knowledge and epistemologies of Poincaré and Einstein. They both considered the same aspect of Kant's theory of knowledge to be important: the role of organizing principles. To Poincaré the potpourri of sensations is organized into concepts by means of two synthetic a priori organizing principles, whereas to Einstein organizing principles or concepts are freely created and are not unalterable. Einstein agreed with Poincaré's criterion for what constitutes objective knowledge and with Poincaré's notion of the order in which objects of Poincaré's real external world are created; that is, the first to be created is the concept of the solid body, then followed by space. Poincaré's theory of knowledge is rooted in sensationism, and thus his epistemology is rooted in empiricism. For Poincaré there is a well-defined direction from empirical facts to the hypotheses of a scientific theory; however, the inductive route is broken where hypotheses are invented. Up until 1904 Poincaré considered that the world is all that is presented to us, and that a quantity is metaphysical if it is not accessible to direct measurement. However, to reconcile Lorentz's theory of the electron with his theory of knowledge (i.e., to maintain an ether that was real but not directly observable), Poincaré was forced to modify his philosophical position toward the metaphysical. However, Einstein's notion of an essential abyss between perceptions or empirical data and concepts or axioms freed him from too much dependence on empirical data. To Einstein the true physical reality lies in an external world that exists independently of the scientist as investigator.

Demonstrating that Poincaré's view of the origins of geometry is part of his theory of knowledge makes it clear why he insisted on only three-dimensional Euclidean geometry for constructing the real external world from the world of sensations. This approach helps us also to understand why he maintained that the statements constituting a scientific theory—especially the ones on relative motion—should be classified according to whether they pertain to geometry, classical mechanics, or the physical sciences. Consistent with their theories of knowledge, both Einstein and Poincaré were relativists. Poincaré's epistemology prevented him from raising his principle of relativity to a convention. Einstein's principle of relativity of 1905 was a concept and belonged to a low-level theory. The mathematical use of both principles of relativity was parallel.

In order to further set the prevailing views of philosophy of science at the turn of the twentieth century and their allied notions of axiomatization, visual thinking, and the quest for unification within either the mechanical or electromagnetic world-pictures, I turn next to Ludwig Boltzmann.

NOTES

1. To some extent this distinction has been previously made by Toulmin (1972), who discusses an epistemics and a philosophical epistemology, and by McMullin (1970).

2. Poincaré demonstrated that this principle is not analytic a priori because it is irreducible to the principle of contradiction. Nor can the principle of induction be obtained from experiment because it permits passage from the finite to the infinite.

3. Poincaré's analysis of the transition from the physical to the mathematical continuum goes thus (see 1902a, 1905b): Call D those displacements that are so small that our sensations cannot distinguish between them. Suppose that we cannot distinguish $9D$ from $10D$, nor $10D$ from $11D$, but that we can distinguish $9D$ from $11D$, that is, $9D = 10D$, $10D = 11D$, $9D < 11D$. The above formula is Poincaré's famous statement of the physical continuum. But this result is "repugnant to reason, because it corresponds to none of the models which we carry about in us." We resolve this "intolerable disagreement with the law of contradiction" (1894) by replacing the small but finite physical displacements that are discussed by a rough group of displacements with the continuous displacements of the continuous Lie groups, thereby inventing the mathematical continuum. Thus, as was the case for the origins of geometry, the notion of the mathematical continuum is "created entirely by the mind, but it is experiment that has provided the opportunity."

4. Analysis of the connections among von Helmholtz, Marius Sophus Lie, and Poincaré remains to be written. Briefly, in the 1860s von Helmholtz demonstrated that the axioms of Euclidean

geometry could be deduced from the laws of motion of rigid bodies, along with the assumption that a space of any number of dimensions is determined by that number of coordinates; similarly, non-Euclidean geometries are the laws of motion of other sorts of bodies. It was essentially the lack of rigor of von Helmholtz's results that spurred Lie in 1867 to begin developing the subject of continuous groups of transformations. As Lie wrote to Poincaré in a letter of 1882: "(Riemann and) v. Helmholtz proceeded a step further [than Euclid] and assumed that space is a *Zahlen-Mannigfaltigkeit* [manifold or collection of numbers]. *This standpoint is very interesting; however it is not to be considered as definitive*" (italics in original). Thus, instead of studying in an approximate manner how geometries are generated by rotations or translations of rigid bodies, Lie developed the means to analyze rigorously how points in space are transformed into one another through infinitesimal transformations—that is, the subject of continuous groups of transformations. Although Poincaré quickly realized the fundamental importance of Lie's work for both mathematics and philosophy, he disagreed with Lie's own interpretations of the origins of geometry and the notion of space. The reason is that for both von Helmholtz and Lie the matter of the group (i.e., the *Zahlen-Mannigfaltigkeit*, or spatial coordinates) existed prior to the group: "For me, on the contrary, the form exists prior to matter" (Poincaré, 1898a). Thus, continued Poincaré, using the group-theoretical approach the origin of geometry can be analyzed without assuming beforehand the existence of both space and geometry. Poincaré agreed with von Helmholtz that an analysis of the origins of geometry should take account of these perceptions, but disagreed with von Helmholtz's and Lie's beliefs that a Euclidean geometry joined with physical laws is empirically testable (Helmholtz, 1876; Lie, letter of 1882 to Poincaré).

5. Piaget, for example, believes otherwise: "Poincaré had the great merit of foreseeing that the organization of space was related to the formation of the 'group of displacements,' but since he was not a psychologist, he regarded this group as *a priori* instead of seeing it as the product of a gradual formation" (1969).

6. See Poincaré (1902a) for a dictionary between Euclidean and non-Euclidean geometry, "as we would translate a German text with the aid of a German-French dictionary." Assuming that Euclidean geometry is self-consistent, then the researches of Bel-

trami, Cayley, and Klein were interpreted as proving that non-Euclidean geometries were also self-consistent. Poincaré (1902a) asks, "Whence is this certainty derived, and how far is it justified?" He considered this problem "not insoluble." It was resolved in 1900 by Hilbert who reduced the axioms of geometry to those of an arithmetic system; such a system was assumed to be self-consistent. This assumption was in turn shattered in 1931 by the work of Kurt Gödel.

7. Poincaré went on to write that "from the law of conservation of energy Leibnitz deduced that of conservation of momentum without doubt because he assumed that the energy, invariable in the absolute motion, should be invariable in relative motion." This assumption, indigenous for classical mechanics, led Poincaré in (1900c) to invoke the ad hoc hypothesis of a "complementary force" in order to impose the invariability of energy and to save action and reaction in an inertial reference system that contains an emitter of unidirectional radiation. For using Lorentz's local time coordinate and the Galilean spatial transformations, Poincaré essentially came upon the transformation law for the energy of a pulse of radiation accurate to order (v/c), which means that energy is not an invariable quantity. Before special relativity there was some confusion between quantities that are conserved (such as energy and momentum) and quantities that are invariable or invariant (such as electric charge). See Miller (1980, 1981b).

8. After discussing the different forms of energy in addition to mechanical energy that were discovered since Leibnitz, Poincaré added that Leibnitz "enunciated [the law of conservation of energy] as clearly and completely as one could in his time."

9. With the advent of general relativity theory, Calinon's speculations on the physical possibility of non-Euclidean geometries were revived as having been prophetic (e.g., Robertson, 1949). Calinon's essay of (1897) is discussed in Capek (1976). Suffice it to say that only Calinon (1885) deserves further study.

10. Just as he had proceeded with geometry, Poincaré (1902a) asked whether the principles of mechanics are a priori or obtained from experiment. For example, if the principle of inertia were a priori, then the Greeks would not have taken the preferred motion to be circular motion. Nor could this principle have been obtained from experiment because no one has ever actually observed a purely force-free motion (i.e., an inertial reference system).

11. Poincaré (1908a) comments on Hertz's mechanics, which is for-
    mulated in a non-Euclidean geometry, thus:

    > It quite seems, indeed, that it would be possible to translate our
    > physics into the language of geometry of four dimensions. At-
    > tempting such a translation would be giving oneself a great deal of
    > trouble for little profit, and I will content myself with mentioning
    > Hertz's mechanics, in which something of the kind may be seen.
    > Yet it seems that the translation would be less simple than the
    > text, and that it would never lose the appearance of a translation,
    > for the language of three dimensions seems the best suited to the
    > description of our world, even though that description may be
    > made, in case of necessity, in another idiom.

12. For example, compare this statement of Poincaré's conven-
    tionalism with J.O. Wisdom's Poincaré: "Theories, for Poincaré
    did *not* flatly have no truth-value; for they were true by conven-
    tion" (1971; italics in original). Clearly, Wisdom missed the
    point.

13. The modern wave theories of light date from the seventeenth
    century. In the early theories the analogy was made between
    sound and light. Since sound travels at a finite velocity it requires
    a milieu; therefore, so does light. (In 1675 Ole Römer ascer-
    tained that light travels at a very fast but finite velocity.) The
    milieu for light waves was termed the ether. Thus, in the prerel-
    ativistic wave theories, light was depicted as a wave phenome-
    non that propagated through an ether, just like waves propagate
    through water. In the latter part of the nineteenth century ex-
    perimental apparatus became available for measuring the effect
    of the earth's motion through the ether on the propagation of
    light. An effect was expected basically for the following reason:
    From the point of view of an observer on the earth there should
    be an ether wind that causes the direction of a light ray to deviate
    from what it would have been if the earth were at rest in the
    ether. The effect of the ether wind was assumed to be calculable
    from Newton's addition law for velocities. Thus, the velocity of
    light relative to an observer on the moving earth, c_r, should be
    the result of adding to the velocity of light relative to an observer
    at rest in the ether c (taken to be a determined constant that today
    we refer to as the velocity of light in vacuum), and the unknown
    velocity of the earth relative to the ether, v. The experiments
    that were set up to measure the difference between c_r and c were
    called ether-drift experiments. To second-order accuracy in the

quantity (v/c), they failed; that is, they found that $c_r = c$. In other words, to this order of accuracy, optical phenomena occurred on the earth as if the earth were at rest in the ether. A goal of the physics of 1905 was to propose hypotheses to explain these failures. (See the Appendix in Chapter 6 for further discussion of the ether-drift experiments.)

14. To Poincaré the advent of quantum theory posed a threat to the formulation of physics in terms of differential equations (1912d):

> We now wonder not only whether the differential equations of dynamics must be modified, but whether the laws of motion can still be expressed by means of differential equations. And therein would lie the most profound revolution that natural philosophy has experienced since Newton.

Thus Poincaré was reluctant to accept the quantum theory as it was posed in 1912.

15. See also Poincaré (1908a), where he writes that the earth's "absolute velocity . . . has no sense," rather one must mean "its velocity in relation to the ether."

16. Initially, Lorentz's theory could explain systematically only the failure of ether-drift experiments accurate to order (v/c). Thus, Lorentz's seminal paper of (1892a) on his electromagnetic theory did not refer to the (then) only trustworthy experiment accurate to order $(v/c)^2$: the Michelson and Morley experiment of 1887. Michelson and Morley used a half-silvered mirror to split a monochromatic beam of light into two rays that traversed to-and-fro along the two orthogonal arms of equal length of an interferometer. Their goal was to determine the velocity of the earth relative to the ether by measuring the time difference between the journey of the two rays. Within their experimental accuracy, no time difference was measured. From Lorentz's correspondence we know that this experiment was of great concern to him. Later (in 1892b) he indulged in physics of desperation: namely, he proposed the hypothesis of contraction according to which matter is deformed owing to its motion through the ether. The version preferred by Lorentz was that only the dimension parallel to the direction of motion through the ether is contracted, while other dimensions are unchanged. This hypothesis was ad hoc because not only was it proposed to explain only one experiment, but the basis of its plausibility argument was unconnected with Lorentz's electromagnetic theory. For details see Miller (1974, 1981b).

Right from the start Poincaré considered the ad hoc hypothesis of contraction to be a blemish on Lorentz's theory. He wrote (in 1902a):

> Experiments have been made that should have disclosed the terms of the first order; the results were nugatory. Could that have been by chance? No one has admitted this; a general explanation was sought, and Lorentz found it. He showed that the terms of the first order should cancel each other, but not the terms of the second order. Then more exact experiments were made, which were also negative; neither could this be the result of chance. An explanation was necessary, and was forthcoming; they always are; hypotheses are what we lack the least.

Abraham's 1902 theory of the electron is based on Lorentz's electromagnetic theory and so it could explain all effects accurate to order (v/c). However, it could not deal with the null results of ether-drift experiments accurate to second order in (v/c).

17. The reason is that in Lorentz's theory the ether acts on matter with no inverse reaction. Poincaré discussed this problem in print (1900c) and in correspondence with Lorentz (Miller, 1980). It turns out that Poincaré was more concerned with the Lorentz theory's violation of Newton's third law than the ad hoc contraction hypothesis because this law was a convention. In order to preserve Newton's third law Poincaré was willing to introduce into Lorentz's theory several hypotheses, one of which was as ad hoc as the hypotheses of contraction (see Note 16, above). However, owing to Lorentz's extension of his theory in 1904 to the electron itself, in such a way as to explain the failure of all extant ether-drift experiments, as well as Kaufmann's data on high-velocity electrons, Poincaré was willing to discard Newton's third law in electromagnetism because it impeded progress (1904a). In other words (see Figure 1.1), without being empirically disconfirmed, Newton's third law lost its status as a convention because it was no longer "fertile."

18. Rayleigh (1902) and, in an improved version, Brace (1904) sent light in two orthogonal directions through an isotropic crystal at rest in the laboratory. Owing to the hypothesis of contraction the indices of refraction should differ in these two directions. To one part in 10 decimal places Rayleigh found no difference, and Brace pushed the accuracy to one part in 13 decimal places. Trouton and Noble (1903) tried to measure the turning couple on a charged condenser at rest in the laboratory. A turning couple was assumed to be present owing to the motion of the labo-

ratory through the ether. This motion was expected to increase the electrostatic energy density between the plates because the charges on the plates are also the source of convection currents. A measurement of the turning couple would permit determination of the earth's velocity relative to the ether. No turning couple was measured.

19. Poincaré's symmetrical version of the Lorentz transformations is mathematically identical to the one familiar to us today from special relativity:

$$
\left.
\begin{aligned}
x' &= k(x - vt) \\
y' &= y \\
z' &= z \\
t' &= k\left(t - \frac{v}{c^2}x\right) \\
\left(k \right. &= \left. \frac{1}{\sqrt{1 - v^2/c^2}}\right)
\end{aligned}
\right\}
\tag{A}
$$

where in the context of the electromagnetic world-picture, the coordinates (x', y', z', t') and (x, y, z, t) refer to an auxiliary reference system S' and to the physically real, but apparently inaccessible, ether-fixed system S. In Eqs. (A), the quantity v can be interpreted only as the mathematical relative velocity between S' and S. The reason is that by definition S is fixed in the ether. In special relativity, however, both S and S' are physically real inertial reference systems; moreover, the space and time coordinates are given in principle operational definitions so that v is a measurable velocity. Lorentz (in his 1904b) wrote the above transformations differently as:

$$
\left.
\begin{aligned}
x' &= kx_r \\
y' &= y_r \\
z' &= z_r \\
t' &= \frac{t_r}{k} - k\frac{v}{c^2}x_r
\end{aligned}
\right\}
\tag{B}
$$

where the coordinates (x_r, y_r, z_r, t_r) refer to an inertial reference system S_r. The systems S and S_r are related through space and time transformations from Newton's mechanics:

$$x_r = x - vt$$
$$y_r = y$$
$$z_r = z$$
$$t_r = t$$

(C)

where here v is interpreted as the heretofore unmeasurable velocity of the earth relative to the ether. Substitution of Eqs. (C) into Eqs. (B) gives Eqs. (A). In prerelativistic electrodynamics the equations of electromagnetism yielded a value for the velocity of light that is independent of the source's motion when they were expressed relative to the ether. When the equations of electromagnetism are transformed to S_r using Eqs. (C), the velocity of light depends on the relative motion between source and observer. But experiment indicated otherwise. So, Lorentz sought further transformations on the coordinates in S_r in order to express the equations of electromagnetism in a reference system in which the source's velocity relative to the observer did not affect the velocity of light. Starting in 1892 he suggested several transformations (see Chapter 3) until he achieved success in 1904. Poincaré (1905a, 1906) dubbed Eqs. (A) the Lorentz transformations. He expressed them in a symmetric form as compared to Eqs. (B), by eliminating the intermediate inertial reference system S_r. Strictly speaking, however, three coordinate systems were employed in the electromagnetic world-picture: S, S_r, and S'. See Miller (1981b) for details.

20. Thus, contrary to Guidymyn (1977), Poincaré never considered the principle of relativity in the physical sciences to be a convention. Further evidence for this point will be offered in the section, "On Poincaré's Post-1905 Thoughts on Geometry, Classical Mechanics, and the Physical Sciences."

21. This innovative school is discussed in Chapter 6.

22. Einstein's (1907a) contains his first published recollections on the state of physics in 1905. Study of this paper, in conjunction with others of his early papers and correspondence, leads to the following description of his assessment of the state of physics in 1905. For a more detailed analysis see Miller (1981b). His previous research on fluctuation phenomena led him by the end of 1904 to conclude that light has both wave and particle properties, and Lorentz's electromagnetic theory could explain only the

wave mode. His research on Brownian motion led him to conclude that mechanics cannot be the basic theory because in volumes the order of the electron's higher-order derivatives of the acceleration cannot be disregarded. Thus, neither the electromagnetic nor mechanical world-pictures could succeed. But restricting considerations to large volumes and not discussing the constitution of matter, the laws of electromagnetism and mechanics could be used with confidence; this is what he did in the special relativity paper.

23. See the letter of 29 September 1909 of Einstein to Arnold Sommerfeld that is quoted in Stachel (1980). Einstein, however, almost certainly embarked on this path independently of Poincaré. The properties of a rotating disc was a much-discussed problem in 1909. It concerned the notion of a rigid body as defined in classical mechanics in contrast with the deformations proposed by Lorentz and by Einstein [see Miller (1981b), Chapter 7].

24. Einstein's description in "Physics and Reality" of the origin of geometry is analogous to those in his (1921a, b), but more detailed. Here he emphasized the importance of the "concept" of the practically rigid body. In "Physics and Reality" Einstein omitted Poincaré's name altogether in these discussions, and in the "Autobiographical Notes" of 1946 he singled out only Hume and Mach as helpful to his thinking toward the special theory of relativity. In the "Replies to Criticisms," however, he rectified this omission. There Einstein exhibited once again his deep understanding of Poincaré by giving his name to the "nonpositivist" in an imaginary dialogue with the logical empiricist philosopher Hans Reichenbach, who was chastised severely.

25. This letter was first discussed by Holton (1978), and in more detail in Holton (1979).

26. It is apropos here to survey those aspects of Einstein's knowledge of physics, philosophy, and technology in 1905 that are relevant to our analysis. See my (1981b) for a more detailed analysis. From analysis of his published papers during 1901–1907, and his lifelong correspondence, there emerges a portrait of someone who was aware of developments in philosophy, technology, and physics. Biographies that depict him as virtually cut off from the world of science do an injustice to his achievements in 1905. In fact, his awareness of contemporary research renders all the more dazzling how he opted for a course of action so different from that of other physicists in 1905.

In the 1905 special relativity paper entitled "On the Electrodynamics of Moving Bodies," Einstein referred to "unsuccessful attempts to discover any motion of the earth relatively to the 'light medium'." He gave no explicit citations to these ether-drift experiments; in fact, there are no literature citations anywhere in the paper. Then he specialized to the class of experiments accurate to the "first order of small quantities"—that is, accurate to order (v/c), where v is the velocity of the earth relative to the ether and c is the velocity of light in the free ether that is measured by an observer at rest in the ether. For example, Martinus Hoek's 1868 experiment is an ether-drift experiment of first-order accuracy. Among the wider class of "unsuccessful attempts" was the second-order Michelson and Morley 1887 experiment. No serious historian has argued that Einstein was unaware of this experiment before 1905. Then there were the second-order null experiments of Rayleigh (1902) and Brace (1904) to detect double refraction in isotropic crystals. Rayleigh and Brace's experiments were analyzed in Max Abraham's widely-cited paper of 1904 in the *Annalen der Physik* entitled, "On the Theory of Radiation and of Radiation Pressure." Since it is inconceivable that someone would publish in a journal to which he had no access, we can safely conjecture that Einstein had at least perused Abraham's 1904 paper.

Another group of first-order experiments are positive experiments that did not attempt to detect the earth's motion through the ether—for example, observations of stellar aberration and experiments to verify Fresnel's dragging coefficient that were performed by Fizeau (in 1851) and by Michelson and Morley (in 1886). In these experiments v was the velocity of ponderable matter relative to the laboratory.

Einstein's emphasis in the special relativity paper on the first-order experiments supports his later comments that stellar aberration and Fizeau's experiment had been the most influential of the often-cited empirical data to his thinking toward the special relativity theory. In a number of places Einstein later recalled of these data, "They were enough" (e.g., Shankland, 1963). Their explanation turned about a quantity that also explained systematically the failure of first-order ether-drift experiments—namely, Lorentz's local time coordinate. (The 1904 version of the local time coordinate is the quantity t' in Eqs. (A) and (B) of Note 19 above; see Chapter 5, Fig. 5.1, for the 1895 version that was known to Einstein in 1905. At the time he wrote the special relativity paper Einstein had not yet seen Lorentz's work (1904b).) In historical context, Einstein's predilection for first-order data is not surprising. Others also emphasized these data over the 1887 Michelson-Morley data. For example, Lorentz especially framed his electromagnetic theory of 1892 to systematically include Fresnel's dragging coefficient. Second-order data are nowhere mentioned in his seminal 1892 opus, *Maxwell's Electromagnetic Theory and Its Application to Moving Bodies*. In a short sequel publication in (1892b) Lorentz proposed the ad hoc hypothesis of contraction to explain the irksome 1887 Michelson-Morley experiment. Another example is Max Abraham's widely-read book, *Theory of Electricity* (1904c), where he considered Fizeau's 1851 experiment to be critical for deciding between Lorentz's and Hertz's theories of the electrodynamics of moving bodies. Almost certainly Einstein had read Abraham's 1904 book, and we know from a letter to his biographer Carl Seelig that before 1905 Einstein had read Lorentz's 1892 presentation of his new version of Maxwell's theory, as well as Lorentz's 1895 monograph entitled, *Treatise on Electrical and Optical Phenomena in Moving Bodies*. In this treatise Lorentz reviewed available first-order experiments and explained them in a more systematic manner than he had in 1892.

Two of Einstein's more reliable biographers, Philipp Frank (1947) and Carl Seelig (1954), wrote of Einstein's having studied on his own while at the Zurich Polytechnic such master philosopher-scientists as Gustav Kirchhoff, Hermann von Helmholtz, and Ludwig Boltzmann.

27. In a book that, as far as I know, Einstein had not read before writing the special relativity paper, but which may have influenced him somewhat afterward, Mach (1905a) provided a similar definition of a concept to those of Boltzmann and von Helmholtz:

> From sensations and their conjunctions arise concepts, whose aim is to lead us by the shortest and easiest way to sensible ideas that agree best with the sensations. Thus all intellection starts from sense perceptions and returns to them. Our genuine mental workers are these sensible pictures or ideas, while concepts are the organizers and overseers that tell the masses of the former where to go and what to do. In simple tasks, the intellect is in direct touch with the workers, but for larger undertakings it deals with

the directing engineers, who would however be useless if they had not seen to the engagement of reliable workers. The play of ideas relieves even animals from the tyranny of momentary impressions.

This excerpt emphasizes Mach's notions of the economy of thought and the primacy of sense perceptions. Mach's definition of a concept squares with his philosophical view in which theories serve as economical summaries of empirical data.

28. Intuitively one would expect that the relative motion between an observer and a source of light can be calculated according to Newton's physics which is consistent with a principle of relativity. But, as Einstein explained (in his 1917a), on the one hand, intuitively means according to the Newtonian principle of relativity, whose velocity addition law disagrees with the properties of light (see also Appendix to Chapter 5). On the other hand, the two axioms of special relativity are compatible, owing to differences in clock readings between observers in relative motion (i.e., the relativity of time).

29. Two facts are simultaneous if they occur at the same time when measured on clocks that are at relative rest. For example, consider two people who are several miles apart, are at relative rest, and they are equipped with flashlights. Their clocks read the same time. As previously agreed, at three o'clock they turn their flashlights on. These two occurrences are simultaneous. But will they be simultaneous to someone else who is in an airplane that is moving along the line joining the two people with the flashlights? The answer is yes, according to Newton's mechanics which is in agreement with our perceptions (see, e.g., the quote from Calinon in the section, "The Structure of a Scientific Theory"). The answer is no according to special relativity in which simultaneity depends on the motion of a reference system (i.e., simultaneity is a relative quantity).

30. Similarly, Maxwell had emphasized that Newton "did not believe in the direct action of bodies at a distance" (1878a). To support this assertion Maxwell quoted from a letter of 17 January 1693 from Newton to Bentley:

It is inconceivable that inanimate brute matter should, without the mediation of something else which is not material, operate upon and affect other matter without mutual contact, as it must do if gravitation in the sense of Epicurus be essential and inherent in it . . . . That gravity should be innate, inherent, and essential to matter, so that one body can act upon another at a distance, through a vacuum, without the mediation of anything else, by and through which their action and force may be conveyed from one to another, is to me so great an absurdity, that I believe no man, who has in philosophical matters a competent faculty of thinking, can ever fall into it.

In his essay "Ether" (1878b), Maxwell gave several scientific reasons for an ether similar to those of Poincaré—for example, that it supports radiation in transit. But he concluded his essay as follows:

Whatever difficulties we may have in forming a consistent idea of the constitution of the aether, there can be no doubt that the interplanetary and interstellar spaces are not empty, but are occupied by a material substance or body, which is certainly the largest, and probably the most uniform body of which we have any knowledge.

Thus did Maxwell and Newton believe that scientific theories should reflect prescientific concepts.

31. In the "Autobiographical Notes" (1946), Einstein delved deeper into the preference for dynamical theories of matter. He recalled a "wonder" that he experienced as a boy of age four or five when his father showed him a magnetic compass. The behavior of the compass's needle came "into conflict with a world of concepts which is already fixed in us. That this needle behaved in such a determined way did not at all fit into the nature of events, which could find a place in the unconscious world of concepts (effect connected with direct 'touch')." Thus, in analogy with occurrences in the world of sense perceptions, something had to be acting on the compass needle, constraining the needle to point in a particular direction.

CHAPTER 2

On the Origins, Methods, and Legacy of Ludwig Boltzmann's Mechanics

I cannot really imagine any other law of thought than that our pictures should be clearly and unambiguously imaginable.

L. Boltzmann (1897a)

Ludwig Boltzmann (1844–1906) set the foundations for the classical theory of gases. He was one of the masters of theoretical physics that Einstein studied on his own while a student at the Zurich Polytechnic Institute (1896–1900).

Sᴜᴄᴄᴇᴇᴅɪɴɢ ʜɪᴍsᴇʟꜰ ɪɴ 1902 to the Chair of Theoretical Physics at the University of Vienna, Ludwig Boltzmann emphasized in his Inaugural Address that "mechanics is the foundation on which the whole edifice of theoretical physics is built, the root from which all other branches of science spring" (1902). The goal of a mechanical interpretation of nature, that is, a mechanical world-picture,[1] had been a central theme throughout Boltzmann's researches in statistical mechanics[2] and electromagnetism. But, owing to recent developments in electromagnetic theory, Boltzmann had begun to qualify this hope somewhat. For diametrically opposite to the mechanical world-picture was the program that Wilhelm Wien had referred to in 1900 as the electromagnetic world-picture, whose goal was to deduce mechanics from electromagnetism.[3] In 1904 Boltzmann's desire for unification of the sciences led him to endorse that program.

It turned out that although circa 1900 Boltzmann's lectures on and research in mechanics and electromagnetism had little effect on major developments in these disciplines,[4] his foundational analysis of mechanics was highly respected. For example, in the obituary of Boltzmann that H.A. Lorentz wrote (in 1907), he spoke of Boltzmann's "striving for clarity in the foundations and to eliminate each bewildering turbidity of thought." It was Boltzmann's eagerness to purify the foundations of mechanics on which I shall focus. This analysis is best carried out by placing Boltzmann in the matrix of the philosopher-scientists that also included as principals Ernst Mach and Heinrich Hertz.[5] Then will follow a discussion of the possible effect of Boltzmann's (1897a) magnum opus on a student at the Zurich Polytechnic who read it sometime between 1897 and 1900 and whose name was Albert Einstein.

THE ORIGINS OF BOLTZMANN'S MECHANICS
PERSPECTIVE

The core of Boltzmann's foundational analysis of mechanics is contained in what he aptly referred to in his 1897 *Vorlesungen über die Principe der Mechanik* as the "characterization of the method chosen":

> It is precisely the unclarities in the principles of mechanics that seem to me to derive from not starting at once with hypothetical mental pictures but trying to link up with experience from the outset.

Boltzmann referred to "starting at once with hypothetical mental pictures" and then linking up with experience as the "deductive method" (1899a). The inverse method, which he referred to as the "inductive mode of representation," begins with empirical facts from which mental pictures are developed. The "unclarities" that Boltzmann sought to remove concerned notions of time, space, mass, and force; in particular, he wanted to reformulate the foundations of mechanics so that the definitions of these quantities did not "overshoot the mark" (1904, 1905b). By overshooting the mark Boltzmann meant the "frequently uttered requirement that natural science must never go beyond experience should therefore in my view be reformulated thus: never go too far beyond experience and introduce only such abstractions as can soon be tested by experience" (1899a). Some examples of queries Boltzmann gave that overshoot the mark are "why the law of cause and effect itself holds; likewise if we ask why the world exists at all." Hertz had referred to such queries as "illegitimate" because they did not contribute to scientific progress (1894).

Boltzmann had begun to investigate the notion of mental pictures in his 1890 lecture, "On the Significance of Physical Theories," where he wrote that the "task of theory consists in constructing an image of the external world that exists purely internally and must be our guiding star in thought and experiment; that is in completing, as it were, the thinking process and carrying out globally what on a small scale occurs within us whenever we form an idea."

We can conjecture that Boltzmann was impelled to elaborate on his notion of mental pictures by Hertz's 1894 formulation of a deductive system of mechanics, and by the phenomenalistic approach of Mach and Wilhelm Ostwald who sought to investigate phenomena without any mechanical explanations or hypotheses on the constitution of matter, such as atoms.[6] It is apropos, therefore, to survey those aspects of Hertz's mechanics on which Boltzmann focused in his philosophical and physical writings, and which we may consequently assume most influenced his philosophical view.

HERTZ'S MECHANICS

Straightaway in his *Principles of Mechanics,* Hertz distinguished between "images themselves" and their development into a "scientific representation." According to Hertz, somehow as a result of "previous experience" in the world of perceptions, "we form for ourselves internal pictures or symbols of external objects."[7] If we assume a "certain conformity between nature and our thought [that] experience teaches us" (i.e., Hertz assumed a not a priori causal principle), then the necessary consequences of the mental pictures of objects in thought are always the objects' necessary consequences in nature. By "conformity between nature and our thought" Hertz was alluding to what he would introduce subsequently into his mechanics as "laws of internal intuition," on which are based Kantian-like organizing principles that serve as axioms. The internal picture of an object, continued Hertz, could not be unique, but the more appropriate image was the one that portrayed the object in the most economical manner. A "scientific representation of images" required Hertz to frame postulates concerning the images' permissibility, correctness, and appropriateness. Without any ambiguity an image was permissible if it did not contradict the "nature of the mind,"—that is, the laws of internal intuition. The correctness of an image was determined unambiguously by empirical tests. Appropriateness, however, could not be decided unambiguously until the "testing of many images" was accomplished. Hertz, like Boltzmann somewhat later, used the term picture or image to cover also conceptual frameworks such as Newtonian mechanics and energetics. But for Hertz, unlike Boltzmann, mental pictures could be discarded once a theory's axiomatic foundation is set (see also Chapter 3, note 11).

Hertz went on to note Gustav Kirchhoff's (1876) statement that "three independent conceptions are necessary and sufficient for the development of mechanics,"—time, space, and mass. However, continued Hertz, in order to understand matter in motion at least one other fundamental conception was necessary, and hitherto this had been taken to be either force or energy. Hertz's critique of the notions of force and energy revealed the inadequacies in "images" of science that took them to be fundamental—that is, the "customary representation" of Newtonian mechanics, and energetics which replaced force with energy.[8] Hertz paid homage particularly to the Newtonian "image" of mechanics which first pointed out that we could not understand matter in motion by restricting ourselves to what "can be directly observed." Then he continued with a statement that we may conjecture influenced Boltzman's own developing notion of mental pictures:

> We become convinced that the manifold of the actual universe must be greater than the manifold of the universe which is directly revealed to us by our senses.

Thus Hertz considered that he had license to assert that there was "something hidden at work, and yet deny that this something belonged to a special category. We are free to assume that this hidden something is nought else than motion and mass again—motion and mass which differ from the visible ones not in themselves but in relation to us and to our usual means of perception." In this way Hertz introduced another method to supplement time, space, and mass—namely, that in addition to the visible masses of a system there are concealed masses that are linked to the visible ones. According to Hertz, no systems are acted upon by a net external force; rather, every system is a portion of a free system whose remainder consists of concealed masses to which the visible mechanical system is linked. Later in his exposition Hertz admitted forces through the undetermined Lagrange multipliers associated with differential constraint conditions. In an effort to meet criticisms of his introduction of concealed masses, Hertz pointed out that Newtonian mechanics posited the invisible notion of force, and energetics that of energy. Hertz's ultimate goal was to simulate the contiguous actions of the ether by rigid connections among hidden masses. Electromagnetism, in Hertz's opinion, revealed that action-at-a-distance forces were only a "first approximation to the truth [and that in electromagnetism] the decisive battle between these different fundamental assumptions of mechanics will be fought out."[9]

Hertz divided his treatise into Books I and II. Book I develops mechanics as far as possible "independent of experience" because all assertions are "a priori judgments in Kant's sense" that are based on "laws of internal intuition." Consequently, in Book I the notions of time and space are those of our internal intuition (i.e., time is absolute and space is Euclidean), and Hertz introduced mass by means of a series of definitions to establish the material point mass. In Book II the quantities time, space, and mass are connected with experience; thus time is measured by a chronometer, space by means of a Euclidean distance scale, and mass by "weighing." In order to proceed beyond the kinematics developed in Book I, Hertz introduced the "fundamental law" according to which a system subject to the hypothesized rigid internal connections and under no external forces, achieves motions for which the sum of the masses multiplied by the square of the accelerations is a minimum (Gauss's principle of least constraint).[10]

THE FOUNDATIONS OF BOLTZMANN'S MECHANICS

There was much discussion in the scientific and philosophical literature circa 1900 of the grand outlines of Hertz's program for a mechanical foundation for physics, to be carefully developed in the "deductive form" of Euclid's geometry. For example, in 1897 it led Henri Poincaré to extend his own developing epistemology into mechanics. However, neither Poincaré nor Boltzmann were entirely satisfied with Hertz's reformulation of mechanics. In addition to agreeing that Hertz's mechanics was laden with hypotheses, they shared another criticism, as Boltzmann wrote (in 1897a)[11]: "There is only one thing that I find lacking here, namely, the proof that nature can really be represented by this picture." Although Boltzmann considered Hertz's mechanics a "program for the distant future" (1899b), he approved of the basis of Hertz's approach. We can discern its effect on Boltzmann in what he considered to be his "one single dissertation of philosophical intent." Boltzmann aimed this dissertation at a laconic statement made by Mach during a debate on atomism. According to Boltzmann (1903), Mach once said, "I do not believe that atoms exist"; since Mach usually drew on an analysis of sensations in support of this sort of statement, Boltzmann did likewise to support its denial.[12]

Thus in his 1897 essay entitled, "On the Question of the Objective Existence of Processes in Inanimate Nature," Boltzmann delved into an analysis of perceptions in order to set his notion of concepts and mental pictures as securely as he could. In order not to overshoot the mark, he began by assuming that "everyone knows what is meant by perceptions of the senses and impulses of the will." Boltzmann also assumed as a "precondition of intelligence" that there are "constant regularities between" perceptions, and that knowledge could be constructed from perceptions. Basically, continued Boltzmann in a footnote, this is equivalent to assuming a principle of causality which could be taken as a priori or as a result of experience; as we shall see, like Hertz, Boltzmann preferred the latter route. Consider, Boltzmann continued, that a complex of perceptions A, followed by a willful impulse (or impulsive complex) B, leads to a sense perception C, while a willful complex D results in another perception, E. These chains of events impart to us memories (*Erinnerungen*) and world-pictures (*Weltbilder*). Next, consider the more complicated situation in which several perceptual complexes A_1, A_2, A_3, . . . , having T in common, always lead to the same sensation, C. If we find another complex A_x which also contains T, then we would expect C to follow. Boltzmann defined the quantities T as follows:[13] "To construct thought pictures we constantly need designations for what is common to various groups of phenomena, thought-pictures or intellectual operations: we call such designations concepts . . . [From the mental pictures or concepts] we come to thought symbols for those regularities of our perceptual complexes that lead to the idea of matter." Thus Boltzmann defined a quantity, T, to be a concept because it is common to a complex of sensations, or thought-pictures, or intellectual operations, thereby providing an understanding of them. The "realist view," according to Boltzmann, is based on the "synthetic description" of thinking: from perceptions concepts are constructed and then ideas of matter." Boltzmann referred to the inverse view as "idealist," and he characterized it as "most appropriate where what counts is the laying bare of concepts [*Herausschälung der Begriffe*]."[14] The idealist view starts with the "most easily grasped rules for constructing this world-picture [*Weltbild*] without bothering how we subjectively came by these rules: the only justification for the world-picture is then seen in its seeking agreement with facts. What previously came first will now come precisely last."

In 1899 in the first of four lectures delivered at Clark University, Boltzmann discussed his notion of the role of concepts in physical theory and how it interacted with his defence of atomism. Lorentz (1907), for example, considered these lectures to have been Boltzmann's best presentation of his foundational analysis of mechanics. Boltzmann began by criticizing Mach's phenomenalism with its claim that physical theory should represent phenomena without going beyond experience. Yet, continued Boltzmann, Mach's referral of inertial motion to the average velocity of all other masses in the universe, for example, or his attempt to fill Newton's cosmic receptacle with an ether—such proposals went beyond experience. Furthermore, Boltzmann added, even Newton had gone beyond immediate experiences despite his avowed goal. The thrust of these discussions was not only to criticize the phenomenologists but also to demonstrate the necessity of transcending direct experience in order to formulate a scientific theory. Boltzmann (1899b) stated emphatically that "phenomenology therefore ought not to boast that it does not go beyond experience, but merely warn against doing so in excess."

Boltzmann prefaced his foundational analysis of mechanics by rejecting Kirchhoff's assertion that the task of mechanics ought to be restricted to describing phenomena. From his analysis Kirchhoff defined force as merely the algebraic expression of mass multiplied by acceleration. Thus he claimed to have eliminated the metaphysical problems of whether force was a property of mass or vice versa, or whether there was a dualism of the separate constituents of force and mass. Boltzmann (1899a) disagreed with Kirchhoff on the grounds that if "we delve really deeply into the mode of our own thinking, into its mechanism, as I should feel inclined to say, then one should like to deny" Kirchhoff's instrumentalist view of mechanics; rather, Boltzmann continued, we should seek to know the mechanism of phenomena, which occurs beyond perceptions. Boltzmann sought a method for laying the foundations of physical theory that permitted freer reign of the imagination than phenomenalism allowed. He proposed to begin neither in the realm of physical theory, per se, as Kirchhoff had suggested, nor in the realm of perceptions, as Mach subsequently proposed, but in the "mode of our own thinking."

Since "all our ideas and concepts are only internal mental pictures," continued Boltzmann, the foundations could be set by blending the "extremes" of metaphysics and physics because the concepts of mass and force are "only mental pictures whose purpose is to represent phenomena correctly." Armed with this methodology Boltzmann could deal with the concepts of mass and force, while maintaining that the metaphysical problems that concerned Kirchhoff "have no meaning at all." Boltzmann emphasized that whereas "Kirchhoff made no material changes in the old classical mechanics,

. . . Hertz went much further" (Boltzmann, 1899b). In particular, Hertz was the first to state with "special clarity" (Boltzmann, 1899a) the use of mental pictures as the basis of an axiomatic approach to mechanics.

Boltzmann recalled that Hertz's mechanics made physicists aware "that no theory can be objective, actually coinciding with nature, but rather that each theory is only a mental picture of phenomena, related to them as sign to designatum" (1899b). "From this it follows," continued Boltzmann, "that we can never find an absolutely correct theory but rather an image that is as simple as possible and that represents phenomena as accurately as possible."

Boltzmann (1899a) disagreed with Hertz on how we "fix" these pictures. Whereas Hertz took the mental pictures to result from laws of thought that were innate in the Kantian sense, Boltzmann considered laws of thought as evolving structures that only seemed to be innate because of their slow evolution in the Darwinian sense. In his 1904 essay, "On Statistical Mechanics," Boltzmann elaborated on this point:[15]

> Our innate laws of thought are indeed the prerequisite for complex experience, but they were not so for the simplest living beings. There they developed slowly, but simple experiences were enough to generate them. They were then bequeathed to more highly organized beings. This explains why they contain synthetic judgments that were acquired by our ancestors but are for us innate and therefore *a priori,* from which it follows that these laws are powerfully compelling but not that they are infallible.

For the most part Boltzmann had only unkind things to say about Kant. For example, to have called the "laws of thought a priori [was] no more than a logical howler of Kant's . . . according to Darwin's theory this howler is perfectly explicable. Only what is certain has become hereditary; what is incorrect has been dropped" (1905b).

Boltzmann agreed with some of Hertz's mental pictures, for example, point masses and coordinate systems. But whereas Hertz demanded that the constructed pictures obey innate laws of thought, Boltzmann required only "that they represent experience simply and appropriately throughout so that this in turn provides precisely the test for the correctness of these laws" (1899a). Consequently, as Boltzmann wrote in his 1897 mechanics book, he considered that a picture of mechanics constructed with the goal of "offering the most definite indications for the purpose of calculation [and] free of vague concepts," was what Hertz meant by requiring that the picture be unambiguous, and thereby coincide with the laws of thought. Boltzmann continued with a statement that captured the style of visual thinking of many German-speaking scientists and engineers circa 1900: "I cannot really imagine any other law of thought than that our pictures should be clearly and unambiguously imaginable" (1897a).

Like Hertz, Boltzmann emphasized that "our ideas of things are never identical with the nature of things. Ideas are only pictures. . . ." At first we "operate only with mental abstractions" and facts of experience are taken into account "only later, after complete exposition of the picture" (1899a). An advantage of the deductive method is that it does not "mix external experience forced on us with internal pictures arbitrarily chosen." However, continued Boltzmann, "it is a genuine mistake of the deductive method that it leaves invisible the path on which the picture in question was reached." His concern here was for the development of abstract notions such as point masses from lower-level internal pictures that had resulted from laws of thought established through the millenia of evolution. We can conjecture that Boltzmann's reason for accepting the "genuine mistake" of an abyss between lower- and higher-level (i.e., more abstract) pictures was the criterion of somehow being able to connect successfully the abstract notions with experience. In later years Einstein would refer to a similarly made distinction between concepts and sense perceptions as the "metaphysical 'original sin',," for which success in formulating descriptions of the "world of immediate perceptions" (1949) was the only justification.

THE METHODS OF BOLTZMANN'S MECHANICS

Boltzmann's deductive account of mechanics differed fundamentally from Hertz's owing to Boltzmann's insistence that mental pictures "represent experience simply and appropriately throughout"—that is, Boltzmann's mental pictures are abstractions from objects actually perceived. Boltzmann began with mental pictures of point masses. Having learned from Hertz that no theory could be completely "objective," Boltzmann replaced Hertz's concealed masses and constraints with the subjectivity of his own mechanics, the notion of point masses interacting through unspecified central forces. These central forces could be chosen so that the "imagined motion of the imagined material point goes over into a true image of actual phenomena" (1899a). As in Book II of Hertz's 1894 *Principles of Mechanics,* Boltzmann defined experiential time as measured on a "perfectly

immutable chronometer" and space as measured relative to a reference system that is "of course nothing real" and which is one of our
"mental pictures" (1897a). The notion of central forces permitted
Boltzmann to determine all masses by "facts of experience" and one
arbitrary mass.[16] Having thus set the foundations Boltzmann went
on in his *Lectures on Mechanics* to develop the entire subject matter of
classical mechanics.

For Boltzmann a bonus of the deductive method was that it rendered superfluous such problems as the "nature of matter, mass and
force" and avoided such problems as absolute space and absolute
motion (1899a). Although these problems could also be circum-
vented by starting with the inductive method, linking up with experience at the outset, "we must determine from experience the
concept of the body's immediate surroundings whose state has influenced its motion." But this brought into the discussion contiguous action, which, wrote Boltzmann, "however *a priori* it may seem
to some, still goes completely beyond the facts and to date remains
well beyond what can be elaborated in detail." If we assume that
contiguous action was the basis of physics, he continued, then we are
left open to an "error that we have laid to the charge of Hertz's mode
of representation: either we should have to invent quite arbitrary
special hypotheses for the way in which contiguous action operates,
or make do with vague general notions about it all." Among the
"quite arbitrary general hypotheses" for an ether that Boltzmann had
in mind was very likely the Lorentz contraction hypothesis which
until 1904 was judged to be ad hoc.

AFTERMATH

By October 1900 Boltzmann realized that despite the impact of mechanical explanation over much of science, "oddly enough it has lost
some ground in its most central field, namely theoretical physics"
(1900). The program of an electromagnetic world-picture was proposed formally by Wilhelm Wien in the December 1900 *Lorentz
Festschrift,* to which Boltzmann had also contributed. Efforts at a
mechanical world-picture paled compared to the advances of those
who subscribed to Lorentz's theory of electromagnetism which took
Maxwell's equations to be axiomatic. Furthermore, the basic submicroscopic constituent of Lorentz's theory had been found empirically in 1897: the electron. In 1904 Boltzmann recalled the "fierce
struggle" (*heftige Kampf*) over whether the emanations in cathode
ray tubes were particles or waves (1905b). It is interesting that

though Walter Kaufmann in Germany and J.J. Thomson in England possessed similar data in 1897 on the emanations in vacuum tubes (in fact, in some ways Kaufmann's were better), the antiatom bias in Germany deterred Kaufmann from concluding that cathode rays were composed of negatively charged submicroscopic particles (see Miller, 1981b).

In suggesting an electromagnetic foundation for physics in 1900, Wien acknowledged the efforts of Maxwell, William Thomson, and Boltzmann toward deducing Maxwell's equations from mechanics and toward pursuing the suggestive analogies between electromagnetism and hydrodynamics and elasticity. Wien went on to mention the difficulty of pursuing Hertz's program to its end as a foundation for electromagnetism, although he hoped that a future electromagnetic foundation for mechanics would measure up to the logical structure of Hertz's developed mechanics.

In (1902–1903) Max Abraham, in Göttingen, supplemented the Maxwell-Lorentz equations with hypotheses concerning the shape and motion of a moving electron and, with certain approximations, succeeded in deducing Newton's second law from Lorentz's force equation. His result for the nonconstant mass of the electron was consistent with available data, and Abraham declared that the goal of an electromagnetic world-picture had been attained. It was Hertz to whom Abraham paid homage. Abraham noted that, although the goal of an electromagnetic world-picture was diametrically opposite to Hertz's, their intents were similar. Abraham quoted from Hertz's mechanics to the effect that rigid connections could be approached only approximately in mechanical systems and so, as Hertz had written in 1894, "we are compelled to seek the ultimate connections in the world of atoms, and they are unknown to us." This was not the case in Abraham's theory of the electron. There, Hertz's rigid connections found their place in a kinematical constraint in which the electron moved like a rigid sphere. In the electromagnetic world-picture, forces were auxiliary concepts (*Hilfsbegriffe*) because they could be traced back to the electron's own velocity and electromagnetic fields. Since the kinetic and potential energies had their source in the electron's own electromagnetic fields, the electromagnetic world-picture went one better than Hertz's mechanics by eliminating even kinetic energy as a fundamental quantity.

Boltzmann welcomed the new developments. In Part II of his lectures on mechanics, published in 1904, there appeared a passage that Boltzmann recalled having written seven years before: "Above all, if one wants to avoid the picture of material points, one should not later introduce them into mechanics after all, but one should start from individuals or elements of different construction, with properties that can be described as clearly as those of material points." After Abraham's theory of the electron was published, Boltzmann could write in Part II of his mechanics lectures: "What I then expected after centuries or even millenia, the half has happened in seven years." And he was especially elated as to how it had happened because, as he wrote, "the ray of hope for a non-mechanical explanation of nature came not from energetics or phenomenology, but from an atomic theory . . . the modern theory of electrons."[17] The electromagnetic world-picture fitted also Boltzmann's view of ultimate explanation of matter in motion where "its simplest fundamental concepts and laws will doubtless remain just as inexplicable as those of mechanics for the mechanical world-picture."

A LEGACY OF BOLTZMANN'S MECHANICS

In the section, "Visual Thinking and Concepts," in Chapter 1, I began to explore Boltzmann's effect on Einstein's thinking toward the 1905 special relativity theory. There, evidence was presented about why we may reasonably assume that Einstein read Volume I of Boltzmann's *Lectures on Mechanics,* and why Boltzmann's "characterization of the method" to set the foundations of mechanics, with emphasis on "mental pictures," held great attraction for someone who thought predominantly in the visual mode. Perhaps before 1905 Einstein had also read Boltzmann's 1897 essay, "On the Question of the Objective Existence of Processes in Inanimate Nature," in which Boltzmann probed the formation of concepts from sense perceptions. Before 1905 Einstein had read at least the introduction to Hertz's *Principles of Mechanics,* which could only have impressed him, as would Boltzmann's mechanics, with the power of an approach to physical theory that emphasized axioms and mental pictures over empirical data. It cannot be overstated that the emphasis on visual thinking among German-speaking scientists and engineers circa 1900 was widespread. Yet in 1905 it was Einstein who combined visual thinking with Gedanken experiments and quasiaesthetic notions with dazzling results. "With the help of certain (imaginary) physical experiments," wrote Einstein in the 1905 relativity paper, he proposed to develop new definitions of time and simultaneity. Einstein chose to develop his theory using a combination of synthetic and deductive methods. His use of concepts ranged from mental pictures, such as ideal measuring rods and clocks or point masses for electrons, to simultaneity and the two basic postulates of his new perspective—the Newtonian principle of relativity applicable to both mechanics and electromagnetism, and the principle that the velocity of light is a determined constant that is independent of the emitter's motion. The two principles functioned for Einstein as Kantian-like organizing principles or concepts that were, however, not unalterable. Thus we may say that Einstein's approach to the foundations of physical theory was in part a combination of Hertz's neo-Kantian framework and Boltzmann's Darwinian view of basic principles. Einstein modified somewhat the realist and deductive schemes of Boltzmann and Hertz by blending together at the outset a suitable mixture of the results of Gedanken experiments (such as magnet and conductor in relative inertial motion), empirical data (such as the results of ether-drift experiments), philosophy (such as an investigation of the notions of time and simultaneity), and theoretical physics. All this led Einstein to conclude that the problems of absolute motion and especially of time could not be dodged by adroit mental play with imaginary reference systems and ideal chronometers, as had been the strategy of Boltzmann and Hertz.

I found the parallel to be striking between Boltzmann's 1897 developments of mechanics and parts of Einstein's kinematics in the 1905 relativity paper. For example, Einstein (1905d) introduced the coordinate system as "three rigid material lines, perpendicular to one another, and originating from a point," and he defined the position of a material point relative to this system by means of rigid measuring rods. In the context of the physics of 1905, which strove for a unified field-theoretical description of matter in motion, the kinematical part of the relativity paper could only have appeared to be inelegant, pedantic, and beside the point. And indeed it was for most physicists. The frontier of physics concerned problems on the nature of the electron, radiation reaction forces, whether particles could travel faster than light, and by mid-1905 Poincaré introduced into electron physics group-theoretical methods and four-dimensional spaces. Did not everyone, therefore, understand what a coordinate system was, and what was meant by position relative to this system? Einstein thought not; and besides, by mid-1905 he realized that neither mechanics, nor electromagnetism, nor thermodynamics could serve as the basis for all of physics.

Perhaps in 1905 Einstein thought privately what he would express in a later publication (1917a) when he attempted to "present the main ideas [of relativity theory] on the whole, in the sequence and connection in which they actually originated": "In the interest of clearness, it appeared that I should repeat myself frequently, without paying the slightest attention to the elegance of the presentation. I adhered scrupulously to the precept of that brilliant theoretical physicist L. Boltzmann, according to whom matters of elegance ought to be left to the tailor and to the cobbler."[18]

The relation of Einstein to Boltzmann and Hertz was best described by a man who had been a student of Boltzmann and became one of Einstein's biographers. Philipp Frank (1947) wrote that from their works Einstein "learned how one builds up the mathematical framework and then with its help constructs the edifice of physics."

In Volume II of his *Vorlesungen über die Principe der Mechanik* (1904) Boltzman wrote, with the intellectual honesty for which he was so admired, that should the mechanical world-picture not succeed, he would be satisfied if his book "contributed to the successful construction of another still more comprehensive and clearer world-picture." Although Einstein never achieved the relativistic world-picture, Boltzmann's foundational analysis may well have aided him in formulating a theory that would set the course of twentieth-century physics and philosophy. That would have satisfied Boltzmann.

CONCLUSION

Ludwig Boltzmann's analysis of fundamental notions in mechanics sets him into the mainsteam of visual thinking in German-speaking countries in the late nineteenth and early twentieth centuries. Although Boltzmann's interest in fundamental matters of physics had always included philosophy, the occasion for his formal entry into philosophy was to defend his notion that atoms are not merely auxiliary quantities against Ernst Mach's phenomenolistic view. Comparison between these two philosophers, both of whom emphasized the importance of sense perceptions, is revealing of their different opinions about the emphasis on mental pictures in setting the foundations of physical theory. For example, whereas they agreed that sense perceptions are the origins of mental pictures, Boltzmann could point to situations where fundamental analysis should start with mental pictures instead of empirical data.

Boltzmann's scientific research determined his unique blend of Mach's phenomenalism and Hertz's idealism, with little inverse reaction from philosophy to science. It is reasonable to conjecture that Boltzmann's philosophical position, and emphasis on mental pictures, may well have affected a budding philosopher-scientist who, by 1905, found himself to be in a similar transitory state between phenomenalism and idealism—namely, Albert Einstein, to whose invention of the special theory of relativity we turn next.

NOTES

(A version of this chapter was presented as a lecture to the International Conference on Ludwig Boltzmann, University of Vienna, 5–8 September 1981, and published in R. Sexl and J. Blackmore (eds.), *Internationale Tagung anlässlich des 75. Jahrestages seines Todes, 5–8 September 1981: Ausgewählte Abhandlungen* (Wiesbaden: Vieweg, 1982). I acknowledge the permission of Vieweg to reprint portions of this material.)

1. Among other historical discussions of Boltzmann's mechanics are Broda (1955), Dugas (1959), and Klein (1970a, 1972).
2. Among the analyses of Boltzmann's researches on statistical mechanics are Brush (1976), Klein (1970a), and Kuhn (1978).
3. Boltzmann's writings on electromagnetism are discussed in Broda (1955) and Klein (1972).
4. For example, in the seminal paper on his version of the Maxwell-Hertz electromagnetic theory, H.A. Lorentz (1892a) gave a footnote citation to Boltzmann's (1891) lectures on electromagnetism. Lorentz wrote that only after having written the treatise had he become aware of Boltzmann's book, whose "principal aim is the mechanical explication inaugurated by Maxwell . . . the problems we had in mind were not the same" (1892a). Lorentz's intellectual debts were to Hertz's reformulation of Maxwell's theory which assumed Maxwell's equations to be axioms, and to Henri Poincaré's 1888 and 1890 lectures at the Sorbonne (Poincaré, 1901). Lorentz formulated an electrodynamics and optics of moving bodies whose principal goal in 1892 was to deduce Fresnel's dragging coefficient. Boltzmann, however, treated only problems concerning the properties of bulk electric and magnetic matter, as well as circuit theory.

     M.J. Klein has aptly written that "an unwary reader glancing through the pages of Boltzmann's book on electromagnetic theory and noting the diagrams might imagine that he had picked up a treatise on the design of engineering mechanisms by mistake" (1972). A widely cited example of Boltzmann's use of mechanics was his doubly cyclic "mechanical analogy" for the mutual induction of two electrical circuits (see figure 2.1). The

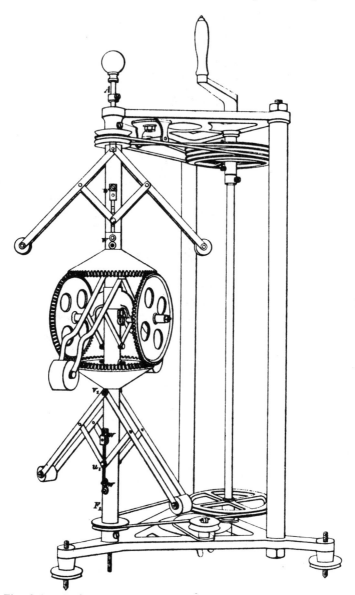

Fig. 2.1.   Boltzmann's ''mechanical analogy'' for the mutual induction of two electrical circuits from his (1891).

mechanical model was constructed and consists basically of two pairs of beveled gears with centrifugal governors. Arnold Sommerfeld, who had heard Boltzmann lecture on electromagnetic theory in 1891 at Munich, many years later in his own text on mechanics utilized Boltzmann's "mechanical analogy for self induction in an exercise on the differential of an automobile to which it is similar in its essential features" (1952).

Boltzmann's method in the second volume of his *Vorlesungen über Maxwells Theorie* for deriving Maxwell's equations differed fundamentally from that in the first volume. He dispensed with pictures of mechanical analogies, emphasizing instead the necessity to distinguish between a mechanical foundation and pictures for illustrating certain consequences of this foundation. Boltzmann proceeded by assuming an ether whose unknown internal motions generated what Faraday had referred to as the "electrotonic state," and which Boltzmann designated by the three components (F, G, H); in modern notation the electrotonic state is related to the vector potential $\vec{A}$. Then Boltzmann defined the "velocity" with which the electrotonic state changed as $\vec{E} = \partial\vec{A}/\partial t$, where in place of Boltzmann's notation I have used $\vec{E}$ to denote that this "velocity" is related to the electric field. Boltzmann next defined the kinetic energy T to be proportional to the magnitude squared of the electric field and the potential energy V to be proportional to the magnitude squared of the magnetic field which he had defined to be $\vec{B} = \vec{\nabla} \times \vec{A}$. Setting up the Lagrangian as $T - V$, Boltzmann used Hamilton's principle in conjunction with Gauss' law for true electricity, to deduce the Maxwell equations for the time variations of the electric and magnetic fields. In summary, Boltzmann's method in the second volume of his lectures on electromagnetism for deriving Maxwell's equations was to set up formal analogies between mechanical and electromagnetic quantities. He went on to claim that this procedure could serve also to sharpen Hertz's method which took Maxwell's equations as axiomatic. Later in the book, Boltzmann developed magnetic phenomena in a manner analogous to electric phenomena, and so he came naturally to Hertz's expression for true magnetism in which $\vec{\nabla} \cdot \vec{B} = m_{\text{true}}$, where m_{true} is the density of true magnetism. But, continued Boltzmann, he had recently shown that every known characteristic of a conducting magnet could be deduced without the notion of true magnetism (1893b). Thus,

Boltzmann wrote in his book, although the Hertzian viewpoint was "without doubt correct" it provided "scarcely a clear insight into the inner connections." In order to avoid the notion of true magnetism, Boltzmann proceeded to investigate such magnetic phenomena as the interaction between two current-carrying solenoids by returning to the *"mechanische Anschauung"* [mechanical intuition] of the doubly cyclic mechanical analogy described above from Volume I.

Subsequent to publication of Volume II, at the 1893 *Versammlung deutscher Naturforscher und Ärzte* at Nürnberg, Boltzmann noted Maxwell's statement to the effect that "infinitely many mechanical hypotheses are possible" (1893c). We note that despite the title of Boltzmann's work (1893c), Lorentz's 1892 theory of electromagnetism was nowhere mentioned, most likely because in Lorentz's theory the Maxwell-Hertz equations are axioms. Boltzmann was aware of Lorentz's theory because he had cited it in the Volume II of his lectures on electromagnetism. Boltzmann went on to divide the possible mechanical theories of electromagnetism into two groups: analogies of electromagnetism with elasticity and hydrodynamics, and mechanical models such as the one proposed originally by Maxwell in 1861–1862 in which the ether was simulated with vortices and rolling particles; he then surveyed the state of the art. The latter approach was the one that appealed to Boltzmann because he considered it to be the surest guideline toward a successful theory. He emphatically reminded physicists "not to permit to grow cold the zeal for the search for mechanical theories that are of great value for the intuition [*Anschauung*] as well as for the discovery of new facts; what this method has proven already is that Maxwell found his fundamental equations through his first mechanical picture."

Poincaré had already emphasized in 1888 that *"Maxwell did not give a mechanical explanation of electricity and magnetism; rather, he restricted himself to demonstrating its possibility"* (italics in original, 1901). Poincaré went on to give a brilliant proof that we may conjecture affected Lorentz's thinking toward formulating his own electromagnetic theory of 1892; namely, Poincaré proved that any system satisfying a principle of least action was open to infinitely many mechanical explanations. Consequently, in Poincaré's view the quest for a mechanical foundation for electromagnetism was futile. Boltzmann thought otherwise, for although the models and pictures could be eventually removed, as had been the case for Maxwell and Hertz, the formal mechanical analogies would remain.

In (1892a) Lorentz used variational principles in a manner similar to Boltzmann's in 1893. However, Lorentz's intent differed from Boltzmann's because Lorentz assumed the fundamental electromagnetic field equations to be axioms. Lorentz sought only to show that the electromagnetic field formalism could be put into a form resembling D'Alembert's principle. This being the case, then his theory agreed with the conservation of energy. In his subsequent major writings on electromagnetic theory, Lorentz dispensed entirely with variational principles of mechanics and, in fact, had only critical comments on a mechanical world-picture. Thus, incorrect is McCormmach's (1970) statement that Lorentz proposed "six hypotheses sufficient for a mechanical derivation of the field equations" [see also Miller (1973) on Lorentz's supposed mechanical models].

5. Other writings on Boltzmann's philosophical viewpoint are Blackmore (1972), Broda (1955), Dugas (1959), and Hiebert (1980).

6. For example, see Blackmore (1972) for evidence in support of the conjecture for Mach and Ostwald. The case for Hertz is developed in this chapter.

7. There is some difference of opinion over Hertz's phrase *"innere Scheinbilder oder Symbole."* In his useful Introduction to Hertz's book, R.S. Cohen notes that Jones and Walley inappropriately translated this phrase as "images or symbols." Cohen renders it as "subjective illusory images" or "inner phantoms." In my opinion the former translation takes poetic license with Hertz's prose; the latter conforms too closely to the dictionary meaning of *das Scheinbild* (illusion, phantom, chimera), while omitting the words "or symbols"; and neither rendering is fully consistent with the passage's context. I concur with Braithwaite's (1968) translation as "internal pictures or symbols" since it squares with the thrust of Hertz's development of mental pictures of objects in the world of perceptions into their scientific representations. Had Hertz the opportunity for revisions (he wrote the book during the last three years of his life, while terminally ill), he most probably would have chosen better terminology for this key point.

8. On the Newtonian image, Hertz was basically critical of the

conception of force which was described by Newton's third law
in a sense different from the other two laws. He deftly illustrated
this point using the description by Newtonian physics of the
motion of a stone attached to a string that is swung around in a
circle. The second law describes the motion to be caused by the
force exerted on the stone by the string. But Newton's third law
requires that the stone exerts an equal and opposite force on the
hand, which is the "centrifugal force" (Hertz implicitly assumed
that the string is massless—see Poincaré (1897) for a more complete critique of Newton's "image"). The centrifugal force is,
however, just the "inertia of the stone." Thus, continued Hertz,
have we not taken the "effect of inertia twice into account—first
as mass, secondly as force?" Hertz deemed his criticisms of the
second image, energetics, to be devastating: whereas kinetic energy could be associated with a system's basic properties of mass
and velocity, potential energy did "not lend itself at all well to
any definition which ascribes to it the properties of substance."
Besides the fact that the potential energy need not be positive, it
is not even uniquely defined owing to the possibility of adding a
constant to it. Furthermore, the situation could not be alleviated
through recourse to force, because force had no place in energetics as a basic quantity. Next, emphasized Hertz, Hamilton's
principle of least action is not applicable to systems with nonholonomic constraints.

9. Hertz assumed that the Newtonian image could be reformulated
   so as to agree with his requirements of permissibility and appropriateness. Consequently, the struggle between the Newtonian
   action-at-a-distance formulation and Hertz's image would be
   decided by which one could better explain empirical data.

10. The fundamental law includes holonomic and nonholonomic
    constraints. Hertz took kinetic energy to be a fundamental quan-
    tity arising from the motion of a system's visible masses, while
    potential energy has its origin in the motion of the system's
    concealed masses. Thus, in Hertz's mechanics potential energy is
    kinetic in origin.

11. In 1898 Boltzmann noted the complications that arise for treat-
    ing with Hertz's mechanics the simple case of elastic impact
    between two spheres; consequently, with this case in mind
    Boltzmann concluded that "in spite of all its philosophical
    beauty and completeness any meaning [of Hertz's mechanics] for
    physics would be very reduced" (1898).

12. For example, Mach had written [(1889) p. 463; pp. 588–589 of English version]: "Atoms cannot be perceived by the senses; like all substances they are things of thought. Furthermore, the atoms are invested with properties that absolutely contradict the attributes hitherto observed in bodies."

13. As an example of a concept Boltzmann gave Faraday's explanation of Arago's experiment by means of induction currents. Arago demonstrated the effect of a magnet on a nearby rotating copper wheel. In this case, according to Boltzmann, T is the notion of an induction current; the set of experiments A_1, A_2, A_3, . . . concern relative motion between magnet and conductor, and Arago's experiment would be the A_x which is explained because it has the part T in common with the other experiments.

14. The realist view, continued Boltzmann, compares the assertion "that one could never imagine how the mental could be represented by the physical let alone the interaction of atoms with the opinion of an uneducated person who says the sun could not be 93 million miles from the earth, since he cannot imagine it."

15. Boltzmann much admired Darwin's work. He described it as the "most splendid mechanical theory in the field of biology. . . . This undertakes to explain the whole multiplicity of plants and the animal kingdom from the purely mechanical principle of heredity, which like all mechanical principles of genesis remains of course obscure" (1905b). As Merz wrote of the nineteenth century, "Germany may be said to have produced *Darwinismus* in this century as France created *Newtonianisme* in the last" (1965).

16. Boltzmann noted that his definition of mass was "due to Mach" [(1889), pp. 203–204; pp. 286–287 of English version]. By experience Boltzmann meant, for example, that there are only positive masses, and that we can measure acceleration (assuming an absoluteness of time).

17. But in his 1904 paper "On Statistical Mechanics" Boltzmann was careful to point out that despite the many successes of the "atomistic theory of the whole of electricity [it] cannot resolve the question as to the limited or infinite divisibility of matter." This problem could be best probed by examining the "foundation of concepts itself . . . The question whether matter is atomistically constituted or continuous therefore reduces to the question: Which represents the observed properties of matter most accurately, the properties on the assumption of an extremely

large finite number of particles, or the limit of the properties if the number grows infinitey large?" Although phrasing the problem thus, continued Boltzmann, does not reply to the old philosophical problem of whether matter is continuous or not, at least it offered a means to avoid the corresponding Kantian antinomy, which earlier in this essay Boltzmann had criticized as merely obfuscating the problem through a logic that relies excessively on "so-called laws of thought." Boltzmann's subsequent recourse to an analysis of concepts involved turning to the statistical interpretation of the "so-called second law of the mechanical theory of heat." In particular, he turned to the statement that a closed system tends toward a state for which its entropy is at a maximum. This law, despite its idealization, when combined with other laws of thermodynamics "always gives the right results." Consequently, concluded Boltzman, further "reflection and research" using statistical mechanics could decide the issue in favor of matter being discontinuous.

18. This opinion of Boltzmann was widely held; for example, Arnold Sommerfeld described Boltzmann's method of setting the foundations of mechanics thus (1944): "There is the standpoint of the axiomatists, who shun no torrent of details in order to clarify fundamental concepts. . . . Compared with the systematic demonstrations of sublime simplicity of Lagrange's mécanique celeste or Hertz's mechanics, Boltzmann's method requires hard work and little elegance. But, as he was supposed to have said, elegance is good for the tailor."

PART II

On Aesthetics, Visualizability, and the Transformation of Scientific Concepts

CHAPTER 3

The Special Theory of Relativity: Einstein's Response to the Physics and Philosophy of 1905

There is no inductive method which could lead to the fundamental concepts of physics. Failure to understand this fact constituted the basic philosophical error of so many investigators of the nineteenth century.

A. Einstein (1936)

I think that only daring speculation can lead us further and not accumulation of facts.

Letter of A. Einstein to M. Besso of 8 October 1952 (Einstein, 1972)

Einstein at the first summit conference of physicists, the 1911 Solvay Conference. Seated (left to right): W. Nernst, L. Brillouin, E. Solvay, H.A. Lorentz, E. Warburg, J. Perrin, W. Wien, Mme. M. Curie, H. Poincaré. Standing (left to right): R. Goldschmidt, M. Planck, H. Rubens, A. Sommerfeld, F. Lindemann, M. de Broglie, M. Knudsen, F. Hasenöhrl, G. Hostelt, E. Herzen, J. Jeans, E. Rutherford, H. Kammerlingh Onnes, A. Einstein, P. Langevin. (Courtesy of the Institut International de Physique Solvay and the AIP Niels Bohr Library.)

IMAGINE THAT YOU are on the editorial board of a prestigious physics journal and that you receive a paper that is unorthodox in style and format. Its title has little to do with most of its content; it has no citations to current literature; a significant portion of its first half seems to be philosophical banter on the nature of certain basic physical concepts taken for granted by everyone; the only experiment explicitly discussed could be explained adequately using current physical theory and is not considered to be of fundamental importance. Yet, with a minimum of mathematics, the little-known author deduces exactly a result that has heretofore required several drastic approximations. Furthermore, you are struck by certain of the author's general principles, and you feel that they promise additional simplifications. So you decide to publish the paper. This could well have been the frame of mind of the most eminent theoretical physicist on the Curatorium of the *Annalen der Physik*, Max Planck, when he received from the editor's office Albert Einstein's 1905 paper "On the Electrodynamics of Moving Bodies"—the relativity paper.[1]

The kind of title Einstein had given his paper customarily signaled a discussion of the properties of moving bulk matter, either magnetic or dielectric. Einstein analyzed neither of these topics. In fact, the paper's first quarter contains a philosophical analysis of the notions of time and length. The paper's second half dispatches quickly certain problems of such fundamental importance that they generally rate separate papers—for example, the characteristics of radiation reflected from a moving mirror—and he concludes with certain results from the dynamics of electrons that generally appear at the beginning of papers in which electrons are discussed. The only experiment developed in detail is at the paper's beginning and concerns the generation of current in a closed circuit as a result of the circuit's motion relative to a magnet, that is, electromagnetic induction.

The phenomenon of electromagnetic induction had ushered the Western world into the age of technology because it is fundamental to electrical dynamos. Everyone knew dynamos worked, but there remained fundamental problems concerning their operation. This chapter discusses the connection that Einstein realized in 1905 between problems concerning huge electrical dynamos, radiation, moving electrical bulk media, the dynamics of electrons, and the nature of space and time. In order to set the stage for Einstein's bold aproach to the physics of 1905, let us review the treatment of these topics by scientists and philosophers of whom Einstein has acknowledged he was aware before 1905.[2]

THE NATURE OF SPACE AND TIME

In his *Science of Mechanics* (1883) the philosopher-scientist Ernst Mach leveled a devastating critique at the Newtonian notions of absolute space and time. According to Newton, absolute space was the ultimate receptacle in which all phenomena occurred, and absolute time flowed independent of the motion of clocks. Mach considered these notions to be "metaphysical obscurities" because they were unavailable to our sense perceptions. Consequently, Mach disagreed with Immanuel Kant, who by 1781 had elevated Newton's notions of absolute space and time to knowledge that we possessed before all else, that is, a priori intuitions. According to Kant, these intuitions serve as basic organizing principles that enable our minds to construct knowledge from the potpourri of sense perceptions. Thus, for example, we are driven irresistibly toward a three-dimensional Euclidean geometry and the law of causality, and then to such higher-order organizing principles as Newton's physics. Although the discovery of non-Euclidean geometries in 1827 had dealt the Kantian view a serious blow, Kant's emphasis on the role of a priori organizing principles was nevertheless considered important to an understanding of how exact laws of nature are possible. A priori organizing principles played an important role in the neo-Kantian frameworks of such influential philosopher-scientists as Hermann von Helmholtz, Heinrich Hertz, and Henri Poincaré, whose writings impressed Einstein no less than did Mach's "incorruptible skepticism" (Einstein, 1946).

Although Mach and Poincaré probed the relation between time and sense perceptions, in their work there remained an absoluteness of time because there was no reason for it to depend on motion. But Mach and Poincaré insisted on replacing motion relative to Newton's absolute space with motion relative to the distant stars or, even better, motion relative to the substance that electrodynamicists assumed to fill Newton's cosmic receptacle—the ether. This brings us to electromagnetism.

ELECTROMAGNETIC THEORY

Newton's mechanics of 1687 had unified terrestrial and extraterrestrial phenomena. The next great synthesis occurred not quite two hundred years later, when James Clerk Maxwell unified electromagnetism and optics. Whereas in the Newtonian mechanics disturbances propagated instantaneously through empty space, in Maxwell's theory disturbances propagated at a large but finite velocity through an ether, like ripples in a pond. In 1892 there appeared the result of over two decades of elaborations and purifications of Maxwell's theory—the electromagnetic theory of that master of theoretical physics, Hendrik Antoon Lorentz.

Lorentz (1892a) assumed that the sources of the electromagnetic fields were as yet undiscovered electrons, which moved about in an all-pervasive, absolutely resting ether. The five fundamental equations of Lorentz's theory are:[3]

$$\vec{\nabla} \times \vec{E} = -\frac{1}{c}\frac{\partial \vec{B}}{\partial t} \tag{1}$$

$$\vec{\nabla} \times \vec{B} = \frac{1}{c}\frac{\partial \vec{E}}{\partial t} + \frac{4\pi}{c}\rho\vec{v} \qquad \text{Maxwell-Lorentz} \tag{2}$$
$$\text{Equations}$$

$$\vec{\nabla} \cdot \vec{E} = 4\pi\rho \tag{3}$$

$$\vec{\nabla} \cdot \vec{B} = 0 \tag{4}$$

$$\vec{F} = \rho\vec{E} + \rho\frac{\vec{v}}{c} \times \vec{B} \qquad \text{Lorentz Force equation} \tag{5}$$

where $\vec{E}$ and $\vec{B}$ are the electric and magnetic fields, respectively, and ρ is the electron's volume density of charge. Since Lorentz's fundamental equations are written relative to a reference system at rest in the ether, which we shall call S, then c is the velocity of light measured in S, and $\vec{v}$ is the electron's velocity relative to S. The Maxwell-Lorentz equations possess the property expected of a wave theory of light, namely, that relative to S the velocity of light is independent of the source's motion and is always c. But this may not necessarily be the result of measuring the velocity of light in a reference system moving with a uniform linear velocity relative to the ether, that is, in an inertial reference system. Therefore, the reference systems in the ether are preferred reference systems. To be sure, despite much effort, experiments had not revealed that the earth's motion through the ether had any effect on optical or electromagnetic phenomena.

Concerning the velocity of light, Newtonian mechanics predicted that the velocity of light emitted from a moving source should differ from the velocity of the light emitted from a source at rest by the amount of the source's velocity; consequently, the velocity of light c' from a source moving with velocity v is given by Newton's law for the addition of velocities,

$$\vec{c}' = \vec{c} + \vec{v}. \tag{6}$$

On the other hand, according to the wave theory of light, the quantity c' measured by an observer at rest in the ether is

$$\vec{c}' = \vec{c} \tag{7}$$

and Lorentz's equations agreed with this requirement. But the effect of the ether on the measuring apparatus was expected to yield a result in agreement with Eq. (6), where c' is the velocity of the light relative to the earth and v is the ether's velocity relative to the earth. However, experiments accurate to second-order in the ratio v/c, where v is the velocity of a body that is moving relative to the ether led to Eq. (7). To this order of accuracy, optical and electromagnetic phenomena occurred on the moving earth as if the earth were at rest in the ether. Therefore, to second-order accuracy in v/c, Newtonian mechanics and electromagnetism are inconsistent with optical phenomena occurring in inertial reference systems.

In an 1895 monograph titled *Treatise on a Theory of Electrical and Optical Phenomena in Moving Bodies,* Lorentz responded fully to the failure of the first-order experiments to detect any effects of the earth's motion on optical and electromagnetic phenomena; these experiments were called ether-drift experiments. For regions of the ether that are free of matter, or within neutral matter that is neither magnetic nor dielectric, the Lorentz equations in the ether-fixed reference system are the set of equations (S) (see Fig. 3.1). Applying the modified space and time transformations to (S), Lorentz obtained their analogues in the inertial reference system S_r.

We can appreciate Lorentz's achievement at a glance because to first-order accuracy in the quantity v/c, the Maxwell-Lorentz equations have the same form in the inertial system S_r as in the ether-fixed system S, and thus the same physical laws pertain to both these reference systems; in other words, to this order of accuracy neither optical nor electromagnetic experiments could reveal the motion of the system S_r. Lorentz called this stunning and desirable result the "theorem of corresponding states." It rested on the hypothesis of the mathematical "local time coordinate" t_L; the real or physical time was still Newton's absolute time. Hence, to order v/c, the velocity of light in S_r was the same as in S, that is, $c' = c$. To this order of accuracy, then, Lorentz's theorem of corresponding states removed

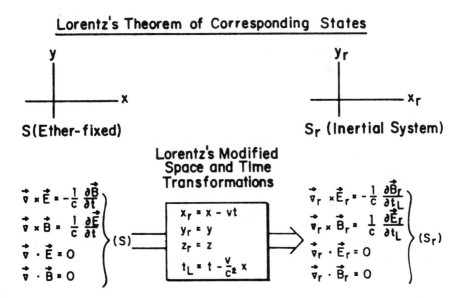

Lorentz's Theorem of Corresponding States

Fig. 3.1. Lorentz's modified space and time transformations contain the local time coordinate, t_L, and Lorentz referred to the electromagnetic field quantities $\vec{E}_r = \vec{E} + \vec{v}/c \times \vec{B}$ and $\vec{B}_r = \vec{B} - \vec{v}/c \times \vec{E}$ as "new" vectors.

the inconsistency between Newton's prediction and that of electromagnetic theory in favor of electromagnetic theory.

But Lorentz had not yet explained the only reliable experiment accurate to second-order accuracy in v/c, namely, the 1887 experiment of Albert A. Michelson and Edward Williams Morley, in which light had been found to take the same time to race back and forth along each of two orthogonal rods of equal length that were at rest on the moving earth. He discussed this experiment in the final chapter of the 1895 treatise, which contained two other experiments that, as the chapter's title indicated, could "not be explained without further ado." In order to explain the Michelson-Morley experiment, Lorentz indulged in a physics of desperation (1892b). From Newton's law for the addition of velocities, which the theorem of corresponding states was supposed to have obviated, he proposed the hypothesis that the dimensions of the rod in the direction of the earth's motion contracted by an amount $\sqrt{(1 - v^2/c^2)}$. In short, Lorentz's contraction hypothesis was admittedly ad hoc. This blemish on Lorentz's theory was emphasized in the philosophic-scientific criticism of Henri Poincaré. Nevertheless, Poincaré was impressed with Lorentz's theorem of corresponding states because it could explain systematically a number of experiments.

So successful had been Newton's physics that many scientists had attempted to reduce all of physical theory to it; that is, they pursued a mechanical world-picture. For example, they attempted to simulate the contiguous actions of the ether with increasingly complex mechanical models. But these attempts paled before the successes of Lorentz's theory. Thus, in 1900 Wilhelm Wien suggested the "possibility of an electromagnetic foundation for mechanics," that is, pursuance of an electromagnetic world-picture based on Lorentz's electromagnetic theory. A far-reaching implication of this program was that the electron's mass originated in its own electromagnetic field and should therefore be a velocity-dependent quantity. From studying the behavior of fast electrons that had been injected transversely into parallel electric and magnetic fields, Walter Kaufmann gave data for a dependence of the electron's mass on its velocity that increased without limit as the electron's velocity approached that of light. Kaufmann's colleague at Göttingen, Max Abraham, developed the first field-theoretical description of an elementary particle. Depending on whether his rigid-sphere electron experienced a force transverse or parallel to its motion, and with certain severe restrictions placed on the electron's acceleration, Abraham predicted that it had transverse (m_T) and longitudinal (m_L) masses (1902a,b; 1903):

$$m_T = \frac{m_0^e}{2\beta^3}\left[(1 + \beta^2)\log\left(\frac{1 + \beta}{1 - \beta}\right) - 2\beta\right] \tag{8}$$

$$m_L = \frac{m_0^e}{\beta^3}\left[\frac{2\beta}{1 - \beta^2} - \log\left(\frac{1 + \beta}{1 - \beta}\right)\right] \tag{9}$$

where $m_0^e = e^2/2Rc^2$ is the electron's electrostatic (that is, rest) mass and $\beta = v/c$. The transverse mass of Abraham's theory agreed with Kaufmann's data, and the goal of an electromagnetic world-picture appeared to be within reach. However, Abraham's theory offered no explanation for the Michelson-Morley experiment, and by 1904 it was in violent disagreement with the new optical experiments of Lord Rayleigh (1902) and D.B. Brace (1904) which were accurate to second-order in v/c.

Prompted by the new second-order data and by Kaufmann's measurements, as well as by Poincaré's criticisms, Lorentz proposed his own theory of the electron in which the contraction hypothesis was deemed no longer to be ad hoc, because it became one of several hypotheses that could explain more than one experiment. Lorentz's electron can be likened to a balloon smeared with a uniform distribution of charge. While at rest, Lorentz's electron is assumed to be a sphere; but moving, it undergoes a Lorentz contraction, and its mass becomes a two-component quantity (1904b):

$$m_T = \frac{4}{3}\frac{m_0^e}{\sqrt{1 - \beta^2}} \tag{10}$$

$$m_L = \frac{4}{3}\frac{m_0^e}{(1 - \beta^2)^{3/2}} \tag{11}$$

where $m_0^e = e^2/2Rc^2$ is the electron's electrostatic (that is, rest) mass and $\beta = v/c$.

Lorentz's m_T agreed with Kaufmann's data as well as did Abraham's. But Abraham immediately leveled a severe fundamental criticism at Lorentz's theory (1904a): Lorentz's deformable electron was unstable because it would explode under the enormous repulsive forces among its constituent parts. Among the Lorentz-Poincaré correspondence that I recently discovered, I determined that Poincaré had recognized this problem independently and then cracked it with his unmatched arsenal of mathematics.[4] Poincaré's resulting papers were the ultimate effort toward an electromagnetic world-picture based on Lorentz's electromagnetic theory. They included such advanced notions of mathematics as group-theoretical methods and four-dimensional spaces. Using a term familiar from fundamental studies in geometry, Poincaré renamed Lorentz's 1904 theorem of corresponding states that embraced all extant data—and, it was *hoped,* future ones as well—the "principle of relativity" (1904a). Einstein, we know, had not encountered Poincaré's 1905 version of Lorentz's theory of the electron when he wrote the relativity paper.

To summarize, by 1905 physicists believed that electromagnetic theory was proceeding in the correct direction. Many of them felt sure that with a little more tinkering, Lorentz's theory of the electron could serve as the cornerstone for a unified field-theoretical view of nature. Lorentz's was a dynamical theory that *explained* such effects as the presumed contraction of length, the observed variation of mass with velocity, and the fact that the measured velocity of light always turned out to be the same—all explained as resulting from the interaction of electrons with the ether. The stage was set for a great new era in science to emerge from what everyone considered to be the cutting edge of scientific research. But, as we shall see, this turned out not to be the case. We move next to an area of science and engineering whose basic problems were deemed unimportant for progress in basic physical theory: the area of electromagnetic induction. In German-speaking countries problems in this area combined technology and basic research. They received a particularly interesting treatment because, as the intellectual historian J.T. Merz has written, the "German man of science was a philosopher" (1965).

ELECTRICAL DYNAMOS

In 1831 Michael Faraday discovered that relative motion between a wire loop and a magnet produces a current in the wire. Faraday interpreted this result as follows: the magnet affects the wire loop through its lines of force, which emanate from the magnet's north pole and enter through its south pole; consequently, relative motion between the loop and the magnet results in the loop's cutting the lines of force. Faraday's law states that the rate at which the lines of force are cut determines the strength of the current induced; furthermore, the direction and magnitude of the induced current depend on only the relative velocity between the loop and magnet.

But Faraday's interpretation of electromagnetic induction differed when circuit and magnet were rotating relative to each other. An apparatus of the sort in Fig. 3.2 was important to Faraday because by

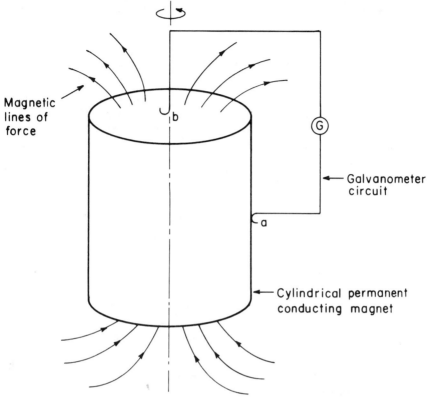

Fig. 3.2. An example of a unipolar dynamo.

1851 he had convinced himself that lines of force participated in the magnet's linear motion but not in its rotation. In Fig. 3.2 the wire loop makes sliding contact with the rotating magnet's periphery and touches one of the magnet's poles; hence, this sort of electromagnetic dynamo became known as a unipolar dynamo. Faraday's interpretation of unipolar induction is: if the loop rotates counterclockwise, a current appears in it owing to its cutting the lines of force. If the loop remains at rest, and the magnet rotates clockwise, a current of the same magnitude and direction appears in the loop; the current was thought to originate in the magnet owing to the magnet's rotating through its lines of force. When magnet and loop turn together in the same direction, there is no net current in the loop because the loop's current is canceled by the magnet's internal current. Thus, two different explanations were required for the current induced in the loop, depending on whether the loop or the magnet rotated.[5] Clearly, the experimental data were more easily understood in terms of relative motion between the wire loop and the magnet with its co-moving lines of force.

Representing the magnetic field by lines of force found fertile ground in the German-speaking countries. Although by the end of the nineteenth century most British scientists considered lines of force to be useful chiefly for pedagogy, the Kantian philosophical position of scientists and engineers in German-speaking countries led them to consider lines of force as a fundamental *Anschauung*. In this context *Anschauung* refers to the intuition through pictures formed in the mind's eye from previous visualizations of physical processes in the world of perceptions; *Anschauung* is superior to viewing merely with senses. In short, lines of force were seen everywhere. There ensued a controversy in the German-speaking scientific-engineering community on the merits of the *Anschauungen* of Faraday versus the rotating-line view, and experiments were offered to distinguish between them. One engineer emphasized that this controversy was "not an academic moot point" for two chief reasons. First, there was intense research and development toward unipolar direct-current and alternating-current generators owing to the relative simplicity of these machines vis-à-vis multipolar machines. Second, calculating the effect of a rotating magnet on stationary armature coils required knowing whether the magnetic field lines remained stationary; this problem appeared also in multipolar machines with stationary field magnets (Weber, 1895).

On the theoretical side, the problem of how current arises in unipolar induction led some physicists to question the universal validity of Faraday's law of induction. Faraday's law was taken to be valid for closed wire circuits. The problem was how Faraday's law could explain the current that arose in the open wire in Figure 3.2 as a result of slipping between the magnet's surface and the wire. This was a complicated boundary-value problem because it concerned assumptions on how electromagnetic quantities varied in the transition layer (that is, the surface of slip) between the spinning magnet and the ether. The widely-read texts of August Föppl (1894) and the 1904 edition rewritten by Max Abraham emphasized the relation between the laws of mechanics and electromagnetism in the explanation of electromagnetic induction: the laws of mechanics are used to discuss the motion of circuit and magnet; the current in the loop depends on only the relative velocity between loop and magnet, and this result agrees with the principle of relative motion from mechanics. This principle asserts that physical phenomena occur as a result of the relative motion of material bodies and that the laws of mechanics are the same for all inertial systems. Abraham stressed that experiments supported the principle of relative motion for Lorentz's theory only to first-order accuracy in v/c. Whereas Föppl's use of complicated notions of surfaces of slip allowed him to express only faith in the validity of Faraday's law, Abraham avoided any problems of unipolar induction by adhering strictly to Hertz's electromagnetic theory, in which the principle of relative motion held exactly. Then he calculated the current for the one case of the loop turning where Faraday's law could be applied with its line-cutting interpretation; the inverse case followed by replacing v with $-v$ in Faraday's law.

In Abraham's view, only the induced current, and not the lines of force, was measurable; that is, the equations of electromagnetic theory dictated what was to be observed, and lines of force were merely auxiliary quantities. In (1904a) Lorentz enlarged his electromagnetic theory to describe moving bulk magnetic and dielectric matter, and his results agreed with empirical data better than did Hertz's. Lorentz's extension required his postulating three different sorts of electrons: conduction electrons responsible for electric current; polarization electrons that produce dielectric properties; and magnetization electrons to explain magnetic properties of matter. Using the local time coordinate, Lorentz showed that the effect of a moving magnet on a stationary current loop—that is, the electric field due to the moving magnet—had its origin in the moving magnet developing dielectric properties. But this explanation for electromagnetic induction involved a complicated intermingling of two different kinds of electrons in a substance that was assumed to possess only conduction and magnetization electrons. Lorentz himself found this explanation puzzling (1910).

Besides appending the hypotheses of additional electrons to an already overburdened superstructure, Lorentz by the end of 1904 was still unable to explain the stability of his electron. Then there was Max Planck's 1900 explanation for the radiation from hot substances, which brings us to the final ingredient necessary for describing Einstein's unique view of the physics of 1905—problems concerning radiation.

RADIATION

Since the late nineteenth century, the light emitted from a cavity within a hot substance had fascinated scientists because its characteristics are independent of the substance's constitution. To his horror,

Planck found that the successful empirical formula he offered to describe cavity radiation required that energy be exchanged in only discrete amounts between the cavity radiation and the hot substance's constituent electrons (1900, 1901).[6] Yet, according to electromagnetic theory, processes involving radiation should be continuous. Furthermore, if only certain discontinuous exchanges of energy were allowed, then the constituent electrons' states of motion were similarly restricted. Consequently, Planck's formula violated both mechanics and electromagnetism. Einstein recalled (1946) that in 1904 this situation was a "second fundamental crisis" (in addition to the instability of Lorentz's electron). He was alone in this assessment.

EINSTEIN'S VIEW OF PHYSICAL THEORY

Einstein's opinion of Planck's theory of cavity radiation received confirmation from the application of his 1904 results on the behavior of atoms in gases—that is, statistical mechanics—to Planck's radiation formula. He found that light exhibited particulate properties that cannot be described by Lorentz's theory (see Chapter 1, Note 22). Thus, current electromagnetic theory was insufficient for discussing the nature of the electron. In addition, his research on particles in solution (Brownian motion) convinced him of the insufficiency of thermodynamics and mechanics in microscopic volumes. These two results led Einstein to conclude that scientists of 1905 were "out of [their] depth" (1923); neither the mechanical nor the electromagnetic world-picture could succeed. Einstein's first two papers in the now famous volume 17 of *Annalen der Physik* explored the consequences of particles of light, that is, light quanta (1905b) and Brownian motion (1905c). In the third paper of that 1905 trilogy— the relativity paper—Einstein again dealt with the characteristics of light.[7]

Einstein's studies of the great philosopher-scientists von Helmholtz, Hertz, and Poincaré had shown him the power of a fundamental analysis within a neo-Kantian framework. But when it came to what to analyze and where to begin anew, he was in a quandary. Theories based on assumptions concerning the constitution of matter were inadequate for a fundamental analysis. At this point, Einstein was aided by his predilection for visual thinking, which was reinforced when he encountered the writings of Ludwig Boltzmann and von Helmholtz, in particular, at the Zurich Polytechnic Institute, or Eidgenössische Technische Hochschule (ETH) in Zurich. Before

Einstein, however, no one had combined visual thinking so effectively with the thought-experiment, that is, an experiment capable of being performed in the mind.

In 1895 Einstein had conceived of a thought experiment that, after ten years of obstinate pondering, finally revealed the "germ of the special relativity theory."★ The essence of this experiment is: (1) Current physics asserts that an observer who is moving alongside a light wave whose source is in the ether should be able to discern the effects of his motion by, for example, measuring the velocity of light. (2) To Einstein it was "intuitively clear" that the laws of optics could not depend on the state of the observer's motion. Statements (1) and (2) are mutually contradictory, and to Einstein this thought-experiment contained a paradox (1946).

In their own ways, Lorentz and Poincaré were also attempting to resolve this paradox—by one degree of accuracy at a time, that is— by proposing, as we saw, hypotheses such as the local time and the ad hoc contraction of moving bodies. But Einstein's 1904 results on the nature of light had convinced him that theories of matter could lead neither to exact explanations for ether-drift experiments nor to a consistent world-picture.

I turn next to the basic scientific-philosophic considerations that led Einstein to discover that the key to the paradox lay, as he said later, in the "axiom of the absolute character of time, viz., of simultaneity, [which] unrecognizedly was anchored in the unconscious" (1946). My research, based on Einstein's writings, has resulted in the following reconstruction of Einstein's thinking toward this momentous invention (1981b). Of all the ether-drift experiments that had been performed, those accurate to first order in v/c were explainable by Lorentz's theorem of corresponding states, which was based on the set of modified space and time transformations that included the mathematical local time coordinate (see Fig. 3.3b). Although the Maxwell-Lorentz equations remained unchanged under the modified space and time transformations of Fig. 3.3b, the same was not true of the equations of mechanics that were transformed between inertial reference systems according to the Galilean transformations (Fig. 3.3a). Thus, according to Lorentz's modified transformations, the laws of mechanics were not the same in every inertial reference system. But this result violated Newton's exact principle of relative

---

★*N.B.* See the appendix to this chapter for further development of this experiment.

Galilean Transformations	**Lorentz's Modified Space and Time Transformations**
$x_r = x - vt$	$x_r = x - vt$
$y_r = y$	$y_r = y$
$z_r = z$	$z_r = z$
$t_r = t$	$t_L = t - \dfrac{v}{c^2}\, x$
(a)	**(b)**

Fig. 3.3.   (a) The coordinates (x_r, y_r, z_r, t_r) and (x, y, z, t) refer to the two *inertial* reference systems S_r and S. (b) The coordinates $(x_r, y_r, z_r, t_r = t)$ and (x, y, z, t) refer to the inertial reference system S_r and to the ether-fixed reference system S; and t_L is the mathematical "local time coordinate."

motion, which states that no mechanical experiment could reveal an inertial system's motion. The mathematical statement of Newton's principle of relative motion lies in the Galilean transformations of Fig. 3.3a, where for mechanics the reference systems S_r and S are *both* inertial reference systems. Consequently, the transformation rules for the laws of mechanics and of electromagnetism depended on two different notions of time—one physical and the other mathematical—contrary to Lorentz's goal of unification. Thus, whereas most scientists in 1905 considered the tension between mechanics and electromagnetism to be rooted in the inability of mechanics to explain the measured velocity of light, Einstein delved deeper and found that current physics rendered mechanics and electromagnetism *incompatible.*

From the texts of Föppl and Abraham, Einstein could have learned well the intimate connection between mechanics and electromagnetism for interpreting electromagnetic induction. Faraday's law was basic also to Lorentz's electromagnetic theory, where the local time made it possible to calculate the moving magnet's effect on either a resting conductor or the open circuit in unipolar induction. But, as we recall, this calculation rested on certain special assumptions on the constitution of matter. Moreover, electromagnetic theory explained electromagnetic induction in two different ways, depending on whether the conductor or the magnet was moving, even though the physically measurable effect was a function of their relative velocity alone: when the conductor moved relative to the magnet, a current flowed owing to a force on the conductor's electrons; in the inverse case, a current flowed owing to an electric field at the conductor's site. For Einstein, two explanations for an effect that depended on only the relative velocity between magnet and conductor was more than a shortcoming of Lorentz's theory; it was, as he wrote in the relativity paper, an asymmetry that was "not inherent in the phenomena." Einstein found the theoretical situation in electromagnetic induction so "unbearable," as he recalled in 1919 (Holton, 1973g), that he focused on the necessity for an equivalence of viewpoints between observers on the wire loop and on the magnet, rather than on the source of the moving magnet's effect on the wire. He turned the fashionable programmatic research efforts of physics sideways by enlarging Newton's principle of relative motion to treat mechanics and electromagnetism on an equal footing, instead of attempting to reduce one to the other. In the relativity paper, he referred to this widened version of Newton's principle of relative motion as the "principle of relativity." Since, as we saw, the Galilean transformations could not explain the velocity of light, the modified transformations of Lorentz, with their apparently ubiquitous local time coordinate, would have to play a role in relating phenomena between reference systems. From Lorentz's modified transformations Einstein deduced a new result for the addition of velocities:

$$w_r = \frac{w - v}{1 - vw/c^2} \qquad (12)$$

where w_r is the velocity of a moving point relative to S_r and w is its velocity relative to S. For the case of $w = c$, then $w_r = c' = c$, instead of Newton's addition law of velocities, $c' = c - v$. Thus Einstein could have realized that, to first order in v/c, the addition law for velocities from Lorentz's modified transformations produced a result that agreed with the intuition of his thought-experimenter.[8] Since Lorentz's modified transformation differed from the Galilean transformation only in the local time coordinate, Einstein asked himself whether the local time might be the physical "time" (1907b)? But this step required asserting that the times in inertial reference systems differed because the local time coordinate depends on their relative velocity. Yet the absoluteness of time had always been accepted. Furthermore, the thought-experimenter's intuition demanded an examination of the mathematical relation between Newton's principle of relative motion and Lorentz's theorem of corresponding states. After all, the spatial coordinates of Lorentz's modified transformation equations of 1895 were mathematically the same as the ones in the Galilean transformations, and the local time coordinate had been invented for use in electromagnetic theory. Einstein's imposing a Newtonian unity upon Lorentz's modified transformation equations meant also his asserting the equivalence of the reference systems S and S_r. This was a big step, for it meant rejecting Lorentz's ether, and with it the dynamical interpretations of an enormously successful and, for the most part, satisfying theory. To his surprise, Einstein had found that the notion of time was both the central point and the Achilles' heel of the electrodynamics of moving bodies.

For aid in analyzing the nature of time, Einstein recalled that he benefited from the "critical reasoning [in] David Hume's and Ernst Mach's philosophical writings." In a letter of 6 January 1948 to his old friend Besso, Einstein wrote that "Hume had a greater effect on me" than Mach (Einstein, 1972). In his 1944 paper entitled, "Remarks on Bertrand Russell's Theory of Knowledge," Einstein provided a discussion of Hume's insights that enable me to offer the following conjecture. [We know that Einstein had read Hume's *Treatise of Human Nature* during 1902–1904 at the informal study group called the "Olympia Academy" (Einstein, 1956; Seelig, 1954).] My conjecture is that Hume's analysis of sense perceptions offered strong evidence that exact laws of nature could not be induced from empirical data.

Hume's analyses of the limits imposed by sense perceptions on notions of causality and of time enabled Einstein to realize that the high value of the velocity of light, compared with the other velocities we encounter daily, had prevented our appreciating that "the absolute character of time, viz., of simultaneity, unrecognizedly was anchored in the unconscious." (Einstein 1946). Poincaré's pregnant statement in *Science and Hypothesis* to the effect that we have no direct intuition of the simultaneity of two distant events may also have been helpful here.

Einstein's fundamental analysis of physical theory went far beyond science as it is normally conceived: from an analysis of electromagnetic induction into an analysis of sensations, and then into an analysis of thinking itself. He concluded that the customary sensation-based notions of time and simultaneity resulted in a physics burdened with asymmetries, unobservable quantities, and ad hoc hypotheses. Thus prevented from lapsing into a dogmatic slumber, Einstein took recourse in the neo-Kantian view that was predicated on the usefulness of organizing principles such as the second law of thermodynamics. He enlarged Newton's principle of a relativity to include Lorentz's theory, and then he raised to axioms this principle and the basic principle of every wave theory of light: in the ether-fixed system S the velocity of light is independent of the source's motion and is always c.

Einstein's two principles do not attempt to explain anything—for example, why the measured velocity of light always turns out to be c. Whereas Lorentz's and Poincaré's principle of relativity was based on a theory of matter that was heavily dependent on experimental data and was contingent on negative results of future ether-drift experiments, Einstein's was independent of assumptions concerning electrons and was an axiom. As Einstein wrote (in 1907b), the principle of relativity was the basis of a theory that specified the form that laws of physics should assume in order to be used again to investigate the constitution of matter; it was a theory of principle. In short, Einstein moved boldly counter to the prevailing currents of theoretical physics by resolving problems in a Gordian manner, that is, by formulating a view of physics in which certain problems do not occur, a view in which the 1895 paradox becomes a mere fiction. Consequently, Einstein's relativity paper looked insignificant alongside the papers of Abraham, Lorentz, and Poincaré, which used the most au courant methods of mathematical physics in order to derive what Einstein took to be axiomatic.

Straightway in the relativity paper, Einstein emphasized that the basic problems confronting physical theory concerned not the constitution of matter but understanding the equivalence of viewpoints between moving observers. For this purpose, he began with the simplest thought experiment illustrating the problems of electromagnetic induction: a magnet and conducting loop in relative inertial motion. No mathematics was necessary to demonstrate that Maxwell's electrodynamics led observers on the wire loop and the magnet to different interpretations for the physically measurable effect, the current induced in the conductor. Thus, Faraday's law was universally valid, though it had been misinterpreted.

In the second paragraph of the relativity paper Einstein mentioned a facet of the results of ether-drift experiments accurate to the first order in v/c (which he left unnamed) that may have been glimpsed only by Abraham (1904c), but within the context of the electromagnetic world-picture: namely, the laws of electrodynamics and mechanics were valid in inertial reference systems to first-order accuracy in v/c. Einstein's masterstroke was to link the experiment of magnet and conductor, which he had just explained, to the ether-drift experiments. He reasoned that electromagnetic induction depends on the laws of mechanics and of electromagnetism, which cover optics also; he then "conjectured" that Newton's principle of relative motion covers these three disciplines to order v/c. He boldly continued: "We will raise this conjecture (whose content will from now on be referred to as the 'Principle of Relativity') to the status of a postulate." Employing the principle of relativity and the principle governing the velocity of light, Einstein went on to develop a view of physics so powerful that the only mention he made of problems involving unipolar induction was to dismiss them as "meaningless." In the relativity paper the Lorentz contraction and most of the other hypotheses of Lorentz's theory surfaced only as "secondary consequences" (1907b), and it was unnecessary for Einstein to review every extant ether–drift experiment, because in his view their results were ab initio a foregone conclusion.

With further simple thought experiments the first part of the relativity paper demonstrated how imprecise were current notions of time and length, and the relative nature of these notions emerged. Most readers of 1905, however, emphasized the end of the paper, where Einstein deduced exactly a formula for the electron's transverse mass that was almost Lorentz's, and so they interpreted Einstein's work as a valuable generalization of Lorentz's theory of the electron.[9] Soon Kaufmann himself pointed out that these two equations should be the same, and he was one of the first to use the name "Lorentz-Einstein theory" (1906). But what did Einstein have to say in the relativity paper about the data in Kaufmann's 1902 and 1903 papers? Nothing, because it is reasonable to conjecture that Einstein knew well that they disagreed with his prediction for the electron's transverse mass (Miller, 1981b). Einstein considered this prediction to be of secondary importance anyway. In a (1907b) paper reviewing the status of the principle of relativity, Einstein discussed Kaufmann's data from late 1905, which, as Kaufmann had written in no uncertain terms, were "*not consistent with the Lorentz-Einstein fundamental assumption*" (1906). Einstein acutely emphasized that the "systematic deviation" between the Lorentz-Einstein predictions and the data could indicate a hitherto "unnoticed source of error"; consequently, Einstein called for further experiments. Then he dismissed Kaufmann's data, because in his "opinion" they supported theories that did not "embrace a greater complex of phenomena," that is, the electron theories of Abraham and A.H., Bucherer, which could not explain optical experiments accurate to second order in v/c.

A letter among Poincaré's correspondence, which I discovered in

Paris, permits us to compare Einstein's bold defense of the principle of relativity in the face of disconfirming data with Lorentz's own response to the same data (see Fig. 3.4). On 8 March 1906, Lorentz wrote to Poincaré: "Unfortunately my hypothesis of the flattening of electrons is in contradiction with Kaufmann's new results, and I must abandon it. I am, therefore, at the end of my Latin. It seems to me impossible to establish a theory that demands the complete absence of an influence of translation on the phenomena of electricity and optics." What a remarkable confession, and what a clear-cut case of falsification. After all those years of work Lorentz was willing to abandon his theory on the basis of a report of a single experiment.

Einstein's intuition served him well here, for in fact, Kaufmann's data were incorrect. In 1908 the Lorentz-Einstein prediction for the electron's transverse mass was considered to have been vindicated by the new data of Bucherer (1908). Bucherer's results were not immediately accepted by the entire scientific community, but Poincaré and Lorentz did not call for additional tests. Incidentally, three decades later it was proved that Bucherer's data were also incorrect.[10] Undoubtedly having the Kaufmann episode of 1907 in mind, and perhaps also the fate of Bucherer's data, Einstein in 1946 described

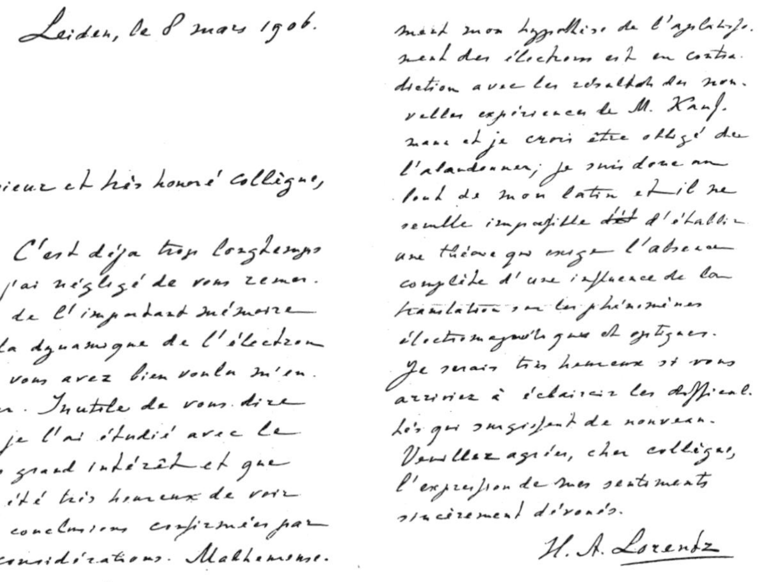

Leiden, le 8 mars 1906.

Monsieur et très honoré Collègue,

C'est déjà trop longtemps que j'ai négligé de vous remercier de l'important mémoire sur la dynamique de l'électron que vous avez bien voulu m'envoyer. Inutile de vous dire que je l'ai étudié avec le plus grand intérêt et que j'ai été très heureux de voir mes conclusions confirmées par vos considérations. Malheureusement mon hypothèse de l'aplatissement des électrons est en contradiction avec les résultats des nouvelles expériences de M. Kaufmann et je crois être obligé de l'abandonner; je suis donc au bout de mon latin et il me semble impossible d'établir une théorie qui exige l'absence complète d'une influence de la translation sur les phénomènes électromagnétiques et optiques. Je serais très heureux si vous arriviez à éclaircir les difficultés qui surgissent de nouveau. Veuillez agréer, cher collègue, l'expression de mes sentiments sincèrement dévoués.

H. A. Lorentz

Fig. 3.4. Letter from H. A. Lorentz to H. Poincaré, 8 March 1906. (Courtesy of the Estate of Henri Poincaré.)

one of two criteria for assessing a scientific theory: "The first point of view is obvious: the theory must not contradict empirical facts. However evident this demand may in the first place appear, its application turns out to be quite delicate" (1946).

CONCLUSION

Although by mid-1905 high-velocity data dominated basic physical theory, Einstein needed only familiar low-velocity data, such as electromagnetic induction, in conjunction with the data of the thought-experimenter, to bring physics into the twentieth century.

Due principally to the fundamental analyses that served to clarify the concepts of length and time in special relativity as compared to Lorentz's theory of the electron, in 1911 most physicists realized the difference between Lorentz's theory of the electron and special relativity. Coincidentally with this demarcation in 1911, Arnold Sommerfeld declared the frontier of physics to be problems concerning the nature of light. Two years later the frontier was redefined by Niels Bohr to be the physics of the atom.

Chapter 4 traces how by degrees physicists lost their perceptual and linguistic anchors to the world we live in because again and again time-honored notions of physics that were abstracted from this realm failed. More than ever before, for scientists raised and working in the German cultural milieu, fundamental problems turned about the perceptual-linguistic Kantian notion of mental imagery that was introduced in this chapter with the term *Anschauung*. This culturally-based mode of mental imagery played a central role in developments in German science and, in certain cases, in German electrical engineering.

It is important to distinguish between the uses of mental imagery by British and German physicists. Many British physicists, such as James Clerk Maxwell, utilized mechanical models and mental pictures in the initial development of a theory. In the theory's formalization the models were either discarded or used strictly for explanatory purposes. For example, compare Maxwell's earlier papers on electromagnetism of 1861–1862 that are based on mechanical models of the ether, with his paper of 1865 in which the equations of electromagnetism are introduced without mechanical models (Maxwell, 1861, 1862, 1865). Conversely for many scientists and engineers trained in the German cultural milieu, mental images became an intrinsic part of electromagnetic theory. For example, the controversy over whether magnetic lines of force rotate with their source magnet was detrimental to the design of a large scale unipolar dynamo in Germany (Miller, 1981a).[11]

Chapter 4 is a case study in which there emerges from the scientific literature and correspondence the individual scientist's struggle to maintain and then to modify in a dazzling way this heretofore successful realist style of visual thinking. The results of this struggle set the basis for the present-day conception of what constitutes physical reality.

NOTES

(A version of this chapter was presented as a lecture at the Einstein Centennial Symposium, The Israel Academy of Sciences and Humanities, Jerusalem, 14–23 March 1979, and published in G. Holton and Y. Elkana (eds.), *Albert Einstein and His Time: Jerusalem Centenary Symposium* (Princeton: Princeton University Press, 1982). I acknowledge the permission of Princeton University Press to reprint portions of this material.)

1. As far as we know, the editorial policy of the *Annalen* was that an author's initial contributions were scrutinized by either the editor (in 1905, Paul Drude) or a member of the Curatorium; subsequent papers were published with little or no refereeing. Einstein had appeared in print in the *Annalen* five times by 1905, so his relativity paper was probably accepted on receipt. I thank Dr. Allan Needell for this information, at which he arrived as a result of studying the correspondence between Planck and Wilhelm Wien.

2. For details see my (1981b) and Holton (1973).

3. Instead of using Lorentz's original units, I take the liberty to write Eqs. (1)–(5) in absolute Gaussian (cgs) units, whose usefulness was emphasized first by Max Abraham in 1902 and 1903. Abraham was also the first to use the nomenclature "Maxwell-Lorentz equations," which appears in Einstein's relativity paper of 1905.

4. These three letters from Poincaré to Lorentz are analyzed in my (1980, 1981b).

5. For a discussion of Faraday's experiments with rotating magnets and of the treatment of unipolar induction in electrical engineering and physics up to the present, see my (1981a). My (1981b), especially Chapter 3, focuses on the influence of unipolar induction on Einstein's thinking toward the special theory of relativity.

6. For discussions of Planck's work see Klein (1962), Jammer (1966), and Kuhn (1978).

7. In addition, Einstein wrote a Ph.D. dissertation (1905a) and published a fourth paper in which he developed the equivalence of mass and energy (1905e).

8. The velocity addition law in Eq. (12) turned out to be valid to all orders in (v/c). In fact, Eq. (12) is not unique to the special theory of relativity because it remains the same even if Eqs. (A) in Note 19 of Chapter 1 were multiplied by any function of k.

9. According to Kaufmann (1906), in 1905 Einstein had avoided the severe approximations on the motion of electrons that was required for deriving Newton's second law from Lorentz's force equation, that is, the quasistationary approximation. Rather, continued Kaufmann, Einstein replaced Lorentz's assumptions on how electrons interact with the ether with purely phenomenological assumptions concerning how clocks are synchronized using light signals.

10. Zahn and Spees (1938) demonstrated that the resolution of Bucherer's velocity filters was inadequate to distinguish between the various competing electron theories. In fact, this shortcoming pervaded every other experimental determination of the electron's mass in the period 1908–1915. For details see Miller (1981b).

11. Thus was Pierre Duhem's connection of the style of thinking of French scientists and German scientists not completely correct. Recall that in his classic *Aim and Structure of Physical Theory,* Duhem described the British scientist as seeking mechanical models in order to understand phenomena, while French and German scientists seek abstract mathematical representations. Duhem chose Hertz as the typical German scientist. But Hertz happened to be one of those German scientists who advocated discarding mental pictures once a theory's axiomatic foundation is set. For example, for Hertz Maxwell's theory was "Maxwell's system of equations" (1893). Thus, Hertz did not consider magnetic lines of force to be a fundamental *Anschauung* (Miller, 1981a). See also the discussion of Hertz's mechanics in Chapter 2.

APPENDIX: EINSTEIN'S GEDANKEN EXPERIMENT OF 1895

In 1895 Einstein conceived of a Gedanken, or thought experiment that, as he recalled in 1946, contained the "germ of the special relativity theory." Owing to the importance of this thought experiment I shall develop it according to Einstein's own description in his 1946 "Autobiographical Notes," while also adhering to the sequence of steps that he set out in his (1917a) book.

What is involved in Einstein's thought experiment (Fig. 3.5) is the picture of a moving observer and the intuitions of light as a wave and of catching up with the point on the light wave. As the Gedanken experimenter catches up with the point on the light wave, he measures different velocities of light in his laboratory. In fact, according to the physics of 1905, the Gedanken experimenter could even achieve the velocity of light ($v = c$), in which case he would see the light behave as a standing wave—that is, moving up and down in place just like a rope that has one end tied to a wall while the other end is driven up and down. But to Einstein (1946) it was "intuitively clear" that the laws of optics could not depend on the state of the observer's motion. Thus, the Gedanken experiment posed the following paradox, whose resolution required that Einstein move beyond the intuition that is developed from sense preceptions:

1. According to contemporaneous physics the Gedanken experimenter should be able to catch up with a point on a light wave.
2. According to Einstein's "intuition" the laws of optics should be independent of the observer's motion—that is, independent of v. Mathematically speaking the horns of this dilemma are: where c_r

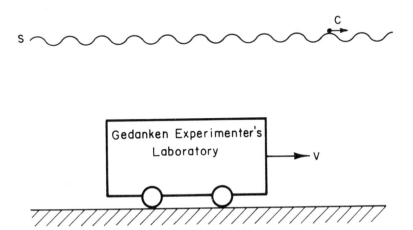

Fig. 3.5. A schematic of Einstein's Gedanken experiment of 1895. The source S of a light wave is at rest in the ether. A point on the light wave moves with a velocity c relative to S. The Gedanken experimenter's laboratory moves with the velocity v relative to the ether.

is the velocity of light relative to the Gedanken experimenter (so, when $v = c$, then $c_r = 0$):

(a) $c_r = c - v$ —according to contemporaneous physics
(b) $c_r = c$ (always)—according to Einstein's "intuition"

Clearly, c_r cannot satisfy both of these conditions, and that is why the Gedanken experiment posed a paradox for Einstein.

CHAPTER 4

Redefining Visualizability

Gedanken ohne Inhalt sind leer, Anschauungen ohne Begriffe sind blind.*

I. Kant (1781)

The German man of science was a philosopher.

J.T. Merz (1965)

Jeder Philosoph hat eben seinen eigenen Kant.**

A. Einstein (Seelig, 1954)

---

*Thoughts without content are empty, intuitions without concepts are blind.
**Every philosopher has precisely his own Kant.

Most of the physicists discussed in this book are in the group portraits from the Solvay Conferences of 1911 (see p. 100), 1927 (above), and 1933 (next page). (Courtesy of AIP Niels Bohr Library.) Standing (left to right): A. Piccard, E. Henriot, P. Ehrenfest, Ed. Herzen, Th. De Donder, E. Schrödinger, E. Verschaffelt, W. Pauli, W. Heisenberg, R.H. Fowler, L. Brillouin. First two rows (left to right): P. Debye, I. Langmuir, M. Knudsen, M. Planck, W.L. Bragg, Mme. Curie, H.A. Kramers, H.A. Lorentz, P.A.M. Dirac, A. Einstein, A.H. Compton, P. Langevin, L. de Broglie, Ch. E. Guye, M. Born, C.T.R. Wilson, N. Bohr, O.W. Richardson.

Sitting (left to right): E. Schrödinger, Madame I. Joliot, N. Bohr, A. Joffé, Madame Curie, P. Langevin, O. W. Richardson, Lord Rutherford, Th. De Donder, M. de Broglie, L. de Broglie, Lise Meitner, and J. Chadwick. Standing (left to right): E. Henriot, F. Perrin, F. Joliot, W. Heisenberg, H.A. Kramers, E. Stahel, E. Fermi, E.T.S. Walton, P.A.M. Dirac, P. Debye, N.F. Mott, B. Cabrera, G. Gamow, W. Bothe, P.M.S. Blackett (at back), M.S. Rosenblum, J. Errera, E. Bauer, W. Pauli, J.E. Verschaffelt, M. Cosyns (at back), E. Herzen, J.D. Cockcroft, C.D. Ellis, R. Peierls, A. Piccard, E.O. Lawrence, and L. Rosenfeld.

IN SETTING THE BASIS for his philosophical system in the monumental *Critique of Pure Reason*, Immanuel Kant demarcated between intuition (*Anschauung*) and sensation (*Empfindung*). *Anschauung* is the intuition, or knowledge, that results from the immediate apprehension of an independently real object, that is, a concept in Kant's philosophy. Whereas etymologically *Anschauung* refers only to visual sensations, Kant extended its applicability to cover any sort of perception. Kant's goal was to develop the notion of the *reine Anschauung* of space and time, and for this purpose the term "pure sensations" would not do.

Owing to the complexity of Kant's philosophy the term *Anschauung* became virtually untranslatable into English, and in any language it had to be understood in the context in which Kant's philosophy was discussed. However, the German *Anschaulichkeit*, and the adjective *anschaulich*, connote visualization or intuition through mechanical models. Some definitions of *anschaulich* from German-language philosophical dictionaries are "the immediately given . . . the readily graspable in the *Anschauung*" (Eisler, 1922), the antithesis of conceptual abstraction" (Eisler, 1927). Thus *Anschaulichkeit* can be interpreted to be less abstract than *Anschauung*.

Anschaulichkeit is a property of the object itself, while the *Anschauung* of an object results from the cognitive act of knowing the object.

Prior to the intense research in atomic physics generated by Niels Bohr's 1913 atomic theory, physicists had dealt with physical systems that with some justification were assumed to be amenable to their perceptions, and for which the space and time pictures of classical physics were applicable. There was general agreement on the sorts of models available for representing a theory adequately—for example, the use of oscillator ions to model dispersion. These pictures were abstracted from previous visualizations of objects in the world of perceptions. Another example of this sort of thinking in nineteenth- and early twentieth-century physics and engineering in German-speaking countries is the notion of magnetic lines of force. From the disposition of iron filings on a piece of paper in the vicinity of a magnet, magnetic lines of force were elevated to an *Anschauung*. The German-speaking physicists and engineers discussed Faraday's *Anschauung* in which magnetic lines of force did not rotate with their source magnet.

Allied with visual thinking in classical physics was the realist view, which provided pictures of submicroscopic systems as continuously developing in space and time. These pictures in turn were associated with the strong notion of causality in which causality and determinism were equivalent and both notions were linked to the conservation laws of energy and momentum.[1]

Central to the ensuing analysis is the difference between the notions of picture and image, which were usually referred to with the German word *Bild*. In 1923 quantum theory had to abandon the intuitive pictures (*anschauliche Bilder*) from the classical theories of mechanics and electromagnetism because they were insufficient and misleading. Loss of visualizability turned out to be an essential prerequisite for Werner Heisenberg's formulation of the new quantum mechanics in 1925. In the period 1925–1927, however, the need for some sort of visualization became paramount. This chapter traces how Heisenberg's 1925 formulation of the quantum mechanics and his attempts in 1926–1927 to interpret the formalism led him to invert the Kantian usage of *Anschauung* and *Anschaulichkeit* and then to use this redefinition as a guiding theme for his research in the quantum theory of fields and nuclear physics. The *Anschauungen* assumed the role of quantities associated with classical physics and thus with close links to the world of perceptions; *Anschaulichkeiten* were promoted to ever higher realms of abstraction. Thus while, loosely speaking, the *anschauliche Bilder* of prequantum physical theory may be referred to as models, this was not the case for the *Anschaulichkeiten* of post-1925 developments. In 1927 Heisenberg began to move toward permitting the mathematical formalism to reveal the properties of the atomic entities themselves; that is, What is *Anschaulichkeit?* This development occurred in the highly charged emotional atmosphere of quantum theory evolving through denial of pictures and images and then the subjective exchanges in the published literature between Heisenberg and Erwin Schrödinger. Nothing less was at stake than the interpretation of physical reality in a realm beyond our perceptions. Examination of Heisenberg's correspondence and published papers reveals a fascinating scenario of a philosophical viewpoint tempered by scientific research.

Consistent with the philosophical-scientific meanings of *Anschauung* and *Anschaulichkeit* that have been discussed thus far, I shall render these terms as follows: *Anschauung* as "intuition," *Anschaulichkeit* as "visualizability," and the adjective *anschaulich* as "intuitive."

BACKGROUND: THE PERIOD 1913–1923

A remarkable and alluring result of Bohr's atomic theory is the demonstration that the atom is a small planetary system . . . the thought that the laws of the macrocosmos in the small reflect the terrestrial world obviously exercises a great magic on mankind's mind; indeed its form is rooted in the superstition (which is as old as the history of thought) that the destiny of men could be read from the stars. The astrological mysticism has disappeared from science, but what remains is the endeavor toward the knowledge of the unity of the laws of the world.

M. Born (1923)

Alas, continued Born, the "possibility of considering the atom as a planetary system has its limits. The agreement is only in the simplest case [the hydrogen atom]." The honeymoon of the Bohr theory was over. Let us survey its successes up to the critical year of 1923. The continuity that had been the hallmark of classical physics was absent in Bohr's 1913 atomic theory owing to Bohr's introduction of a "quantity foreign to the laws of classical electrodynamics, i.e., Planck's constant," which served both to set the scale of atoms and to account for the emission of "*homogeneous* radiation" (1913a; italics in original).[2] In Bohr's opinion, whereas classical mechanics could be

Niels Bohr (1885–1962) is shown here in the early 1920's. In September 1911 Arnold Sommerfeld declared that the special theory of relativity "was in the safe possession of physicists" and that the new frontier of physics was defined by Planck's energy quantum and Einstein's light quantum. Two years later the frontier was redefined by Bohr's theory of the atom in which the atom is depicted as a miniature Copernican solar system. By 1923 Bohr's theory was beset by severe fundamental problems. Yet one of its basic statements—the correspondence principle—provided the pathway in 1925 to the new theory of the atom. Bohr's keen instinct for basic problems, and his philosophical bent, were essential toward clarification of the new theory. He was an inspiration to the generation of physicists that struggled with atomic theory from 1913 through the 1930's. Werner Heisenberg, the inventor of the new atomic theory, recalled that he had "learned so infinitely much from Bohr" (1967). (Courtesy of AIP Niels Bohr Library)

used in a suitably quantized form for calculating the stationary states of atoms, the discontinuities inherent in atomic processes rendered classical electrodynamics "inadequate." It was the symbols taken from ordinary mechanics that permitted the atom in its stationary states to be visualizable as a miniature Copernican system (see Fig. 4.1). However, the transition of an electron between its "waiting places" was unvisualizable (1913b); the electron in transit behaved much like the Cheshire cat. Thus Bohr did not deem his "atomic model" as anywhere near completion, for its "considerations conflict with the admirably coherent set of conceptions which have been rightly termed the classical theory of electrodynamics."

Bohr in (1918) proposed a mathematical method for describing the unvisualizable transition process, namely, the *A* and *B* coefficients that Einstein had considered to be a "weakness" in his (1916c; 1917b) quantum theory of radiation. The A and B coefficients are the probabilities for an atom to make transitions that are spontaneous or induced by external radiation, respectively. Einstein assumed the statistical laws for atomic transitions to be like that of radioactivity. So the A and B coefficients reflected one's ignorance of the mechanism for atomic transitions.[3] On the other hand, for Bohr probability was just the prescription for dealing with unvisualizable quantum jumps whose dynamics, that is, causes, were unknown. Although the withering away of the picture of the Bohr atom was presaged by its lack of success in dealing with the constitution of atomic systems more complex than that of the hydrogen atom, the problem of dispersion drastically altered Bohr's model of the atom.[4] The reason is that the response of bound electrons to radiation could be correlated neither with their simple motion in Keplerian stationary states nor with their representation as Planckian oscillators. To Bohr neither the three-body problem nor the anomalous Zeeman effect was among the "fundamental difficulties" confronting the atomic theory; rather, as he explained in (1923), these "fundamental difficulties" had as their common denominator the explanation of the interaction of light with atoms.

THE PICTURE OF LIGHT QUANTA

In (1913a) Bohr had described the radiation emitted from an atom as "*homogeneous* radiation" and not as light quanta.

As M.J. Klein has written, circa 1910 "such men as Max Planck and H.A. Lorentz had only sharply critical things to say about Einstein's light quanta" because light quanta could not account for inter-

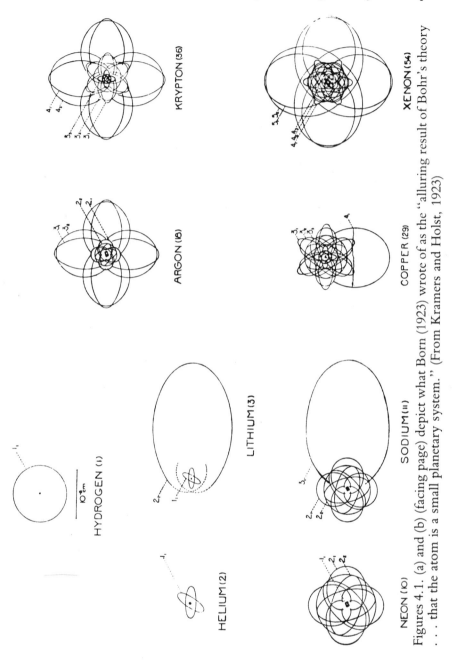

Figures 4.1. (a) and (b) (facing page) depict what Born (1923) wrote of as the "alluring result of Bohr's theory . . . that the atom is a small planetary system." (From Kramers and Holst, 1923)

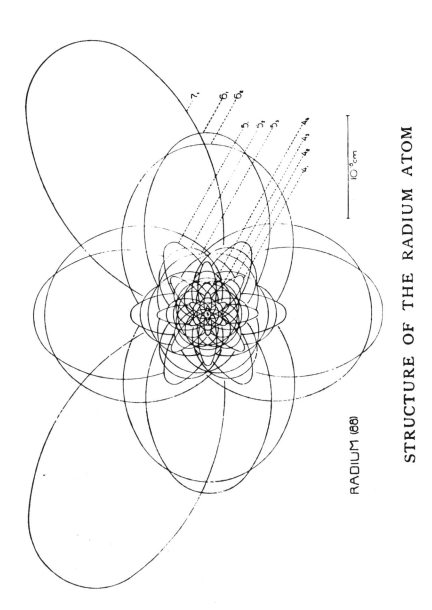

RADIUM (88)

STRUCTURE OF THE RADIUM ATOM

ference and diffraction (1970b). Yet by 1916 the usefulness of light quanta for discussing the photoelectric effect had led some physicists to entertain their reality seriously, and to be deeply concerned over the conundrums involved in reconciling two apparently disparate concepts of light as a particle and as a wave. In (1916) Owen W. Richardson expressed the conceptual situation as follows:

> It is difficult, in fact it is not too much to say that at present it appears to be impossible, to reconcile the divergent claims of the photoelectric and interference groups of phenomena. The same energy of the radiation behaves as though it possessed at the same time the opposite properties of extension and localization. At present there seems no obvious escape from the conclusion that the ordinary formulation of the geometrical propagation involves a logical contradiction, and it may be that it is impossible consistently to describe the spatial distribution of radiation in terms of three-dimensional geometry.

During 1913–1920 Bohr did not publish on the nature of light. His approach to emission and absorption of light by atoms was via a set of limiting procedures for extending the classical wave theory of light into the atomic domain, which in (1920) he referred to as the "correspondence principle." In his widely read treatise *Atomic Structure and Spectral Lines* (1922), Arnold Sommerfeld wrote of the "magic wand," that is, the correspondence principle, that "it is indeed astonishing how much of the wave theory still remains even in spectroscopic processes of a decidedly quantum character." But Sommerfeld considered Bohr's correspondence principle to be only a stopgap, and he wrote in the same vein as had Richardson that "modern physics is thus for the present confronted with irreconcilable contradictions and must frankly confess its non-liquet."

In lectures of (1920) and (1921a) Bohr broached the basic conceptual problem concerning light quanta, namely, the "insurmountable difficulties from the point of view of optical interference." In (1923), in the face of the dissolution of the Copernican atom, he confronted the problem of "forming a constant picture [*Bild*] of phenomena." The principal point was reconciling the discontinuities of atomic physics with classical electrodynamics. One approach that Bohr suggested involved the light quantum and maintaining conservation of energy and momentum in individual processes. But this was an unsatisfactory solution, continued Bohr, because the "picture [*Bild*] of light quanta precludes explaining interference." Yet the undeniable usefulness of the light quantum hypothesis for explaining certain

phenomena led Bohr to conclude that a contradiction-free description of atomic processes could not be arrived at by "use of conceptions borrowed from classical electrodynamics." Since in classical
physics the conservation laws of energy and momentum were linked
with a continuous space-time description, then, continued Bohr,
these laws may "not possess unlimited validity." Bohr's guide in the
atomic domain would be the correspondence principle, upon which
rested his second method for resolving "fundamental difficulties";
namely, Bohr proposed the "coupling mechanism," according to
which atoms responded to incident light like an ensemble of harmonic oscillators whose frequencies were those of the possible
atomic transitions. This method permitted Bohr to renounce the
"so-called hypothesis of light quanta." In 1923 Rudolf Ladenburg
and Fritz Reiche proposed a mathematical formulation for the coupling mechanism by using the correspondence principle to translate
the classical equation for dispersion into a form that could be interpreted "formally" as the atom responding to radiation like "*Ersatzoszillatoren*" (1923).[5]

Their work permitted Bohr (in 1924b) the means to avoid interpreting the results of Arthur Holly Compton's experiments on the
scattering of X-rays from atoms in terms of light quanta (1923).[6] In
order to reach this goal Bohr resorted to combining the most extreme consequence of the first method of 1923 (renouncing energy
conservation) with the *Ersatzoszillatoren*, which he referred to in the
1924 Bohr, Kramers, and Slater paper as "virtual oscillators." Of
these oscillators, wrote Bohr, "Such a picture [*Bild*] was used by
Ladenburg," where Bohr did not use the term "picture" to mean
"visualization." For how could one visualize a planetary electron in a
stationary state as represented by as many oscillators as there are
transitions to and from this state? He meant the term "picture" to
refer to the interpretation of the mathematical framework. Whereas
the picture of the Copernican atom had been imposed on the 1913–
1923 theory owing to Bohr's use of the language of "ordinary mechanics," the image of the virtual oscillators was synonymous with
the scheme of the 1924 Bohr, Kramers, and Slater version of the
Bohr atomic theory.

VISUALIZABILITY LOST

Out with the pictures went also the conservation laws of energy and
momentum because there were as many virtual oscillators as spectral
lines, and the virtual oscillators representing the free electron need

not be correlated spatially with the atomic entity. Furthermore the virtual oscillators emitted a field that carried only the probability of inducing atomic transitions. Consequently, for example, the virtual radiation field of one atom could induce an upward atomic transition in another atom without the source atom undergoing the corresponding downward transition. Thus in order to preserve the wave concept of radiation Bohr abandoned "any attempt at a causal connection between transitions in distant atoms," as well as jettisoning visualization and momentum and energy conservation. As support for this radical move to resolve the tension between the thema-antithema couple of particle and wave in terms of the wave mode, Bohr noted the paradoxes involved in the wave-particle duality that "perhaps for the first time [were] clearly expressed by O.W. Richardson" (in Richardson, 1916). While subsequent work on dispersion by Born, Heisenberg, and Hendrik Kramers utilized the virtual oscillators, neither the violations of energy nor momentum were well received.

Einstein, for example consistent with his realist view, which embraced an objective physical reality that existed independently of the experimenter, did "not want to be forced into abandoning strict causality," as he wrote to Born on 29 April 1924 (Born, 1971).

On 6 December 1924 Pauli wrote to Sommerfeld that concerning the "problem in your book, 'wave theory and particle theory,' [your] 'frank non liquet' is a thousand times preferable to me than the well-constructed artificial apparent solution of the problem by Bohr, Kramers, and Slater (even if the experiment of Geiger and Bothe should verify this theory)" (Pauli, 1979). Pauli welcomed Bothe and Geiger's 1925 experimental disproof of the Bohr, Kramers, and Slater theory. In a letter to Kramers of 2 July 1925 Pauli wrote that "you move in an entirely wrong direction: The concept of energy is not to be modified, but the concepts of motion and force" (Pauli, 1979). Pauli's unsuccessful work on the three-body problem and his researches on the anomalous Zeeman effect had led him to distrust atomic models. He set down his strong opinion on this point in a letter of 12 December 1924 to Bohr (Pauli, 1979): "I believe that the energy and momentum values of the stationary states are somewhat more real than the 'orbits.' The (still unattained) goal must be to deduce these and all other physically real, observable characteristics of the stationary states from the (determined) quantum numbers and quantum-theoretical laws. We do not want, however, to clap the atom into the chains of our bias (to which in my opinion belongs the assumption of the existence of electron orbits in the usual kinematic meaning), but, on the contrary, we must adjust our concepts to experience." If this was to be the route to the new quantum mechanics, then what about *Anschaulichkeit*? Pauli had referred to this point earlier in the letter under discussion in the course of an acid critique of the pictures of atomic models that would appear in the book of Kramers and H. Holst entitled, *The Atom and the Bohr Theory of Its Structure* (1923): "Thus I consider for certain—despite our good friend Kramers and his colorful picture books—and the children they listen to it with pleasure. Even though the demand of these children for *Anschaulichkeit* is in part legitimate and healthy, still this demand should never count in physics as an argument for retaining systems of concepts. Once the systems of concepts are settled, then will *Anschaulichkeit* be regained."[7] Exactly how this was to be accomplished Pauli knew not. The editors of the Pauli correspondence rightly note that Bohr almost certainly showed this letter to Heisenberg, who was at Copenhagen working on the problem of dispersion. Whereas for Pauli the "virtualization of physics"[8] was anathema because of its association with mechanical models, for Heisenberg it would be the key to renouncing visualizability and redefining *Anschaulichkeit*. Pauli's suggestion would offer Heisenberg support for his new work on dispersion.

To a qualitative 1925 discussion of collision processes using the Bohr, Kramers, and Slater theory, Bohr appended a note added in proof that recognized the theory's refutation by the experiment of Bothe and Geiger (1925a). Bohr interpreted the results of Bothe and Geiger as forcing on the "picture a corpuscular transmission of light," but not definitely deciding between the wave and particle theories. Whereas in 1923 Bohr had linked the light quantum with the classical billiard ball picture of collisions, when confronted in 1925 with the light quantum's reality he accepted the validity of the conservation laws of energy and momentum, but witheld the possibility of visualization of individual processes. For he added the complication of the light quantum's "unavoidable fluctuations in time making it even more difficult to use intuitive pictures [*anschauliche Bilder*]" to analyze collision processes and the structure of the atom. And if this were not enough, continued Bohr, there were Louis de Broglie's (1924) proposal of a wave-particle duality of matter and Einstein's (1924–1925) results on the indistinguishability of particles in an ideal quantum gas. Not only had the bound electron lost its visualizability in the classical meaning of this term, but free electrons had lost their individuality.

Loss of visualizability had paid off, however, for as Bohr wrote in a paper of December 1925 in *Nature* (1925b), "Quite recently Heisenberg, who has especially emphasized these difficulties, has taken a step probably of fundamental importance by formulating the problems of the quantum theory in a novel way by which the difficulties attached to the use of mechanical pictures may, it is hoped, be avoided." Bohr referred to Heisenberg's (1925b) recent formulation of the new quantum mechanics. When Bohr had presented the lecture "Atomic theory and mechanics" on 30 August 1925 he had been unaware of Heisenberg's quantum mechanics paper. The published version of Bohr's lecture contained a footnote to the effect that the published text "had been notably influenced" by Heisenberg's recent work. Except for the final section entitled, "The development of a rational quantum mechanics," it is difficult to discern how Heisenberg's paper effected changes in Bohr's 30 August 1925 lecture. Bohr had begun the 1925 lecture by emphasizing the "essential failure of pictures in space and time," but in the final section he wrote of only "a limitation of our usual means of visualization." For by December 1925 Bohr was aware of the rapid developments of Heisenberg's original formulation by Born and Pascual Jordan (1925), in which the equations of Hamiltonian mechanics could be utilized when suitably reinterpreted. Thus as had been the case in 1913, the pictures from the world of perceptions could reenter the atomic domain via the symbols of "rational mechanics." Meanwhile, concluded Bohr, "It will at first seem deplorable that in atomic problems we have apparently met with such a limitation of our usual means of visualization," and mathematics will have to serve as the guide. In a note added in proof to George Uhlenbeck and Samuel Goudsmit's 1926 *Nature* article proposing an essentially nonvisualizable fourth degree of freedom for the electron, namely, spin, Bohr wrote that electron spin "opens up to a very hopeful prospect of our being able to account more extensively for the properties of elements by means of mechanical models, at least in the qualitative way characteristic of applications of the correspondence principle." Bohr's attitude toward electron spin is reasonable because most likely he considered it as obviating the unvisualizable *"unmechanische Zwang"* that he had proposed in 1923 as a means to account for the anomalous Zeeman effect in alkali atoms. In fact, the title of the German version of Uhlenbeck and Goudsmit's paper in *Die Naturwissenschaften* is "Replacement of the hypothesis of the *unmechanischen Zwang* by a requirement on the internal behavior of each individual electron" (1925).[9]

THE NEW QUANTUM MECHANICS

A wonderful combination of profound intuition and formal virtuosity
inspired Heisenberg to conceptions of striking brilliance.

L. Rosenfeld (1967)

Among the major figures of quantum physics in 1923–1925 Heisenberg was the only one who published in detail on every aspect of fundamental problems—calculation of stationary states for three-body problems, the anomalous Zeeman effect, and dispersion. He recalled that although mechanical models were useful, the line of papers from Ladenburg, Bohr-Kramers-Slater, Heisenberg-Kramers, and "my paper" (*AHQP*: 19 February 1963) ("On an application of the correspondence principle to the problem of the polarization of fluorescent light" (1925a)) indicated the path away from "cheap solutions" (*AHQP*: 13 February 1963).[10] The Bohr-Kramers-Slater paper, continued Heisenberg, indicated the "kind of price" required for progress, and in "my paper" Heisenberg drove the notion of virtual oscillators to its inevitable conclusion—complete freedom from planetary orbits.

The occasion for Heisenberg's writing "my paper" was the publication in 1923–1924 by Robert W. Wood and A. Ellett of data on the polarization of the resonance radiation resulting from the scattering of light by mercury and sodium vapors (1923, 1924). It seemed as if certain transitions induced by absorption of radiation could be squelched by appropriately orienting a weak external uniform magnetic field. For example, for the case of mercury, whereas an external uniform magnetic field of one gauss oriented along the direction of polarization of the incident radiation resulted in an unpolarized scattered beam, for no external field the polarization of the scattered beam was 90 percent. The appearance of such rich and perplexing data is comparable today to the discovery of a new elementary particle, and the response from the physics community of 1924 was similar; namely, they proposed a plethora of models. Bohr's overall 1924 criticism of the schemes proposed was that they considered the bound electron to be either a Planckian oscillator or in a Keplerian orbit (1924c). This situation is indicative of the resistance by many physicists to the use of virtual oscillators.[11] The motion of an electron in a degenerate state, Bohr emphasized, could not be associated with any particular stationary state, and hence the electron could not be uniquely assigned quantum numbers. In the spirit of the Bohr, Kramers, and Slater theory, Bohr suggested that bound electrons

Werner Heisenberg (1901–1976) is shown here in 1926, one year after he had invented the new atomic theory—quantum mechanics. In 1926 he embarked on a remarkable series of extensions of the quantum mechanics that set the foundations for molecular physics, quantum theory of magnetism, nuclear physics, and elementary-particle theory. (Courtesy of AIP Niels Bohr Library)

responded to incident radiation like an ensemble of virtual oscillators. Whereas for the nondegenerate case Bohr, as he had in the Bohr, Kramers, and Slater paper, wrote of the possibility of only a "formal connection" between the bound electron and an ensemble of virtual oscillators, for the degenerate case he attributed to the virtual oscillators some measure of physical reality. "In contrast to the nondegenerate system we must be prepared to find the behavior of degenerate atoms, so far as concerns radiation, is not connected with motion in particular states, but requires further specification through the virtual oscillators." The problem was how to apply this notion to the degenerate case; in a footnote Bohr alerted the reader to a paper in which Heisenberg had succeeded.

Heisenberg's (1925a) paper was his first publication on the problem of dispersion. The association of the virtual oscillators through the correspondence principle with transition processes led Heisenberg to write, "Therein are the grounds that the virtual oscillators are connected with the electron's motion in stationary states only in a symbolic manner . . . [thus] the virtual oscillator approach provides a greater degree of freedom than the notion of electrons in stationary states." Heisenberg took advantage of this point to make a "broadening and sharpening of the correspondence principle," thereby arriving at "clear-cut conclusions as concerns intensity and polarization." The principal problem Heisenberg treated was the degenerate case from the data of Wood and Ellett, and his results agreed adequately with these data. Heisenberg recalled that this paper showed the "necessity for detachment from intuitive models" (1967).

In a subsequent paper on dispersion Kramers and Heisenberg referred to Heisenberg's 1925 paper as "an attempt to set up a theory of observations on the basis of the correspondence principle" (1925). The first steps toward the new quantum mechanics had been taken. The results in Heisenberg's 1925 paper reinforces what he wrote to Pauli on 9 July 1925, after having suitably reinterpreted the expansion of the bound electron's position using the model of an electron performing its periodic oscillations as an anharmonic oscillator: "It is my genuine conviction that an interpretation of the Rydberg formula in the case of circular and elliptical orbits of classical geometry has not the least physical meaning . . . . The conception of the orbit, which we are unable to observe, is wiped out completely and replaced in a qualified manner" (Pauli, 1979). Thus by 9 July 1925 Heisenberg had relinquished not only the picture of Keplerian stationary states but also the picture of the bound electron as a localized entity. The virtual oscillator representation of the bound electron permitted him to formulate a theory, as he wrote in the seminal paper on the matrix mechanics, "founded exclusively on relations between quantities which in principle are observable [rather than on rules that] lack an intuitive [*anschaulich*] physical foundation unless one still returns to the hope" that the position of an electron in an atom could be observed (1925b). He characterized the unvisualizable electron by its radiation and the probabilities for transitions, that is, the coefficients in the virtual oscillator expansion. Heisenberg's overall approach was consistent with that suggested earlier by Pauli, and on 27 July 1925 Pauli wrote to Kramers that "I have welcomed with joy Heisenberg's *Ansatz*" (Pauli, 1979).

In a tour de force of early 1926 Pauli deduced the Balmer formula from Heisenberg's theory, and emphasized at the outset that "Heisenberg's form of the quantum theory completely avoids a mechanical-kinematical representation [*Veranschaulichung*] of the notion of electrons in the stationary states of the atom" (1926).

Although the renunciation of a picture of the bound electron was a necessary prerequisite to the discovery of the new quantum mechanics, nevertheless the lack of *Veranschaulichung* or *Anschaulichkeit* or an intuitive [*anschaulich*] interpretation was of great concern to Bohr, Born, and Heisenberg, and this concern emerges from their scientific papers of the period 1925–1927. For example, in the important paper of Born, Jordan, and Heisenberg of late 1925, Heisenberg wrote in the introduction that the present theory labored "under the disadvantage that there can be no directly intuitive [*anschaulich*] geometrical interpretation because the motion of electrons cannot be described in terms of the familiar concepts of space and time" (1926a). Heisenberg continued, "In the further development of the theory, an important task will lie in the closer investigation of the nature of this correspondence between classical and quantum mechanics and in the manner in which symbolic quantum geometry goes over into intuitive classical geometry [*anschaulich klassische Geometrie*]." In the section entitled, "The Zeeman effect," probably also written by Heisenberg, the notion of planetary stationary states arose when he wrote of the inability of the new quantum mechanics to resolve the problem of the anomalous Zeeman effect as perhaps due to the result of an "intimate connection between the innermost and outermost orbits . . . "; however, Heisenberg hoped that the recent hypothesis of an electron spin by Uhlenbeck and Goudsmit might provide an alternative route. Yet this fourth degree of freedom for the electron could not be visualized because a point on a spinning electron of finite extent could move faster than light.

VISUALIZABILITY REGAINED

My theory was inspired by L. de Broglie, *Ann. de Physique* (10) *3*, p. 22, 1925 (Thèses, Paris, 1924) and by short but incomplete remarks by A. Einstein, *Berl. Ber.* (1925) pp. 9ff. No genetic relation whatever with Heisenberg is known to me. I knew of this theory, of course, but felt discouraged not to say repelled, by the methods of transcendental algebra, which appeared very difficult to me and by the lack of visualizability [*Anschaulichkeit*].

E. Schrödinger (1926a)

The more I reflect on the physical portion of Schrödinger's theory the more disgusting I find it. Just imagine the rotating electron whose charge is distributed over the entire space with axes in 4 or 5 dimensions. What Schrödinger writes on the visualizability [*Anschaulichkeit*] of his theory . . . I consider trash. The great accomplishment of Schrödinger's theory is the calculation of matrix elements.

W. Heisenberg (letter of 8 June 1926 to W. Pauli, in Pauli, 1979)

In a more objective tone one of Schrödinger's principal criticisms against the quantum mechanics was that in his opinion it was "extraordinarily difficult" to approach processes such as collision phenomena with a "theory of knowledge" in which we "suppress intuition [*Anschauung*] and operate only with abstract concepts such as transition probabilities, energy levels, and the like." For although, he continued, there may exist "things" that cannot be comprehended by our "forms of thought," and hence do not have a space-and-time description, "from the philosophic point of view" Schrödinger was sure that "the structure of the atom" did not belong to this set of things (1926a).[12]

Schrödinger preferred a continuum-based theory over the "true discontinuum theory" of Heisenberg. In fact, Schrödinger pushed his proof of the mathematical equivalence of the wave and quantum mechanics to the conclusion congenial to his own viewpoint— when discussing atomic theories one "could properly also use the singular."

But what sort of picture did Schrödinger offer? He preferred none at all to the miniature Copernican atom, and in this sense the purely positivistic standpoint of the quantum mechanics was preferable because of "its complete lack of visualizability"; however, this conflicted with Schrödinger's philosophic viewpoint. Schrödinger based his visual representation of bound and free electrons on comparison of the classical electrodynamics of an electron with the wave function from the Schrödinger equation. Wave mechanics took the electron in a hydrogen atom to be represented as a distribution of electricity around the nucleus. However, Schrödinger's (1926b) proof of the localization of a free electron represented as a wave packet was shown by Heisenberg to be invalid (1927); rather, the wave packet did not remain localized.

Schrödinger (1926a) went on to emphasize that his visual representation was unsuitable for systems containing two or more electrons because the wave function may be represented in a space of $3N$ dimensions, where N is the number of particles. Hence, vis-à-vis Heisenberg, Schrödinger took energy levels and transition probabilities to be abstract quantities. We see emerging here personal taste for an aesthetic-philosophical sort as to the meaning of abstract quantities and thus of *Anschaulichkeit* too.[13]

In summary, whereas no adequate atomic theory existed as of July 1925, by mid-1926 there were two seemingly dissimilar theories. Heisenberg's quantum mechanics was supposed to be a "true discontinuum theory." Although a corpuscular-based theory, it renounced any visualization of the bound corpuscle itself. However, its mathematical apparatus based on matrices was unfamiliar to most physicists and also difficult to apply. On the other hand, there was Schrödinger's wave mechanics, which was a continuum theory focusing entirely on matter as waves, offering a visual representation of atomic phenomena and accounting for discrete spectral lines. Its more familiar mathematical apparatus based on partial differential equations set the stage for a calculational breakthrough. Although the final experimental verification of the wave-particle duality of matter would not appear until 1927, many already subscribed to it. The wave mechanics delighted the more continuum-based portion of the physics community, who were intent on preserving pictures and clinging to a classical realism. For example, on 26 April 1926 Einstein wrote to Schrödinger, "I am convinced that you have made a decisive advance with your formulation of the quantum condition, just as I am equally convinced that the Heisenberg-Born route is off the track" (Schrödinger, 1963).[14] On 27 May 1926 the venerable H.A. Lorentz wrote to Schrödinger that "if I had to choose between your wave mechanics and the matrix mechanics, I would give preference to the former, owing to its greater visualizability [*Anschaulichkeit*]" (Schrödinger, 1963).[15]

Heisenberg expressed his sharp opinion of Schrödinger's wave mechanics in the 8 June 1926 letter to Pauli quoted from in the epigraph to this section. And, as Heisenberg wrote in his 1926 paper "Many-body problem and resonance in quantum mechanics," it was for "expediency" in calculations that he used Schrödinger's wave functions. In this paper Heisenberg exploited the mathematical equivalence of the two theories with the caveat that Schrödinger's "intuitive pictures [*anschauliche Bilder*]" should not be imposed on the quantum theory. For, continued Heisenberg, as Schrödinger himself had written, such pictures led to conceptual difficulties in treating many-body problems. Schrödinger's reason was that both theories used formulas from classical mechanics that considered atomic objects to be point particles, but this was "no longer permissible" because atomic objects were "actually extended states of vibration that penetrate into one another" (1926a). Heisenberg, on the other hand, avoided such conundrums for he had renounced visualizability of atomic processes. Thus Heisenberg could declare Schrödinger's opinion of many-body atomic problems to be illegitimate, much as Heinrich Hertz in (1894) had judged certain problems that he took to be detrimental to the advance of electromagnetic theory, for example, the nature of electric charge. In the 1926 paper *treating* the many-body problem Heisenberg wrote that we must set limitations "on the discussion of the intuition problem [*Anschauungsfrage*] owing to situations in which the wave representation is more constrained"—for example, "the notion of the spinning electron," which resisted perception or intuition through pictures. Furthermore, Heisenberg continued, Schrödinger's method was not a consistent wave theory of matter in the sense of de Broglie's because while de Broglie's waves were in a space of three dimensions, Schrödinger's were not and yet borrowed "concepts from the corpuscular theory." On the other hand, the result of the program of "de Broglie and Einstein" would be a "comprehensive description of atomic phenomena in our customary concepts of space and time." Yet Heisenberg was skeptical of such a wave theory of matter because "the hope of the continuum theorists" was for corpuscles to emerge as "singularities of the metrical structure of space," and such a theory would be formulated in a noneuclidean space in which there could be no description in terms of our customary concepts of space and time. Thus, according to Heisenberg, since any route in atomic theory led to loss of visualizability, physicists should maintain only the quantum mechanics that arrived at this point naturally, that is, via the correspondence principle.

The tension between the quantum and wave mechanics increased with the appearance of Born's quantum theory of scattering in the latter part of 1926. Born's analysis of data from the scattering of electrons from hydrogen atoms convinced him of the need for a quantum-theoretical description of scattering consistent with the conservation of energy and momentum; yet, Born writes, neither scattering problems nor transitions in atoms can be "understood by the quantum mechanics in its present form" (1926b)—here, under quantum mechanics, Born includes *both* Heisenberg's and Schrödinger's mechanics. His reasons are: Heisenberg's quantum mechanics denies an "exact representation of the processes in space and time" (1926c); Schrödinger's wave mechanics denies visualization of phenomena with more than one particle. In Born's view treating problems concerning scattering and transitions requires the "construction of new concepts," and his vehicle would be Schrödinger's version because it allows the use of the "conventional ideas of space and time in which events take place in a completely normal manner," that is, the possibility of visualizability.

One new concept that Born proposed was rooted in some unpublished speculations of Einstein, namely, that light quanta were guided by wave fields that carried only probability, thereby providing the means for discussing interference and diffraction using light quanta. Born boldly assumed the "complete analogy" between a light quantum and an electron in order to postulate the interpretation that the "de Broglie-Schrödinger waves" (the wave function in a three-dimensional space) was the "guiding field" for the electron.

Owing to the mathematical characteristics of the wave function, Born attributed physical reality to its magnitude squared. For example, Born continued (1926b), in a scattering process the magnitude squared of the coefficient of the outgoing wave function is the probability that a particle incident originally along the z-axis is scattered into a solid angle, $d\Omega$.[16] This view led Born "to be prepared to give up determinism in the world of atoms." For although the carrier of probability, that is, the wave function, developed causally, all final states were probable (although in general not equally probable) that were consistent with the conservation laws of energy and momentum. In Born's view, quantum mechanics distinguished between causality and determinism. Furthermore, since the quantum "jump itself" in an atom "def[ies] all attempts to visualize it," then for atomic transition processes, the notion of causality was meaningless, and one was left only with the quantum-mechanically determinate" (1927).

Heisenberg was enraged over Born's use of the Schrödinger theory and his assessment of the quantum mechanics. In fact, from the beginning Heisenberg had been no less outspoken than Schrödinger, for soon after the appearance of wave mechanics he referred to it as "disgusting" in a letter to Pauli. Heisenberg recalled that "Schrödinger tried to push us back into a language in which we had to describe nature by *'anschauliche Methoden'* . . . . Therefore I was so upset about the Schrödinger development in spite of its enormous success" (*AHQP*: 22 February 1963). Then came Born's paper in which "he went over to the Schrödinger theory."

Heisenberg described these developments as very disturbing to his "actual psychological situation at that time"; namely, that quantum mechanics was a complete and thus closed system. Born, on the other hand, had assessed it as incomplete and introduced a new hypothesis using Schrödinger's wave mechanics. It was in response to this highly charged emotional atmosphere that Heisenberg wrote his paper of (1926c), "Fluctuation Phenomena and Quantum Mechanics." He remembered that although this paper received very little attention, "For myself it was a very important paper."[17] Indeed, it is a paper written by an angry man in which Born's theory of scattering is not cited and Schrödinger is sharply criticized. There Heisenberg demonstrated that a probability interpretation emerges naturally from the quantum mechanics and can be understood only if there are quantum jumps. At the conclusion he comes down firmly in favor only of a corpuscular viewpoint. This becomes crystal clear in Heisenberg's important review paper of 1926 "Quantum Mechanics" where once again Born is not mentioned and Schrödinger is soundly criticized.

Heisenberg's first epistemological analysis of the quantum theory, entitled "Quantenmechanik" (1926b) was written at Copenhagen, where during the latter part of 1926 through the spring of 1927 he and Bohr struggled toward a satisfactory physical interpretation of the new theory. The fundamental problems on which Heisenberg focused in the 1926 paper enable us to glimpse the central points of their struggle, even though it expressed more Heisenberg's own view.

"According to our customary intuition [*gewöhnliche Anschauung*]," Heisenberg wrote, "that is, the application of the usual space and time concepts, space and matter are imagined to be in the last analysis divisible in principle into arbitrarily small parts." However, he continued, this turned out not to be the case for the following reasons: empirical evidence supported by theoretical calculations for atomic

structure; and owing to fluctuation phenomena processes occurring
in small regions of space and time exhibit "a typically discontinuous
element." In the past we attributed to electrons the "same sort of
reality as the objects of our daily world; we represented to ourselves
these basic building blocks as extraordinarily small particles of
known charge and mass but unknown internal structure, which
move precisely according to fathomable laws in space and time and
certainly complying with our intuition [*Anschauung*] of the familiar
continuity of the space-time world." Yet, Heisenberg went on, "in
the course of time this representation has proved to be false." For the
"electron and the atom possess not any degree of direct physical
reality as the objects of daily experience." Then there was the
"fecundity of the light quantum hypothesis," as demonstrated by the
photoelectric effect, the Compton effect, and the experiment of
Bothe and Geiger. Nevertheless, he continued, from the beginning,
"in contrast to the material particle we have never attributed the
degree of reality to the light quantum which is due the objects of
daily experience" because the light quantum contradicted the
"known laws of classical optics." Thus in November 1926 Heisenberg was still equivocal on the reality of the light quantum, although
he went on to note that empirical verification was accruing. On the
other hand, Heisenberg continued, perhaps it was, as Einstein had
emphasized, that "conversely the *electron* is due a similar degree of
reality as the light quantum" (italics in original). Heisenberg considered this point as symptomatic of the fundamental problem of
atomic physics: "the understanding of each typically discontinuous
element and its 'kind of reality'."

Recalling the models of the Bohr theory Heisenberg wrote, "The
program of quantum mechanics has above all to free itself from these
intuitive pictures [*anschauliche Bilder*]" (or, as he had written in the
previous paragraph of this essay, "*anschauliche Modelle und Bilder*").The new theory, Heisenberg continued, should deal only with
directly measurable quantities: "The earlier theory had the benefit of
direct visualizability [*Anschaulichkeit*] and the use of accepted physical
principles with the disadvantage that in general it calculated with
relations that in principle were not testable and thereby could lead to
internal contradictions; the new theory ought above all to give up
totally on visualizability [*Anschaulichkeit*]," thereby avoiding any internal contradictions.

Heisenberg's "*first decisive restriction* in the discussion of the reality
of the corpuscle" (italics in original) was to discard notions of the

bound electron's position because this had been necessary in order to formulate the quantum mechanics in the first place. Heisenberg's second restriction on the reality of the corpuscle was a result of Einstein's theory of the ideal quantum quantum gas, namely, that the "individuality of the corpuscle is lost."

Whereas Heisenberg began this essay by demonstrating how nature in the small contradicted our customary *Anschauung*, he concluded with emphasis on the fact that the existing scheme of quantum mechanics contained contradictions of the "intuitive interpretations [*anschauliche Deutung*]" of different phenomena, and this was unsatisfactory. For besides the quantum theory's restrictions on the reality of the corpuscle, there lurked the light corpuscle. Then, after repeated warnings throughout this paper against intuitive interpretations of the quantum mechanics, Heisenberg concluded by reporting from Copenhagen that "hitherto there has been missing in our picture [*Bild*] of the structure of matter any substantial progress toward a contradiction-free intuitive [*anschaulich*] interpretation of experiments that in themselves are contradiction-free."

At Copenhagen Bohr and Heisenberg struggled with the restrictions on physical reality discussed by Heisenberg in "Quantenmechanik" and such attendant paradoxes as, in the diffraction by a double-slit grating of a low-intensity beam of electrons, which slit does a single electron pass through, or does it pass through both slits? (*AHQP*: 25 February 1963). Then there was the "paradox of polarized light," which concerned the interpretation of a polarized light quantum. Their approaches to Gedanken experiments involving these restrictions and paradoxes differed. By late 1926 Bohr had accepted the wave-particle duality of light and matter and, recalled Heisenberg, "wanted to take this dualism [as the] central point." Consequently, Bohr could deal effectively in Gedanken experiments with pictures. On the other hand, Heisenberg persisted in focusing on the corpuscular aspect of matter, for which there was a "consistent mathematical scheme. I just wanted to forget about the wave packets and the waves." In Heisenberg's opinion Schrödinger's 1926 equivalence proof of wave and matrix mechanics was unsatisfactory, and so Heisenberg recalled that he "had always in mind the old theory of Duane," according to which wave phenomena could be obtained from particles through the phase integrals from the old quantum theory.

After reading a preprint of P.A.M. Dirac's transformation theory (1926), on 23 November 1926 Heisenberg wrote Pauli: "I consider

Dirac's work as an extraordinary advance (Pauli, 1979). It then became clear to Heisenberg how "things are related," and he could mathematize Bohr's Gedanken experiments. Nevertheless, Heisenberg continued in the letter to Pauli, he remained focused on the mathematical formalism with its essential discontinuities and unvisualizability: "That the world is continuous I consider more than ever as totally unacceptable. But as soon as it is discontinuous, all words that we apply to the description of facts are so many numbers. What the words 'wave' or 'corpuscle' mean we know not any more."[18] Heisenberg recalled of this period that "we couldn't doubt that [quantum mechanics] was the correct scheme but even then we didn't know how to talk about it . . . [these discussions left us in] a state of almost complete despair.[19]

Requiring a respite from their intense conversations, in February of 1927 Bohr went on a skiing holiday to Norway. During this break Heisenberg (1969) "suddenly remembered my conversation with Einstein [spring of 1926, in Berlin] and particularly his statement, 'It is the theory that decides what we can observe.' I was immediately convinced that the key to the gate that had been closed for so long must be sought here."[20] Although in 1926 Heisenberg had demurred, starting in 1927 he would move beyond Einstein's suggestion. Development of this aspect of Heisenberg's research begins with his letter to Pauli of 23 February 1927, in which Heisenberg described in some detail his realization of the uncertainty relations. Heisenberg wrote of the "intuitive [*anschaulich*] meaning I have made by applying Dirac-Jordan to the suitable mathematically complete [nonrelativistic] quantum mechanics." But what sort of "intuitive meaning" could Heisenberg have meant when previously he had been emphatic on rejecting visualizability in the modes of pictures or images in a scientific theory? Heisenberg had concluded the 1926 essay "Quantenmechanik" on a similar note when he called for a "*Bild*" of matter that offered a contradiction-free "*anschaulich* interpretation" of experiments. He provided the answer in a paper that can be regarded as the sequel to "Quantenmechanik," namely, his 1927 paper "On the intuitive [*anschaulich*] content of the quantum-theoretical kinematics' and mechanics" (1927), where he demarcated boldly the notion of how "we believe that we understand a theory intuitively [*anschaulich*]" from the visualization of atomic processes. Heisenberg was driven to this conclusion because "heretofore, the intuitive [*anschaulich*] interpretation of the quantum mechanics is full of internal contradictions that become apparent in the struggle of opinions concerning the discontinuum-and-continuum theory, waves and corpuscles." The quantum mechanics, therefore, satisfied only one of Heisenberg's 1926 criteria "to be understood intuitively." Taking support from the redefinitions of physical reality in the large required by the special and general theories of relativity, Heisenberg asserted that in the atomic domain a revision of our usual kinematical and mechanical concepts "appears to follow directly from the fundamental equations of the quantum mechanics." Thus, Heisenberg permitted the mathematics of the quantum mechanics to determine the restrictions on such perception-laden symbols as position and momentum. Using the Dirac-Jordan transformation theory he went on to deduce the famous uncertainty relations. For example, since the product of the uncertainties in measurements of an electron's position and momentum is a small but non-zero number (Planck's constant), then the more precise that the electron's position is known, the less precise can its momentum be ascertained. Focusing exclusively on the particle aspect of matter, and misusing the light quantum (e.g., in the γ-ray microscope Gedanken experiment), led Heisenberg to state that ultimately the reason for revising customary concepts of physical reality is rooted in the "typical discontinuities" of atomic processes. Since the uncertainty relations placed limits on the accuracy to which initial conditions could be determined, Heisenberg rejected the causal law from classical mechanics that required visualization and the continuous development of physical systems. Thus, concluded Heisenberg, if we bear in mind the uncertainty principles, we "should no longer regard the quantum mechanics as unintuitive [*unanschaulich*] and abstract."[21] The uncertainty principle paper was, therefore, a turning point in Heisenberg's view of the nature of physical reality. In later years he recalled Pauli's brief reply to the 23 February 1927 letter as "*Es wird Tag in der Quantentheorie*" (1960).

Bohr thought otherwise. During his skiing trip he had concluded that the only way to reestablish some sort of visualization using notions from the world of perceptions was to separate the space-time pictures from the conservation laws and causality. Furthermore, Bohr realized that the wave-particle duality of light and matter, and not the essential discontinuities, lay at the root of the necessity for the redefinition of physical concepts. He emphasized that Heisenberg's concentration on half the picture had led him astray, causing Heisenberg to renounce as misleading any visualizability of the quantum-theoretical formalism. In Bohr's view the wave-particle duality and not the essential discontinuities was the basis of the uncertainty relations. Like Heisenberg, Bohr took recourse to Einstein's special relativity theory as an indicator of the need to change physical concepts, for example, Einstein's investigations of the consequences of the large but not infinite velocity of light, even though our perceptions are not open to such a high velocity. But unlike Heisenberg, Bohr did not at this juncture permit the theory to decide what was observable, as Heisenberg recalled he had learned from a conversation in the spring of 1926 with Einstein. Instead Bohr delved into the atomic regime to analyze the consequences of Planck's constant, whose smallness led to consequences beyond our perceptions, such as linking the wave and particle modes of matter and light. In a lecture delivered on 16 September 1927 at the International Congress of Physics, Lake Como, Italy, Bohr encompassed his viewpoint in what he referred to as the complementarity principle: "The very nature of the quantum theory thus forces us to regard the space-time coordination and the claim of causality, the union of which characterizes the classical physical theories, as complementary but exclusive features of the description, symbolizing the idealization of observation and definition respectively" (1928).[22]

Thus by separating causality from a space-time description Bohr was able to regain, in a suitably restricted form, pictures from the world of perceptions, that is, *Anschaunngsbildern*, whose loss he had lamented in 1925 and for which he was willing to go to the length of welcoming the notion of spin in 1926. Bohr's 1923–1926 essays are filled with visual words such as *Bild* and *Vorstellungen vor Augen*. Consequently the lack of anchoring words with perceptions in 1926, an anchor for which Bohr had emphasized the need in 1913, must indeed have driven him to despair. As Bohr wrote in 1927 on the possibility of further renunciation of visualization when relativity is applied to quantum theory, "Indeed we find ourselves here on the very path taken by Einstein of adapting our modes of perception borrowed from the sensations to the gradually deepening knowledge of the laws of nature. The hindrances met with on this path originate above all in the fact that, so to say, every word in the language refers to our ordinary perception" (1928).

Bohr returned to Copenhagen sometime after 23 February 1927 and pressed his own view relentlessly, even attempting to prevent Heisenberg from publishing the uncertainty principle paper. Heisenberg's only concession was a note added in proof, in which he wrote that Bohr's recent investigations had led him to conclude that observational uncertainties were rooted in the wave-particle duality of matter and not "exclusively on the presence of discontinuities."

By the end of March 1927, owing principally to the mediation of Oskar Klein, relations became once more amicable between Bohr and Heisenberg. An indicator of the warm atmosphere at Copenhagen is in Heisenberg's letter to Pauli of 16 May 1927 in which Heisenberg wrote of Bohr's discussing with him a "general work on the 'conceptual foundations' of the quantum theory written from the viewpoint, 'There are waves and particles' " (Pauli, 1979). Differences of opinion remained between them; for example, in this letter Heisenberg wrote that to him "the discontinuities are the uniquely interesting aspects of quantum theory, and we can never emphasize them enough." In addition, continued Heisenberg, there are "presently between Bohr and myself differences of opinion on the word '*anschaulich*'." Indeed, for Bohr, who was sharpening the notion of complementarity, the mathematical formalism could not dictate what was to be intuitive, particularly a formalism that emphasized discontinuities, that is, particles over waves.

On 31 May 1927 Heisenberg wrote to Pauli, "As you know I deem the Dirac-Jordan theory as better than the wave mechanics (even in the discontinuous correspondence rule form) because the Dirac-Jordan theory is unintuitive [*unanschaulich*] and general and the discontinuities are easier to formulate" (Pauli, 1979). This passage underscores Heisenberg's distinction between *anschaulich*, referring to pictures and images analogous to those from the world of perceptions (as was the case for the wave mechanics), and the classical nonvisualizability of the quantum mechanics with its accent on the discontinuum.

Heisenberg's writings from the period 1928–1929 in conjunction with his reminiscences reveal that he came to understand the symmetry of the "complementarity picture" from the mathematical statements of the complementarity principle—the methods of the quantization of wave fields formulated for the electromagnetic field by Dirac in 1927 and then extended to particles with mass by Jordan, Klein, and Wigner in 1927–1928.[23] This is a form of the quantum theory in which the wave and particle aspects of matter can be transformed mathematically into one another and yet remain mutually exclusive. Dirac wrote the 1927 paper in Copenhagen and thanked "Bohr for his interest and for friendly conversations"; Bohr communicated the paper to the *Proceedings of the Royal Society*. In fact, we can speculate that Bohr's influence on Dirac can be discerned through

Dirac's statement that "there is thus a complete harmony between the wave and particle descriptions." Dirac's results may even have supported Bohr's thinking on complementarity. (Dirac's paper was received 2 February 1927, at least two weeks previous to Bohr's trip to Norway.)

FURTHER DEVELOPMENTS IN VISUALIZABILITY

After Dirac's great paper on the theory of the electron one had the impression that all the fundamental features of atomic physics had been neatly incorporated into the new conceptual structure, and with characteristic eagerness the other pioneers of the atomic world, Heisenberg and Pauli, leaving to lesser fry the polishing off of details, turned to the major remaining task of applying the new methods of quantization to the electromagnetic field. It is difficult to those who did not witness it to imagine the enthusiasm, nay the presumptuousness, which filled our hearts in those days. I shall never forget the terse way in which a friend of mine (now a very eminent figure in the world of physics) expressed his view of our future prospects: "In a couple of years", he said, "we shall have cleared up electrodynamics; another couple of years for the nuclei, and physics will be finished. We shall then turn to biology."

L. Rosenfeld (1967)

Visualizability (*Anschaulichkeit*) had been attained for the atomic regime by being redefined along with the notion of intuitive (*anschaulich*). These notions had undergone transformations in the direction that Heisenberg had indicated in the uncertainty principle paper and gone beyond what Pauli had called for in his 12 December 1924 letter to Bohr: the theory decided what was to be intuitive and visualizable. The mathematics of the quantum mechanics decided the meaning of the theory's symbols and thus the theory's "intuitive content," as well as the notion of visualizability in the atomic realm. This was an important step because in the atomic domain visualization and visualizability are mutually exclusive. Visualization is an act of cognition. So visualization is what Heisenberg had referred to in the 1926 paper "Quantum Mechanics" as the "customary intuition [*Anschauung*]" that could not be extended into the atomic domain. Visualizability concerns the intrinsic properties of elementary particles that may not be open to our perceptions, e.g, electron spin. In classical physics visualization and visualizability are synonymous. But from 1927 through 1932 Heisenberg resisted any imagery of atomic phenomena—that is, for Heisenberg and other quantum physicists such as Pauli, in the atomic domain visualizability did not yet possess a depictive or visual component.

Bohr was at first uncomfortable with this style of thinking. Concerning, for example, the 1929 work of Heisenberg and Pauli on quantum electrodynamics Léon Rosenfeld recalled that Bohr "always regarded with deep suspicion any theory not solidly anchored in some concrete bit of reality, and the difficulties encountered in the attempted extension of quantum theory to electrodynamics seemed to him so remote from any familiar physical situation that he was not easily persuaded to take them seriously."

NUCLEAR PHYSICS: METAPHOR BECOMES PHYSICAL REALITY

But Heisenberg thrived on this mode of thinking, as is clear in his series of papers in 1932–1933 entitled, "Über den Bau der Atomkerne. I," "II," "III." These key papers in the history of science are vintage Heisenberg. As perhaps nowhere else in twentieth-century physics we can glimpse in published form the struggles of a scientist trying to formulate a theory in a situation where basic laws of nature as well as current theory seem to be violated. In this case Heisenberg sought to frame a theory of the nucleus that included the newly discovered neutron. In order to establish a basis for the first modern nuclear theory Heisenberg thought it necessary to spin new ideas on the notions of force and of physical reality itself.

The euphoria recalled by Rosenfeld in the epigraph to the section, "Further Developments in Visualizability," was short-lived owing to several interconnected problems. Besides the difficulty of interpreting the electron's negative energy states, there was Klein's paradox, Dirac's hole theory, the electron's divergent self-energy, and the vexatious problem of explaining the continuous β–ray spectrum from the decay of nuclei. The β–ray spectrum's continuous nature (in velocity) had been known since 1902 when it was considered to be merely an inconvenience for measuring the charge-to-mass ratio of high-velocity electrons (see Miller, 1981b). By 1927 the spectrum's continuity was established as a property of the disintegrating nucleus. Furthermore, since protons and electrons were assumed to comprise nuclei, then N^{14} had the wrong statistics.[24] Thus physicists suggested that nuclear electrons may not obey quantum theory (Dirac's equation included) because their spins are suppressed, among other reasons. Some physicists, principally Bohr, suggested that owing to the nuclear electron's continuous spectrum in energy for a supposedly two-body final state, then energy conservation is invalid in β-decay.[25] In unpublished speculations of 1930, Heisenberg had even flirted with dropping charge conservation in the nucleus. This was too much even for Bohr.[26]

Introducing the neutron into the nucleus, wrote Heisenberg in his first paper on nuclear physics, leads to an "extraordinary simplification for the theory of the atomic nucleus" (1932a). Because in addition to settling the problem of the correct statistics for N^{14}, the neutron permitted Heisenberg the means to relate the problem of β-decay to the form of the attractive force that binds the nucleus. As Heisenberg wrote to Bohr on 20 June 1932 of his work on nuclear physics, "The basic idea is to shift the blame for all principal difficulties onto the neutron and to refine quantum mechanics in the nucleus."[27]

Introducing the neutron into the nucleus, wrote Heisenberg in his first paper on nuclear physics, leads to an "extraordinary simplification for the theory of the atomic nucleus" (1932a). It permitted him to relate the "fundamental difficulty which we face in β-decay," to the form of the attractive force that operates between the neutron and proton in order to bind the nucleus. First he had to decide whether the neutron is a composite particle consisting of proton and electron or a fundamental particle. In either case, Heisenberg assumed that the neutron is a particle with spin one-half in order to provide N^{14} with the correct statistics. The conception of the neutron as a bound state, or as a collapsed hydrogen atom, had been proposed by Rutherford in 1920 and was advocated by the neutron's discoverer, James Chadwick. However, to Heisenberg, a nonelementary neutron would require the electron to obey incorrect statistics, and so "it does not seem appropriate to elaborate further on such a picture [*Bild*]."[28] If under the proper circumstances a fundamental neutron decays into a proton and electron, then the "conservation laws of energy and momentum are probably no longer applicable." At this time Heisenberg had not accepted Pauli's hypothesis of a neutrino, which is a massless particle that permits application of these two conservation laws with the result that the proton, electron, and neutrino can share the energy and momentum in a variety of ways that are characteristic of a three-body final state.[29] Although Heisenberg chose to consider the neutron as a fundamental particle, in the succeeding deliberations he equivocated, owing principally to the problem of where the electrons originated in β-decay. And so, when necessary, he invoked arguments based on conservation of energy—for example, when he discussed the stability of certain nuclei against β-decay. But then, at the paper's conclusion, he suggested that for certain processes such as the scattering of light from nuclei, it is useful to assume that the neutron is a composite particle. In fact, the difficulty of understanding exactly what Heisenberg was attempting to do in 1932 is attested to by the inaccurate ways in which his work was described. For example, Wigner (1933) wrote that Heisenberg discussed the possibility that the "only elementary particles are the proton and electron." In a review article commemorating the fiftieth anniversary of the neutron's discovery, Frank (1982) wrote that the "hypothesis of the neutron-proton model of nuclei . . . was stated almost simultaneously by Iwanenko . . . and Heisenberg." Actually, Heisenberg considered both cases seriously! It is noteworthy that of all the Heisenberg reminiscences on the history of physics in the 1920s and 1930s that I have read, those pertaining to his 1932 paper are the least trustworthy. For example, he recalled: ". . . to keep an electron in the nucleus would be a dreadful affair from the ordinary point of view of quantum theory. Therefore, if actually there were electrons in the nucleus, then probably you would have to change everything again. So I was extremely happy to see that this was not necessary" (*AHQP*: 12 July 1963). We shall see that in the 1932–1933 papers on nuclear physics, Heisenberg tried in every way to include electrons in the nucleus.

The force between a charged proton and a neutral neutron cannot be of the same sort as that between two charged particles and any proposed form must yield the property of saturation. Thus, in the first 1932 paper on nuclear physics, Heisenberg instead drew an "analogy" between the attractive force of a proton and a neutron and the exchange force that is the dominant factor in a molecule's stability. For example, in the solution to the problem of the H_2^+-ion, a strictly quantum-mechanical contribution to the ion's energy can be described metaphorically as the electron's being shared or exchanged back and forth between the two protons at a frequency that is equal to the exchange energy divided by Planck's constant. Mathematically, during the interval when the electron is primarily bound to one proton, its wave function overlaps the other proton. The exchange contribution to the helium atom and the H_2-molecule arises through the indistinguishability of the electrons and so any depiction of this force "should not be taken seriously" (Bethe and Salpeter, 1957). In fact, the inability of the old Bohr theory to account for the H_2^+-ion's stability presaged the theory's fundamental problems (Fig. 4.2). It turned out that the ion's stability is due to

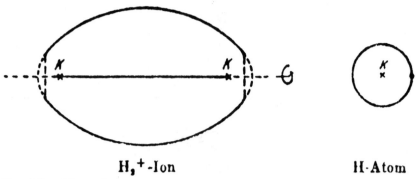

H₂⁺-Ion   **H-Atom**

Fig. 4.2.   These diagrams are from Pauli's (1922) unsuccessful attempt to deduce the characteristics of the H_2^+-ion from Bohr's atomic theory. The letter K denotes the central positive charge. In the "H-Atom" the electron's orbit can be a circle and in the "H_2^+-Ion" it can be an ellipse. (From Pauli, 1922.)

an intrinsic property that is unvisualizable because the exchange force cannot be developed from intuitions constructed from the world of perceptions—that is, our "customary intuitions."

The exchange force was another great invention in quantum theory that Heisenberg had introduced in (1926a) in order to understand the helium atom spectrum.[30] Walter Heitler and Fritz London (1927) used Heisenberg's concept of the "exchange phenomenon" [*Austausch phänomen*] in their theory of homopolar bonding in molecules. Heisenberg (1928) extended the work of Heitler and London toward "clarifying" ferromagnetism." In (1928) Heitler wrote that it is as "yet incomprehensible what exchange in reality means."

As if in response to Heitler, in 1932 Heisenberg extended the exchange force into the nucleus in a way that offered visualizability and thus, in this case, an understanding of what the exchange force "in reality means." Modern nuclear physics and elementary-particle physics began with this passage in Heisenberg's (1932a):

Suppose we bring the neutron and proton to a separation comparable to nuclear dimensions; then, in analogy to the H_2^+-ion, the negative charge will undergo a migration [*Platzwechsel*], whose frequency is given by a function $J(r)/h$ of the separation r between the two particles. The quantity $J(r)$ corresponds to the exchange [*Austausch*] or more correctly migration integral [*Platzwechselintegral*] of molecular theory. The migration can again be made more intuitive [*anschaulich*] by the picture [*Bild*] of electrons that have no spin and follow the rules of Bose statistics. But it is surely more correct to regard the migration integral [*Platzwechselintegral*] $J(r)$ as a fundamental property of the neutron-proton pair without wanting to reduce it to electron motions.

Yet, as we shall see, Heisenberg went on in the (1932–1933) papers to discuss nuclear electrons that "follow the rules of Bose statistics." Heisenberg's change of terminology from "exchange" [*Austausch*] to "migration" [*Platzwechsel*] that I have noticed here for the first time in the literature emphasizes the new concept to follow because he had something else in mind for the neutron-proton force. It is reasonable to conjecture that with this switch from exchange to migration, Heisenberg's visualizability (*Anschaulichkeit*) of the migration (*Platzwechsel*) of the electron in the neutron-proton force is, however unintentionally, the visualizability in Fig. 4.3. In Fig. 4.3 the quantity $J(r)$ is the attractive force between a fundamental nuclear proton and a composite nuclear neutron. The attractive force operates through the "migration" of charge from the neutron to the proton which, capturing the Bose electron, becomes a neutron. For this reason I have rendered *Platzwechsel* as "migration." "Change of place" is inappropriate because in Heisenberg's view the neutron and proton do not merely change places. The metaphor of motion is of the essence here. "Migration" also conveys the meaning of *Platzwechsel* that enabled Heisenberg's vision of the nuclear force to be brought to fruition (in 1935) by Hideki Yukawa. But we are moving ahead of ourselves.

As Heisenberg wrote in (1932b), "the circumstances that the neutron behaves also in many respects like a composite proton and neutron enters only in the considerations for the expressions for the *Platzwechsels* and for β-decay." In order to render the "picture" [*Bild*] of the attractive force more "intuitive" [*anschaulich*], Heisen-

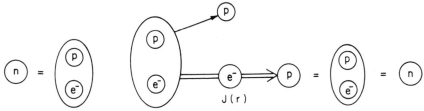

Fig. 4.3. At the left is a composite neutron (*n*) made up of a fundamental proton and a spin zero electron. At the right, according to Heisenberg, the Bose electron from the composite neutron "migrates" over to a nuclear proton that captures it and becomes a composite neutron. The quantity $J(r)$ is the attractive neutron-proton force. Hence, the Heisenberg exchange force is not merely a change of position for neutron and proton.

berg proposed the "migration" of an electron that "has no spin and obeys the rules of Bose statistics." Was this not Heisenberg's reason for rejecting the nonelementarity of the neutron? But was it not also the case for Heisenberg that the theory decided what was *anschaulich*? Heisenberg had taken the first step toward regaining the visual component of visualizability.

Heisenberg's analogy for the attractive force $K(r)$ between two neutrons was the exchange force in the H_2-molecule that arises from the indistinguishability of the two electrons; thus, this force is more conventional than the neutron-proton force. In the subsequent literature the quantities $J(r)$ and $K(r)$ were referred to as "exchange forces"; thus, to a first approximation, Heisenberg assumed that the exchange forces between neutrons and protons are static and central. Since $J(r)$ is a short-range force, the expressions most frequently used for it are

$$J(r) = e^{-r/a} \quad \text{and} \quad J(r) = e^{-(r/a)^2}$$

where a and b are constants that are adjusted to fit empirical data (e.g., Heisenberg, 1933b, and Gamow, 1937).

For considerations other than the *Platzwechsels* and β-decay, continued Heisenberg, the "neutron can be interpreted as a solid elementary building block of the nucleus." Thus, as Heisenberg recalled, in order to describe mathematically the exchange force in which the neutron would be "just a proton without charge . . . I came to this isotopic spin business" (*AHQP*: 12 July 1963). The notion of isotopic spin permitted Heisenberg to describe the nucleus as if it were composed of identical particles of spin one-half that obeyed the Pauli exclusion principle; thus, in Heisenberg's new formalism that includes the notion of migrating electrons, the neutron and proton are two different states of the nuclear particle (or nucleon), depending on whether the nucleon's fifth degree of freedom, or isotopic spin, is $+1$ or -1. Thus, in Heisenberg's theory of the nuclear force, the composite and elementary neutron stand side by side.[31] Rosenfeld's description of Heisenberg in the epigraph to the section, "The New Quantum Mechanics," was well put.[32]

In summary, in order to describe the attractive force between the neutron and proton, Heisenberg extended the metaphorical exchange force from the quantum theory of molecules to an *Anschaulichkeit* (visualizability) with the "migration" of an electron, however "unintentionally," because the migrating electron obeyed the wrong statistics; this problem had been Heisenberg's reason for not preferring the composite neutron in the first place.[33]

The principal problem that confronted Heisenberg in his attempt to formulate a theory of the nucleus in which the β-decay interaction was linked to the neutron-proton force lay in the origin of the electrons in β-decay. According to the uncertainty principle, the neutron could not be a proton-electron bound state because the composite neutron's binding energy would have to be of the order of 137 mc^2 (m is the mass of an electron), which is one hundred times greater than the measured neutron-proton mass difference (1932b). On the other hand, continued Heisenberg, a "meaningful definition of the concept of binding energy is impossible for the electron in the neutron on account of the denial of the energy law for β-decay." Thus he was led to conclude that the fact that "the stability of the neutron cannot be described by the accepted theory permits the separation of quantum mechanics into accessible and inaccessible domains"—that is, the usual quantum mechanics does not apply to nuclear electrons (1933a). Heisenberg went on to emphasize that whereas either the composite or fundamental neutron was adequate for discussing the properties of lighter nuclei, the composite neutron was required for the heavier nuclei which are β-emitters. Another choice for the heavier nuclei was to assume that they are composed of fundamental neutrons, protons, and electrons, whereas ordinary quantum mechanics applies to the fundamental protons and neutrons but not to the electrons. But the nuclear electrons would have to be bound to the neutrons, protons, and α-particles, which gives rise to yet another problem: the nuclear electrons should transfer some of their energy to the α-particles because of the strong binding between them. But the emitted α-particles have a definite energy and yet the nuclear electrons to which the α-particles are tightly bound do not have a definite energy.

Playing all ends against the middle in this theoretical free-for-all offered Heisenberg an "arbitrariness [in seeking] a formulation that will not lead sooner or later to internal difficulties."

Other sorts of exchange forces were proposed that agreed with available data better than Heisenberg's. These forces did not require inappropriate migratory electrons.[34] But Heisenberg's visualizability of a nuclear exchange force with something exchanged turned out to be fruitful. It led to Enrico Fermi's 1934 theory of β-decay whose success, wrote Heisenberg (in 1935), was "proof of the existence of exchange forces" in nuclei.

In late 1933 Enrico Fermi solved the problem of the origin of the β-decay electrons (1934). Fermi took seriously the neutrino hypothesis that Pauli had offered as a "desperate conclusion" for understanding the continuous β-decay spectrum without resorting to non-conservation of energy.[35] Fermi's theory of β-decay is based on Heisenberg's isotopic spin formalism and the techniques of second quantization. In fact, until Fermi's work, research in quantum field theory was of a largely speculative-theoretical sort.[36]

Fermi (1934) wrote that in order to understand β-emission we should seek "a theory of emission of light particles from a nucleus in analogy to the theory of the emission of a light quantum from a decaying atom according to the usual basis of radiation processes." Fermi was able to reap the benefits of Heisenberg's 1932 *Anschaulichkeit* by viewing the exchange process differently from Fig. 4.3. Rather, for β-decay, he considered the exchange to be analogous "to the emission of a light quantum" in a two-level atomic system. Thus, for Fermi, Fig. 4.3 became the diagram in Fig. 4.4.

Here both neutron and proton are considered to be as elementary as the energy levels in an atom. So, according to Fermi, just as the light quantum emitted by an atom in an atomic transition was not present in the atom prior to the transition, neither was the electron present in the nucleus prior to β-decay.

Fermi's Hamiltonian interaction for Fig. 4.4 is

$$H_{\text{interaction}} = g_w \, (\bar{\psi}_p \, \gamma_\mu \, \psi_n) \, (\bar{\psi}_{e^-} \, \gamma_\mu \, \psi_\nu).$$

The precise mathematical meaning of these symbols is unimportant for our purposes because the point I want to make—as did Fermi, although without reference to a diagram like Fig. 4.4—is that they have a depictive connotation: $\psi_\nu(\bar{\psi}_e)$ designates the creation of a neutrino (electron), ψ_n designates the destruction of a neutron; and $\bar{\psi}_p$ the

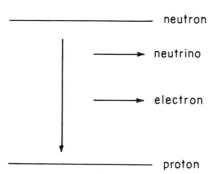

Fig. 4.4.   According to Fermi a neutron becomes a proton with the emission of an electron and a neutrino. The inverse process is the one in which a proton becomes a neutron with the "disappearance of an electron," that is, production of a positron and a neutrino.

creation of a proton (where g_w is the strength of the β-decay process).[37]

In his 1936 book *Anschauliche Quantentheorie*, Jordan wrote elatedly of Fermi's "*anschaulich* thinking of the analogy between γ-radiation and β-radiation."[38] It is reasonable to conjecture that by 1934, as a result of such investigations of the decay schemes of many-level atomic systems—such as that of Victor Weisskopf and Eugene Wigner in 1930—physicists were gaining insights into atomic and then nuclear processes through diagrams like Fig. 4.5.

In (1934) Igor Tamm elaborated and further generalized Fermi's *anschaulich* theory of β-decay. Tamm assumed that the neutron and proton result from the splitting of a degenerate state by the exchange energy that originates in the neutron's emission of the electron and neutrino, or inversely the proton's emission of a positron and neutrino. The splitting of the energy levels is analogous to what occurs in molecules. So far Tamm has offered nothing really new. But next he generalized the mental imagery in Fermi's theory of β-decay as follows: "The role of light particles (ψ-field) [i.e., electrons and neutrinos] providing an interaction between heavy particles corresponds exactly to the role of the photon (electromagnetic field), providing an interaction between electrons . . ." Using suitably modified methods of quantum electrodynamics Tamm went on to show that the interaction energy from Fermi's β-decay theory was too small to explain the binding energies of neutrons and protons in nuclei. Thus, contrary to the hopes of Heisenberg and Fermi, a theory of β-decay could not cover the neutron–proton force as well.

Directly following on Tamm's paper is one by D. Iwanenko that is similar in thrust to Tamm's.[39] Like Tamm, Iwanenko (1934) compared the "interaction *exchange* energy (Heisenberg's *Austausch*) be-

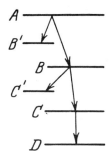

Fig. 4.5. Weisskopf and Wigner's (1930) depiction of atomic transitions. (From Weisskopf and Wigner, 1930.)

tween proton and neutron [to] the birth and absorption of a photon in the case of two electrons" (italics in original). It is noteworthy that in his English-language paper Iwanenko used the German *Austausch* for exchange even though he compared the mechanism of the neutron-proton force to the coulomb interaction mediated by a photon. We can conjecture that Iwanenko, like most other physicists circa 1934, missed the purpose of Heisenberg's change of terminology. One physicist who did not was Hideki Yukawa. In November 1934 Yukawa considered the exchange force between the neutron and proton to be a "migration force describable with a *Platzwechselintegral*" (1935).

Yukawa set out to "modify the theory of Heisenberg and Fermi" by changing the analogy of the field of "light particles" transmitting the neutron-proton force, to the exchange of a single new particle between the neutron and proton. He suggested that the transition of a "neutron state to a proton state" need not always proceed through the emission of a neutron and electron,

> but the energy liberated by the transition is taken up sometimes by another heavy particle, which in turn will be transformed from proton state into neutron state. If the probability of occurrence of the latter process is much larger than that of the former, the interaction between the neutron and proton will be much larger than in the case of Fermi, whereas the probability of emission of light particles is not affected essentially. Now such interaction between the elementary particles can be described by means of a field of force just as the interaction between the charged particles is described by the electromagnetic field.

I have quoted extensively in order to let Yukawa's own transformation of imagery unfold. Yukawa's Hamiltonian for the system of neutron, proton, and the "U-field," that is, the interaction force between neutron and proton, is a straightforward generalization of the one in Heisenberg's 1932 isotopic spin paper[31] where, as Yukawa wrote, "we take for the '*Platzwechselintegral*'

$$J(r) = -g^2 \frac{e^{-\lambda r}}{r},"$$

(instead of the expressions for *J* (*r*) on p. 160) and for the moment Yukawa ignored the interaction between neutrons and the coulomb interaction between the protons. The quantity λ is directly proportional to the mass of the exchanged particle and inversely proportional to the range of the strong interaction force between the neutron and proton. We may conjecture with certainty that Yukawa's explicit use of the German word *Platzwechsel* was to signal that he meant not a metaphorical exchange or *Austausch*, but a real "migration." Yukawa noted that the migrating particles must be Bose particles, which can be either positively or negatively charged, and have a mass about 200 times the electron's mass. He offered the theory with frank reservations because, as he said, "such a quantum with large mass and positive or negative charge has never been found by experiment, [and therefore] the above theory seems to be on a wrong line." He did, however, indicate that the new quanta "may have some bearing on the shower produced by cosmic rays."

A GLIMPSE OF THE MICROCOSM

Diagrams depicting two electrons interacting through the exchange of a photon, or a neutron and proton interacting through the exchange of particles are nowhere present in the scientific literature of the period 1926 to 1943. This was the situation despite the proper verbal descriptions of particles transmitting forces. For example, despite P.A.M. Dirac's quantization of the electromagnetic field (in 1927), the conceptualization of the coulomb force transmitted by a photon played little if any role in subsequent work on electromagnetic processes such as electron–electron scattering. Nor have I found diagrams depicting the transmission of forces by particles in the correspondence that I have thus far studied. The only depictions of β-decay are diagrams like the one in Fig. 4.6. But Fig. 4.6 was, after all, the sort of diagram that was the basis of Fermi's *anschaulich* theory of β-decay.[40]

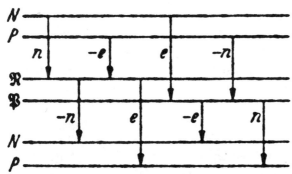

Fig. 4.6.   Depicting β-decay as a "cascadelike" process analogous to the transition between atomic energy levels. Where N (P) denotes a neutron (proton), 𝔑 (𝔓) a "hypothetical stationary state" of a neutron (proton), n (−n) a neutrino (antineutrino), and e (−e) an electron (positron). (From Wentzel, 1936)

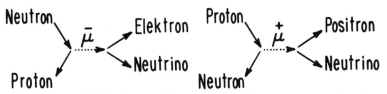

Fig. 4.7. Wentzel's "schemata" for depicting beta–decay as a "two stage process" that is intermediated by a meson. (From Wentzel, 1943—German edition)

In 1943 Gregor Wentzel proposed a "didactic" device for depicting Yukawa's treatment of β-decay as an "indirect" process that is mediated by a virtual meson (Fig. 4.7). These "schemata," wrote Wentzel, depict the "beta-decay process as a two stage transition." Wentzel's Fig. 4.7 depicts the nuclear exchange force that had been first proposed by Heisenberg with no analogy to the photon as an intermediary of the interaction between two electrons. We recall that Iwanenko and Tamm did make the photon analogy, although without any diagram. On the other hand, Yukawa's successful theory of the nuclear exchange force was formulated directly along the lines proposed by Heisenberg, namely, of introducing a *Platzwechselintegral* for describing the nuclear force. It was Yukawa's genial idea to propose a *Platzwechselintegral* that described the actual "migration" of a proper particle—that is, a proper Boson and not a Bose-electron. But Yukawa offered no depiction of this process.

In the 1949 Preface to the English edition of his book, Wentzel wrote that extensive changes were necessary only in the section where the Fig. 4.7 appeared and "here it was easy to modernize the text and to adapt it to the present state of knowledge." Wentzel's book became a valuable stop-gap until the publication of up-to-date books such as that of Bethe and de Hoffmann (1955). But the transition from Wentzel's schematics of 1943 to the Feynman diagrams of 1949 required further transformations of what constitutes physical reality and its accompanying mental imagery.

By 1943 the mental imagery of most physicists had already undergone several transformations. For is it not the case that Fig. 4.7 is another way of "seeing" the energy levels in Fig. 4.6? In turn, Fig. 4.6 is a descendant of the energy level representations that replaced the *customary intuition* of the Copernican atom that classical physics had imposed on the Bohr theory.

I am unable, at this point, to document a direct connection between the line of reasoning initiated by Heisenberg and the one developed by Richard P. Feynman in 1949 into the diagrammatic method. In his Nobel Prize address (1965) Feynman said little about his scientific antecedents. He cited only his work with John A. Wheeler on action-at-a-distance electrodynamics and a (1932) paper by Dirac. Yet the notion of particles other than photons carrying forces was in the literature—for example, Feynman (1949) cited Wentzel's masterful 1943 survey of quantum field theory. At this point I can say only that diagrammatic methods were in the air; not only were there the schematics of Wentzel's kind, but also the notion from Fermi's β-decay paper of letting the mathematics indicate how particles were created and destroyed. The missing ingredient was an intuitive physicist. Feynman fitted the bill. There was, however, one other path that may have led to Feynman's work in 1949. This path was also initiated by Heisenberg.

Elementary-Particle Physics: Visualizability Transformed

While Fermi's theory of β-decay offered some hope for the contemporary quantum theory of fields, there remained such persistent infinities in the interaction between light and matter as in vacuum polarization and in the bound and free electron's self-energies. Amidst the skeptical mood of such major physicists as J. Robert Oppenheimer, Rudolf E. Peierls, and Pauli, Heisenberg continued to realize bold conceptual schemes—for example, in his (1934) paper, "Bemerkungen zur Diracschen Theorie des Positrons," he proposed a method for an *"anschauliche Theorie der Materiewellen."*[41] However, in this method no mental images were involved. The situation that Heisenberg faced in 1934 was similar to the one in 1925 when the old Bohr theory had finally collapsed. Thus, as he had done in 1925, Heisenberg discarded imagery in favor of relying on trustworthy formulae.

His mathematical method was based on Dirac's recently proposed density matrix expressed in the Hartree self-consistent field approximation, in which electrons and positrons can be treated as noninteracting (1934). So, wrote Heisenberg (1934), "there does not enter [into this approximation] the physical quantum-theoretical *unanschaulich* features of the phenomena: the density matrix thus brings about an *anschaulich* correspondence-rule picture [*Bild*] of the actual phenomena—in a similar way as do the classical-mechanical atomic models." The normalization condition for the density matrix, continued Heisenberg, played a role "parallel to the quantum condition of the earlier half-classical theory."[42] After incorporating the conservation laws for charge, energy, and momentum into the new formalism, Heisenberg calculated the functional form for the finite contribution from vacuum polarization. He then offered an intuitive interpretation of vacuum polarization: The dipole moment of an oscillating charge distribution is reduced owing to the vacuum polarization that results from the induced charge in the neighborhood of the oscillating charges. In order to discuss pair creation, pair annihilation, and the self-energy of the photon, Heisenberg extended the procedures of the *anschauliche Theorie der Materiewellen* to the formulation of a *Quantentheorie der Wellenfelder*. In the latter theory the density matrix was expressed in the quantized matter waves of Jordan and Wigner, which were used effectively for the first time in quantum electrodynamics. Heisenberg went on to demonstrate that in the new formalism the photon's self-energy was only logarithmically divergent. This result, in conjunction with the finite value for vacuum polarization, led Heisenberg to be optimistic over the *anschauliche Theorie der Wellenfelder*, which was free of the divergences of previous calculational schemes since it contained the correct "correspondence-rule description of the phenomena." Nevertheless, divergences remained and Heisenberg concluded in the same vein as he had in the 1926 paper, "*Quantenmechanik,*" to the effect that he sought a "contradiction-free unification of quantum theory with the demands that derive from the correspondence to the *anschaulich* field theory."

Heisenberg soon lost faith in the "subtraction physics" because such divergences as in the electron's self-energy remained. Then there were the lack of agreement between experiment and theory as well as the divergences that occurred when the techniques used in the interaction between light and matter, and in nuclear forces and β-decay were applied to cosmic ray phenomena and particle production in high-energy collisions.[43]

In (1938) Heisenberg discussed a notion on which he had been speculating for some time—namely, that since the theory of elementary particles often gave unreliable results for energies above $137mc^2$ (where m is the electron's mass), then there is a "fundamental length" of about 10^{-13} cm below which a new theory is required. (In quantum mechanics there is an inverse relation between energy and length.) Thus, just as classical physics fails for velocities approaching the velocity of light c and for situations where Planck's constant h cannot be neglected, so does quantum field theory fail for processes occurring within the fundamental length. Many physicists rejected

Heisenberg's "fundamental length" and preferred instead to resolve all difficulties through investigating the existing theory of elementary particles. For Heisenberg the main problem was how to move across the boundary into distances less than the fundamental length.

In contrast to most other physicists circa 1938 (or today), Heisenberg excelled in situations where theory was in flux. In 1925, when all else had failed, he had formulated the new quantum mechanics through adroit use of correspondence-rule connection with the old Bohr theory. In 1932 he had proposed the isotopic spin formalism in combination with a new version of an exchange force which was carried by electrons that violated the laws of quantum theory because, in his view, quantum theory may not be valid within the nucleus. These ideas had spurred subsequent developments which led to Fermi's theory of β-decay and Yukawa's theory of nuclear forces, although it turned out that quantum mechanics and conservation of energy did not have to be abandoned for nuclear phenomena. In 1934 he had sought a new version of the contemporary quantum theory of the interaction between light and matter through a correspondence-limiting procedure that avoided troublesome aspects of this theory. The "subtraction physics" was ultimately unsatisfactory, but it offered a ray of hope and its basic theme would survive in the new theory of Feynman. Just as in 1925 when Heisenberg had sought a new theory for the domain in which Planck's constant could not be neglected, and in 1932 when he entertained the possibility that quantum theory failed within the nucleus, so in 1938 he investigated the notion of the "fundamental length."

By 1943 Heisenberg judged the situation in physics to be serious enough to warrant a return to his strategy of 1925 in combination with the "fundamental length" (1943a). He proposed to replace the existing theory of elementary particles with the scattering matrix or S-matrix formalism that was based on only experimentally measurable quantities like the energy and momentum in the initial and final states of a scattering process. He hoped that the S-matrix would provide the means to penetrate interaction distances less than the fundamental length. In Heisenberg's view the crux of the basic problems in elementary particle physics was the starting point for calculations, which was the mathematical expression from classical physics for the energy of the system of particles (i.e., the Hamiltonian interaction) that led to divergences. The S-matrix depended on the interaction Hamiltonian in such a way that Heisenberg was optimistic about avoiding divergence problems. He believed that in the future theory the S-matrix would contain the interaction Hamiltonian in a new way that would go over in some sort of correspondence-rule limit to the mathematical form of the one from classical physics (1943b, 1946). He achieved success for certain idealized situations. Perhaps recalling the problems of interpretation in 1927 for the (then) new quantum theory, he wrote that the next step after the new theory of elementary particles is formulated is to consider its "intuitive contents" [*anschaulichen Inhalt*] (1943b).

Heisenberg's style was first to try to resolve fundamental problems with the existing theory and then to seek a new theory through the method of correspondence-rule limiting procedures. This approach had been dazzlingly successful in 1925, and in 1932, 1934, and 1943 it led to far-reaching results.

After the war, high-precision measurement techniques became available for measuring such small effects as the displacement of energy levels due to the interaction between light and the electrons bound in atoms—that is, the Lamb shift. These experimental results were among the catalysts for theorists to attempt again to cure the infinities in quantum electrodynamics. It turned out that Heisenberg's "subtraction physics" contained the germ of the so-called renormalization program in which infinite terms are grouped together to form by definition measurable quantities like the electron's empirically measured finite mass and charge.[44]

The new theory of the interaction between light and matter came packaged in two formulations that turned out to be equivalent—namely, those of Julian Schwinger and Sin-Itoro Tomonaga on the one hand, and that of Richard P. Feynman on the other. Neither theory required a fundamental length. Schwinger's theory had the aura of mathematical rigor, whereas Feynman's was based on a diagrammatic description which originated in certain mathematical rules whose origin was not rigorous. At first Feynman's formalism was distrusted because it lacked mathematical proofs compared with Schwinger's. In his 1965 Nobel Prize address, Feynman wrote that the absence of rigorous mathematical proofs for his diagrammatic rules caused "the work [to be] criticized, I don't know whether favorably or unfavorably, and the 'method' was called the 'intuitive method'." Detailed proofs turned out to be unnecessary when Freeman J. Dyson in articles written in 1948–1949 established the identity of Feynman's theory with the seemingly more mathematically complete theories of Schwinger and Tomonaga. At this point most of the physics community took up the more intuitive methods of Feynman.

Just as in 1934 when there was the *Anschaulichkeit* for representing atomic decay schemes using energy level diagrams, today Feynman diagrams are *Anschaulichkeit*. In (1950) Heisenberg wrote, "The researches of Schwinger and Dyson on quantum electrodynamics have taught us how we deal with nonlinear reciprocal effect terms, and how the correspondence limit-*anschaulich* representation can be effected after carrying out certain renormalizations. With it has been also established the connection to the S-matrix formalism that had been developed independently of the quantum electrodynamics."

At first I was puzzled as to why Heisenberg singled out Dyson and not Feynman. The reason is very probably that in (1949) Dyson proved that "the Feynman method is essentially a set of rules for the calculation of the elements of Heisenberg's S-matrix . . . ." Thus, in (1950) Heisenberg meant by the "correspondence limit" that the classical physics interaction Hamiltonian can be used if it is properly rewritten according to renormalization prescriptions of the new theory of the interaction of light and matter. And he meant by the "correspondence limit-*anschaulich* representation" what he had referred to in (1943b) as the "intuitive contents" of the new theory that turned out to be the Feynman diagrams. Once again for Heisenberg theory decided what was to be intuitive (*anschaulich*).

Heisenberg's approval of Feynman's diagrams as *anschauliche Methoden* was in accordance with the new meaning of *Anschaulichkeit*. For example, the anthropomorphic or customary version of the coulomb force between two electrons is usually depicted, as any freshman physics text will attest, as in Fig. 4.8a, rather than the appropriate but abstract Fig. 4.8b that is Feynman's diagram.

Fig. 4.8(a). The anthropomorphic or intuitive representation of the repulsive interaction between two electrons.

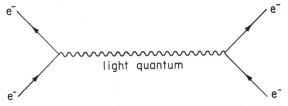

Fig. 4.8(b). The new *Anschaulichkeit* for the coulomb repulsion between two electrons. Two electrons interact by exchanging a light quantum (the wavy line). Today the words exchange and migration are synonymous.

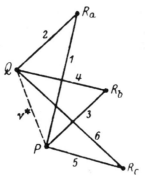

Fig. 4.9(a). A term diagram from the Kramers–Heisenberg paper of 1925, where R_a, R_b, R_c, Q, and P are levels in an atom from which light is scattered. The incident light causes the atom to make transitions from a state P to a state Q via the intermediate states R. The energy difference between the states P and Q is $h\nu^\star$, where the frequency ν of the incident light is much greater than $\nu^\star$. Note that energy need not be conserved in the intermediate transitions. (From Kramers and Heisenberg, 1925)

Fig. 4.9(b). The Feynman diagrams for Fig. 4.9a where $E(E')$ is the energy of the incident (scattered) light, E_P and E_Q are the energies of the atom's initial and final states, and E_R is the energy of possible intermediate states. The atom's trajectory in space-time is designated by the horizontal line.

Basically, the method for construction of Feynman diagrams parallels the one Fermi used in 1934 (Fig. 4.4), which in turn is related to Heisenberg's *Anschaulichkeit* in the isotopic spin paper of 1932, and that in turn derives from the "term diagrams" in the Kramers–Heisenberg paper of 1925 (see Fig. 4.9).

With hindsight Heisenberg wrote that the "term diagrams were like Feynman diagrams nowadays" because the term diagrams were suggested by the mathematics of the scattering process (*AHQP*: 13 February 1963). The Feynman diagrams corresponding to Fig. 4.9a are in Fig. 4.9b. A basic difference between 4.9a and 4.9b is that Fig. 4.9b is a space-time diagram in the special relativistic sense.

Today in quantum theory we use terms taken over directly from the German language that are indicative of the success of Heisenberg's style of doing physics—permitting the theory to decide what is *Anschaulichkeit*. For example, in the extreme there are terms like Schrödinger picture and Heisenberg picture, and interaction picture that have neither picture nor image content; rather, they refer only to symbols. On the other hand, elementary-particle physicists often think in terms of Feynman diagrams which are the new *Anschaulich-keit*.

CONCLUSION

Now, the question to be determined is this: What are the most perfect bodies that can be constructed, four in number, unlike one another, but such that some can be generated out of one another by resolution? If we can hit upon the answer to this, we have the truth concerning the generation of earth and fire and of the bodies which stand as proportionals between them. For we shall concede to no one that there are visible bodies more perfect than these, each corresponding to a single type. We must do our best, then, to construct the four types of body that are most perfect and declare that we have grasped the constitution of these things sufficiently for our purpose.

Plato (circa 350 B.C.)

After a brief period of spiritual and human confusion, caused by a provisional restriction to "*Anschaulichkeit*," a general agreement was reached following the substitution of abstract mathematical symbols, as for instance psi, for concrete pictures. Especially the concrete picture of rotation has been replaced by mathematical characteristics of the representations of the group of rotations in three dimensional space.

W. Pauli (1955)

If we wish to compare the knowledge of modern particle physics with any of the old philosophies, the philosophy of Plato appears to be the most adequate: The particles of modern particle physics are representations of symmetry groups, according to the quantum theory and to that extent they resemble the symmetrical bodies of the Platonic doctrine.

W. Heisenberg (1976)

The principal problem with the *Anschauungen* of the pre-1923 quantum theory was that they were abstractions from the world of perceptions and consequently were encumbered with pictures linked to attendant schemes of causality and conservation laws. A clue to a new meaning of picture and image was that since 1926 Heisenberg used almost exclusively the term *Anschaulichkeit*, and reserved *Anschauung* to denote the pre-1923 state of affairs. For example, Heisenberg began the 1926 paper "Quantenmechanik" by reviewing our "*gewöhnliche Anschauung*" or our "*einfache Anschauung*," which considered matter to be infinitely divisible in principle, and hence was inapplicable to the atomic regime with its essential discontinuities. *Anschaulichkeit* and *anschaulich*, on the other hand, were used by Heisenberg in the uncertainty principle paper to denote the visualizability that was dictated by the quantum mechanics. In the 1924 Bohr, Kramers, and Slater paper, Bohr had used the term *Bild* in this manner.

In summary, in the mechanics of the atom *Anschauung* had to be abandoned. Taking up Bohr's suggestion to redefine the term *Bild*, Heisenberg then began to move beyond *Anschaulichkeit* with stunning results in quantum mechanics and quantum field theory. Yet Pauli persisted in pursuing his 1924 rejection of *Anschaulichkeit*. For example, in one of the epigraphs to this section, Pauli recalled that the unvisualizable fourth degree of freedom for the electron found its place in the representations of the group of rotations in three-dimensional space (Pauli's Whig history notwithstanding). Heisenberg's notion of *Anschaulichkeit* set the stage for Feynman's diagrammatic method, which was the direct descendant of the subtraction physics of the 1934 *anschauliche Theorie*. Bohr's lifelong notion of complementarity, with its emphasis on the measurement process, was a higher form of positivism. Heisenberg, on the other hand, moved his notion of *Anschaulichkeit* away from the sensual Kantian notion of *Anschaulichkeit* and the intellectual Platonic notion of *Anschaulichkeit*. In a statement that supplements nicely the one in the epigraph to this section, Heisenberg wrote, "Our elementary particles can be compared with the regular solids of Plato's *Timaeus*. They are the Archetypes of the Ideas of matter" (1969).

NOTES

(A version of this chapter is in A. Shimony and H. Feshbach (eds.), *Physics as Natural Philosophy* (Cambridge, Massachusetts: MIT Press, 1982). I acknowledge the permission of MIT Press to reprint portions of this material.)

1. By the law of causality as it was understood in classical physics I mean the following: according to the law of causality since initial conditions can be ascertained with in-principle perfect accuracy,

then a system's continuous development in space and time can be traced with in-principle perfect accuracy. Needless to say, there are limitations to the accuracy of measurements even in classical physics, but these limitations were assumed not to be intrinsic to the phenomena. By determinism I mean that a system will occupy every point on its possible trajectory and/or that a system will develop on the trajectory that is required by its initial conditions because every successive state is entailed by the one previous. (See, for example, Poincaré, 1902b, 1912b,d; Cassirer, 1956; Margenau, 1950). Thus, Max Planck devoted many years of his life to seeking other theories for the law governing cavity radiation that were consistent with the continuity of classical physics (see Chapter 5). Planck's radiation law was of great concern to Poincaré. Recalling discussions at the 1911 Solvay Conference, Poincaré wrote that should "discontinuity reign over the physical universe" then determinism would be invalid (1912b, see also, 1912d). Miller (1978) explores principally the relation between the conservation laws of energy and momentum with *Anschaulichkeit* and *Anschauung*. Here I discuss how developments in the quantum theory served as a springboard to go beyond *Anschaulichkeit*.

2. For discussions of Bohr's (1913) papers see Jammer (1966) and Heilbron and Kuhn (1969).

3. For discussions of Einstein's 1916–1917 theory of radiation see Jammer (1966), Klein (1970), and Pais (1982). Although Einstein's researches in the years 1905–1916 are too often regarded as revolutionary, he viewed them as extending classical physics. So it is not surprising that Einstein considered the appearance of probabilities as a "weakness" of his quantum theory of radiation. For example, Ernest Rutherford's law for how many of a large number of atoms will undergo radioactive decay in a certain time period is a statistical law in the sense of classical physics, wherein statistics and probability were interpreted as reflecting our ignorance of the dynamics of individual processes. At the 1911 Solvay Conference, Poincaré emphasized that the laws governing the behavior of individual atoms would be complicated but not statistical—that is, these laws would turn out to be causal laws. This faith, continued Poincaré, would be realized by future analyses of Rutherford's recently proposed model of the atom as a minuscule Copernican system (1912b). For despite the important role of probability in the statistical mechanics of

Boltzmann, physicists clung to the belief that somehow phenomena at every level could be described with a yet-to-be formulated version of classical physics that contained the customary notions of causality and continuity.

4. By 1920 Bohr's highly refined atomic theory had compiled an impressive list of successes—for example, it could account for most of the spectral lines, intensities, and polarizations for the normal Zeeman effect and the linear Stark effect as these processes occurred in the hydrogen atom, as well as accounting reasonably well for the place of elements in the periodic table. Yet by mid-1923 this picture of the atom was withering away principally for the following reasons. The deduction of stationary states for the next simplest elements to the hydrogen atom, for example, the H_2^+-ion, and H_2-molecule and the helium atom, resisted treatment by some of the most able calculators of the day—namely, Born, Pauli, and Heisenberg. The successful methods of Alfred Landé to calculate the structure of the multiplets in the anomalous Zeeman effect defied visualization. Furthermore, Bohr's attempt to formulate a mechanical picture for the anomalous Zeeman effect, as it occurred in the alkali atoms, led him to postulate the nonvisualizable "*unmechanische Zwang,*" which served to distort the atom's core in two different ways.

   Among discussions of the stage of the Bohr atom circa 1923 see Jammer (1966), Forman (1968), Serwer (1977), Miller (1978), and Heilbron (1983).

5. In (1921b) Bohr had introduced the "coupling principle," which was a first approximation to the coupling mechanism. The coupling principle treated radiation and atoms in an enclosure; independently Ladenburg (1921) had a similar idea.

6. For a discussion of Compton's experiments see Stuewer (1975).

7. This passage is the footnote denoted by the asterisk to Pauli's statement that "not only the concept of force but also the kinematic concept of motion from the classical theory shall have to undergo profound modification★ (I have, for that reason, avoided entirely in my work [on the anomalous Zeeman effect] the designation orbit)."

8. From a letter of W. Heisenberg to N. Bohr, 8 January 1925. On deposit at the Center for History of Physics, American Institute of Physics.

9. For further discussions see Jammer (1966), van der Waerden (1960), and Heilbron (1983).

10. For other discussions of this paper see Miller (1978) and Mackinnon (1977). Mackinnon, however, analyzes this paper within a philosophical framework that concerns models and hence underplays the role of changing conceptual frameworks.

11. J.H. Van Vleck considered that the virtual oscillators were "in some ways very artificial" (1924). W.F.G. Swann took them to be a "temporary expedient" (1925). G. Breit preferred the notion of "virtual orbit" that had been suggested to him by Van Vleck (1924). Heisenberg recalled that Sommerfeld was "terribly skeptical" (*AHQP*: 19 February 1963). Pauli vehemently protested the "virtualization of physics" (letter of Heisenberg to Bohr, 12 December 1924, on deposit at the Center for History of Physics, American Institute of Physics).

12. For further discussions of the relation of Schrödinger's wave mechanics to the research of de Broglie and Einstein, see Jammer (1966), Klein (1964), and Miller (1978).

13. Another interesting example of personal taste in these matters is P.A.M. Dirac who wrote in the first (1930) edition of his *Principles of Quantum Mechanics* that the "matrices of Heisenberg's representation fit in very well with the 'anschaulich' forms of the quantum theory in existence before quantum mechanics, in particular with Bohr's theory of the atom." Dirac's point was that the state vectors in the Heisenberg representation are time independent, and so omitting the interaction between atoms and radiation the Hamiltonian's eigenvalues are "Bohr's energy levels."

14. Ten years later Einstein maintained this preference. In his 1936 essay "Physics and Reality," Einstein prefaced his criticisms of the quantum theory by developing the theory's fundamentals, and he wrote, "I shall try here to sketch the line of thought of de Broglie and Schrödinger [and not 'Heisenberg and Dirac'], which lies closer to the physicist's method of thinking . . . ."

15. Lorentz continued by noting that although for more than one particle matrix mechanics had certain advantages, nevertheless the wave mechanics was preferable because the wave equation could discuss individual quantum states while matrices "cannot at all be analyzed into pieces."

16. Pauli (1927) generalized this statement as follows. The quantity $|\psi(x)|^2 dx$ is the probability that a particle is in the interval between x and $x + dx$.

17. In fact, Heisenberg's colleague Jordan rediscovered its results in

his (1927a) and had to add a footnote that he had been unaware of Heisenberg's (1926c).

18. Heisenberg recalled that even after Dirac's transformation theory he refused to concede that there was a dualism, rather that the transformation theory showed how "very flexible" was the matrix mechanics (*AHQP*: 5 July 1963). In (1927b) Jordan published a formulation of transformation theory that was equivalent to Dirac's. Heisenberg recalled (*AHQP*: 28 February 1963) that at first "I was more happy with the Dirac paper than with the Jordan paper . . . because Dirac kept within the spirit of the quantum theory while Jordan, together with Born, went into the spirit of the mathematicians"—that is, Jordan's was an axiomatic approach.

19. See, for example, Heisenberg (1975). Throughout Heisenberg's reminiscences he recalled the "uncertainty," "confusion," and "despair" of the period 1926–1927. See also Heisenberg (1974).

20. In more detail Heisenberg reported Einstein having said that "on principle it is quite wrong to try founding a theory on observable magnitudes alone. In reality the very opposite happens. It is the theory that decides what we can observe." At the time Heisenberg disagreed with Einstein, holding instead to the view that a "good theory must be based on directly observable quantities"; this had been Heisenberg's methodology in formulating the 1925 matrix mechanics.

21. At this point Heisenberg responded in kind to Schrödinger's subjective published opinion of matrix mechanics that had appeared as a footnote in one of Schrödinger's 1926 papers (see the epigraph to the section, "Visualizability Regained"). Thus in a footnote Heisenberg recalled Schrödinger's harsh words. Although, Heisenberg continued, he had praise for the calculational value of Schrödinger's theory, its "*populäre Anschaulichkeit*" had led physicists astray from the "direct path" for the consideration of physical problems.

22. For discussions of the genesis of Bohr's complementarity principle see Holton (1973f), Jammer (1966), and Meyer-Abich (1965).

23. See, for example, Heisenberg (1929) and (*AHQP*: 12 July 1963). The relevant papers are Dirac (1927), Jordan and Klein (1927c), and Jordan and Wigner (1928). See also Heisenberg (1930) where he refers in the preface to the "Kopenhagener Geist der Quantentheorie."

Heisenberg's deep-rooted preference for the matrix mechanics is clear from the text of the Chicago lectures (1930). For whereas in the preface he wrote of the "symmetry of the book with respect to the words 'particle and wave,' " Schrödinger's wave equation did not appear until halfway through the book, where Heisenberg deduced it from Dirac's transformation theory applied to the matrix mechanics. Incidentally, the English-language version of Heisenberg's lectures has to be handled with care, for its omissions and cavalier style does an injustice to Heisenberg's discussions of *Anschaulichkeit*.

24. If the N^{14} nucleus were composed of protons and electrons then it would have 14 protons and seven electrons and so would obey Fermi-Dirac statistics. But N^{14} obeys Bose statistics, which is the case if it were composed of seven protons and seven neutrons, that is, an even number of spin one-half particles.

25. For a survey of the experimental data on nuclear electrons during 1920–1934 see Stuewer (1983).

26. For a discussion of Bohr-Heisenberg correspondence during 1930–1932 see Bromberg (1971).

27. From a letter of W. Heisenberg to N. Bohr 20 June 1932. On deposit at the Center for History of Physics, American Institute of Physics.

28. In 1932 James Chadwick had dismissed this problem with the comment that the only advantage "to suppose that the neutron may be an elementary particle [is] the possibility of explaining the statistics of such nuclei as N^{14}" (1932). In his Nobel Prize address, Chadwick, in turn, dismissed with short shrift the possibility of a composite neutron because it assigns the wrong spin to the electron (1935).

29. Pauli openly discussed the neutrino (so-named subsequently by Enrico Fermi) at a meeting in 1931 at Pasadena, California. See *Structure et Propriétés des Noyaux Atomiques: Rapports et Discussions du Septième Conseil de Physique tenu à Bruxelles du 22 au 29 Octobre 1933 sous les auspices de l'institut International de Physique Solvay* (Paris: Gauthier-Villars, 1934), pp. 324–325. In 1930 Pauli had proposed the neutrino in a letter to "friends and colleagues" (Jensen, 1972). Pauli originally considered a neutrino with a very small mass.

30. Heisenberg's paper "Many-body Problem and Resonance in Quantum Mechanics" was a watershed event in the history of modern physics. In addition to being the root of his subsequent work on nuclear physics, in it Heisenberg also introduced the

notion of intrinsic symmetry into quantum mechanics. This facet of Heisenberg's paper is explored in Miller (1984b).

31. The Hamiltonian that Heisenberg proposed is (1932a):

$$H = \frac{1}{2M} \sum_k p_k^2 - \frac{1}{2} \sum_{k>\ell} J(r_{k\ell})(\rho_k^\zeta \rho_\ell^\zeta + \rho_k^\eta \rho_\ell^\eta)$$

$$- \frac{1}{4} \sum_{k>\ell} K(r_{k\ell}) (1 + \rho_k^\zeta) (1 + \rho_k^\ell)$$

$$+ \frac{1}{4} \sum_{k>\ell} \frac{e^2}{r_{k\ell}} (1 - \rho_k^\zeta) (1 - \rho_\ell^\zeta) - \frac{1}{2} D \sum_k (1 + \rho_k^\zeta)$$

where M is the proton or neutron mass, $r_{k\ell} = |\vec{r}_k - \vec{r}_\ell|$ is of the order of nuclear dimensions or 10^{-13} cm, p_k is the momentum of particle k, and

$$\rho^\xi = \begin{pmatrix} 0 & 1 \\ 1 & 0 \end{pmatrix}, \quad \rho^\eta = \begin{pmatrix} 0 & -1 \\ 1 & 0 \end{pmatrix}, \quad \rho^\zeta = \begin{pmatrix} 1 & 0 \\ 0 & -1 \end{pmatrix}$$

—that is, the ρ's are the Pauli spin matrices, J and K are the exchange forces for neutron-proton pairs and the neutron pairs (he assumed that J is larger than K), e^2/r is the coulomb interaction between proton pairs (he neglected any attractive force between nuclear protons), D is the mass defect between protons and neutrons, and the summations are over all particles in the nucleus. Thus, according to Heisenberg's exchange force the neutron and proton exchange charge but not spin.

32. This situation brings to mind Heisenberg's seminal 1925 paper on the quantum mechanics where he wondered why $xy \neq yx$, where x and y are the bound electron's x and y coordinates, because he did not realize that he had rediscovered matrices. Seminal papers often contain oversights and missed connections—for example, in the 1905 special relativity paper, Einstein missed the mass-energy equivalence which he published later in (1905e).

33. Others at this time also speculated that the nucleus was composed of neutrons and protons. See, for example, the short note of Iwanenko with no details (1932).

34. Majorana (1933) introduced another sort of exchange force in which the neutron and proton are treated as elementary particles and where charge and spin are interchanged. Majorana's force turned out to be more nearly the correct one because it saturated at the α-particle, whereas Heisenberg's saturated at the deuteron which is not stable.

35. Quoted by Jensen (1963) from a letter sent by Pauli to "several friends and colleagues" in December 1930.

36. Jensen (1963) recalled that Fermi's β-decay theory finally convinced Pauli that "there was tangible physics in second quantization."

37. As we know today Fermi's interaction is not quite correct because the β-decay process requires also an axial-vector part.

38. Jordan (1936) defined his notion of an *"anschauliche Verständnis"* as follows: Just as the limiting case of the velocity of light becoming infinite permits correspondence between relativity and classical mechanics, the limiting case of Planck's constant approaching zero permits correspondence between quantum and classical mechanics. Although in both situations important symmetries and simplicities of presentation are lost, one can begin to better understand these two theories. Thus, Jordan continued, "In this way we gain by degrees an *'anschauliche Verständnis'* of the quantum theory." Jordan went on to apply the correspondence principle to such elementary problems as the harmonic oscillator and central force motion. However, continued Jordan, writing of the state of quantum theory in late 1926, the abandoning of this methodology became clear to Bohr and Heisenberg owing to their work on thought experiments: "They demanded the development of an *anschauliche Vorstellung* from the essential regularities of the quantum phenomena, in such a way that without starting from any complicated mathematical apparatus we are capable of surveying the essential features of the processes in concrete experiments . . . . On the process of historical development, these thoughts crystallized certainly first out of the grounds of the quantum and wave mechanics (Heisenberg-Born-Jordan-Dirac; Schrödinger, respectively) and their amplification from the 'statistical transformation theory' (Dirac, Jordan, v. Neumann)." These statements support Heisenberg's reminiscences.

   The new higher level of *Anschaulichkeit* achieved through complementarity became the foundation for a correspondence principle for the development of quantum electrodynamics and nuclear physics. For this reason Jordan considered Fermi's β-decay theory to be *Anschaulichkeit.)*

39. Tamm (1934) wrote that his "friend Iwanenko" had independently and simultaneously realized Fermi's theory of β-decay.

40. The scientific literature supports Wentzel's recollection that Yukawa's "ingenious idea . . . was not received, wherever it became known, with immediate consent or sympathy" (Wentzel, 1960). This situation changed in 1937 when Yukawa's meson was declared to have been discovered in cosmic ray phenomena. However, this meson turned out not to be the nuclear field, or strong interaction meson (See Cassidy, 1981; Galison, 1983).

41. For example, Peierls (1934): "[On the infinities occurring in vacuum polarization] it does not seem probable that this difficulty can be removed without making essential changes in the concepts underlying theory"; Furry and Oppenheimer (1934): "[The infinities occurring in vacuum polarization] require here a more profound change in our notions of space and time." In a similar vein as he had frowned on virtual oscillators ("virtualization of physics"), Pauli (1936) referred to Heisenberg's density matrix scheme as "subtraction physics."

   Although Wentzel (1943) discussed the Dirac-Heisenberg formulation "which seems satisfactory" at some length, he concluded that it "has, however, not led to any improvement regarding the self-energy problem of quantum electrodynamics." With the hindsight of almost three decades, Wentzel recalled (1960) of physics circa 1934 that the "pessimists were wrong this time: a workable though very complicated formulation of hole theory was accomplished through the efforts of Dirac (1934) and Heisenberg (1934)."

   Until postwar developments in quantum electrodynamics that resulted in the renormalization program, Heisenberg's method for calculation was the only viable one. In fact, Heisenberg's work contained the seeds of the renormalization program: Heisenberg defined the density matrix $R_s = R - \frac{1}{2}R_F$, where R is the density matrix for the Dirac hole theory and R_F the density matrix for the distribution in which every energy level is occupied. Heisenberg was able to identify singularities by defining the matrix R_s in a nondiagonal coordinate representation and then expanding in powers of the relative momenta between electrons and positrons. He then subtracted the singularities from R_s

to form the matrix *r*, which he defined to be the physically significant part of the density matrix formalism. The so-called Dirac-Heisenberg subtraction formalism was used to calculate the scattering of light by light [Euler and Kockel (1935) and Heisenberg and Euler (1936)], and the displacement of energy levels in an atom due to vacuum polarization (Uehling, 1935).

42. By the "quantum condition of the earlier half-classical theory" Heisenberg undoubtedly meant what had been referred to as "integral pdq"—∮pdq = nh—which is the quantized phase integral of classical action-angle theory. In 1925 this trustworthy half classical, half quantum quantity was the basis for Heisenberg's move toward the matrix mechanics. In 1934 this function would be performed by the normalization condition for Dirac's density matrix which is $R^2 = R$.

43. For discussions of these aspects of quantum field theory see Cassidy (1981) and Galison (1983).

44. For a survey of quantum field theory during the period 1930–1950 see Wentzel (1960) and Weinberg (1977).

PART III

On the Psychology of Concept Formation and Creative Scientific Thinking

Albert Einstein and Max Wertheimer: A Gestalt Psychologist's View of the Genesis of Special Relativity Theory

What were the decisive steps in the development of Einstein's theory of relativity? Although this is quite a task, I shall try to make them clear to the reader.

M. Wertheimer (1959)

Remark: Professor Max Wertheimer has tried to investigate the distinction between mere associating or combining of reproducible elements and between understanding [*organisches Begreifen*]. You may be interested in the manuscript (not yet published). I cannot judge how far this psychological analysis catches the essential point.

A. Einstein to J. Hadamard, 17 June 1944 (in Hadamard, 1954)

Max Wertheimer (1880–1943) was one of the founders of Gestalt psychology. His colleagueship with Einstein at the University of Berlin led to a life-long friendship. Among the topics they discussed was the process of creative scientific thinking. This photograph shows him in 1939 working at his home in New Rochelle, New York. (Courtesy of the Estate of Max Wertheimer)

THROUGHOUT HIS LIFE Einstein maintained discussions with colleagues in virtually every discipline. His interactions with Freud on scientific matters—the best-known exchange Einstein had with the psychologists—are known not to have been satisfactory (e.g., Jones, 1963). They were constantly at loggerheads over whether psychoanalysis is a science. For example, a draft of Einstein's reply circa 1927 to someone who suggested that Einstein permit himself to be psychoanalyzed reads thus: "I regret that I cannot accede to your request, because I should like very much to remain in the darkness of not having been analyzed" (Hoffmann and Dukas, 1979). Further evidence for Einstein's disagreement with Freud's psychoanalysis can be found in the Einstein Archives.

Einstein's relationship with the Gestalt psychologist Max Wertheimer was quite the opposite. Their common interests in physics, sailing, and music engendered a colleagueship and friendship spanning the period from 1916 to Wertheimer's death in 1943. Gestalt psychology abounds with terminology and concepts that are borrowed from physics, a discipline they considered as the paradigm of science. Each of the famous triumvirate of Gestalt psychology—Wertheimer, Karl Koffka, and Wolfgang Köhler—had a deep inter-

est in physics. Wertheimer in World War I did research for the Aus-
tro-Hungarian army on the construction of devices to detect hidden
artillery batteries by a triangulation technique based upon the reports
of the cannons.[1] Koffka and especially Köhler received training as
physicists. Köhler's *Die physischen Gestalten in Ruhe und im stationären
Zustand,* published in 1920, presented a discussion of a Gestalt physics. Gestalt field theory was formulated in analogy with electromagnetic field theory. Kurt Lewin, a student of Köhler, trained as a
physicist, extended Gestalt field theory into social psychology.[2] An
indicator of how congenial the viewpoint of Gestalt psychology was
to Einstein's own holistic view of science is this excerpt from Einstein's (published) foreword to a proposed collection of Wertheimer's essays (Michael Wertheimer, 1965):[3]

> Behind these essays lies above all an epistemological requirement
> which derives from the Gestalt-psychological point of view: beware of
> trying to understand the whole by arbitrary isolation of the separate
> components or by hazy or forced abstractions.

In 1916, Wertheimer wrote some 27 years later, Einstein and
he began discussing intensely the process of creative thinking, especially concerning Einstein's thought process that may have led
to his discovery in 1905 of the relativity of simultaneity, which is
the cornerstone of the special theory of relativity. At least two decades later, Wertheimer decided to reconstruct these conversations.
The result appeared as Chapter 10 of *Productive Thinking*—a book
Wertheimer meant as a "prolegomena . . . a mere introduction" to
two more detailed volumes on the Gestalt theory of thinking, which
alas were never completed. The chapter on Einstein entitled "Einstein: The thinking that led to the theory of relativity," consists of
two parts. Part I is divided into ten "Acts" of a scenario wherein
Wertheimer reconstructs the "drama"—"the drama developed in a
number of acts"—of Einstein's discovery in 1905 of the relativity of
simultaneity, a drama in which, in Wertheimer's version, the interferometer experiment of Michelson and Morley plays a key role.
Part II contains Wertheimer's explicit Gestalt analysis of this drama.

As preparation for the discussion to follow of Wertheimer's work I
shall sketch in broad strokes his general explanation of "productive
thinking."[4] Facts enter into the field of knowledge, and a problem
situation arises. There is a "desire to get at real understanding" of
these facts, but the relationship of the facts to one another is unclear:
"When one grasps a problem situation its structural features and
requirements set up certain strains, stresses, tensions in the thinker."

Then there enters into the field of knowledge a piece of information that is particularly important, perhaps because it is particularly disconcerting to the Gestalt already held by the mind. This piece of information "may drop in externally" (this will be seen to be the presumed role of the Michelson-Morley experiment), or it may be internal, generated by the mind of the discoverer (as was supposed to be the case for Galileo).[5]

This "region in the field becomes crucial"; thought is focused upon it, yet it "does not become isolated." "In real thinking," says Wertheimer, the "strains and stresses . . . yield vectors in the direction of improvement of the situation, and change it accordingly."[6] A gap in understanding is realized. Filling in the gap, that is, solving the problem, causes the field to be restructured: "A new, a deeper structural view of the situation develops, involving changes in the functional meaning, the grouping, etc., of the items." A "good Gestalt" has been achieved; one has obtained "a whole consistent picture," and moreover one can now look back and better understand the parts.

In the case study of Einstein, Wertheimer asserts that the gap was closed by Einstein's discovery of the invariant quantity that removed the arbitrariness in the problem situation introduced by the subjectivity of the observer. The new Gestalt (relativity physics) is centered about the invariant quantity.

According to Wertheimer the tendency towards a good Gestalt is "almost irresistible." The mind will seek the gap in the field and fill it so that the field becomes reorganized in the simplest and clearest manner that the prevailing conditions allow. This is referred to by Wertheimer as the "*Prägnanz* principle." Wertheimer asserts that Gestalt psychology offers deeper insights into the process of productive thinking than does traditional logic or associationism. Traditional logic provides connections that "are blind or neutral to questions of their specific structural functions in the process." Associationism does not always distinguish clearly "between sensible thought and senseless combinations."

Wertheimer was in the process of formulating a "Gestalt logic" which he outlined in the concluding chapter of *Productive Thinking*: "Dynamics and logic of productive thinking." Wertheimer intended to devote to the Gestalt logic one of the volumes of the proposed triad. The Gestalt logic consists of rules for dealing with the dynamics of thinking, for example, "of grouping, of centering, of reorganization." All of this, of course, could provide only further insight into that non-quantifiable raison d'être for productive thinking—

mankind's "desire," "thirst," indeed "passionate desire," "the craving" in all fields of endeavour for "real understanding . . . for true orientation."

With these preliminaries, let us now turn to Wertheimer's analysis of Einstein's thought. As we shall see below, and as may not be surprising in any case, Part I of Chapter 10 of *Productive Thinking* turns out to be a "reconstruction" according to the Gestalt theory of thinking. This was not unintended. Wertheimer near the start of his book explains that in the ten case studies comprising the content of the book's first ten chapters, the analyses contain the "Gestalt interpretation of thinking." Recently, however, this ruling intention has been overlooked by some historians trying to chronicle the actual sequence of Einstein's thought and by some philosophers of science, in particular by R.S. Shankland,[7] A. Grünbaum,[8] G. Gutting,[9] and K. Schaffner.[10] The misuse of Wertheimer's Chapter 10 raises from a different point of view the interesting and current problem of the legitimacy of "reconstruction" of steps in scientific advance. Shankland, Grünbaum, Gutting and Schaffner (especially the last three, who are philosophers of science) attempt to establish that Einstein's thought leading to the discovery of the special relativity theory can be reduced to and reconstructed in purely logical terms. This reductionistic viewpoint was anathema to both Einstein and Wertheimer; using Wertheimer in support of it does an injustice to the life's work of the great psychologist.

Furthermore, such use of Wertheimer clearly displays a lack of familiarity with the complete book, for at the beginning Wertheimer makes his view against the possibility of reducing thought processes to the rules of traditional logic quite explicit:[11]

> If one tries to describe processes of genuine thinking in terms of formal traditional logic, the result is often unsatisfactory: one has, then, a series of correct operations, but the sense of the process and what was vital, forceful, creative in it seems somehow to have evaporated in the formulations.

On this point Einstein was in complete agreement with Wertheimer.[12] On the other hand, Wertheimer did believe in the possibility of explaining productive thinking from the viewpoint of a Gestalt logic.

Elsewhere I discuss problems concerning the current vogue of "rational reconstruction" of history (1974, 1984a). Here I shall limit discussion to an analysis of Wertheimer's Chapter 10 placed in the context of the entire book, and the results of historical case studies of Einstein's early work towards the special theory of relativity. In addition, my analysis incorporates the results of studying Wertheimer's five manuscripts for his book *Productive Thinking,* as well as a considerable quantity of notes and letters generously placed on deposit at the New York Public Library for scholarly use by Wertheimer's elder son, Mr. Valentin Wertheimer.[13] These materials have been long available to any scholar interested in historical research on the status of Wertheimer's account of the genesis of relativity, but evidently have not been used for this purpose before.

The conclusion which will emerge is that Wertheimer's Chapter 10 is not a historical account but a "reconstruction" of Einstein's thought according to the Gestalt theory of thinking. Thus it may be studied as a pioneering effort in psychobiography—and an adventurous one at that, for we shall see that Wertheimer did not allow his lack of knowledge of the history of physics or of physics itself to impede his work. In fact, in Chapter 9, on "A discovery of Galileo"—another case study in the creative process according to Gestalt theory—Wertheimer openly and honestly notes his lack of expertise concerning history of science: "I shall not attempt here a historical reconstruction. This would require a thorough discussion of much source material—and I am no historian."[14]

THE CHAPTER ON EINSTEIN IN *PRODUCTIVE THINKING*

There exist in the Wertheimer archives at the New York Public Library no notes of the "hours and hours" of conversations that Wertheimer says he and Einstein had in 1916. There are, however, the five manuscript versions of Chapter 10. And in examining them, our attention is first drawn to a footnote by Wertheimer that appears on the first page in the early manuscripts (1 and 2), but *not* in the published book:

(1) A request to the reader: I should like to receive letters telling me how the reader fared with this chapter and what his reactions were after following me to the end. This proposal I address not only to the layman and to those acquainted with the theoretical situation in physics; I am deeply interested in the reactions of teachers of physics with experience in teaching these matters.

Either Wertheimer himself removed this note, or his death before publication made unnecessary its inclusion in the final draft. The importance to a textual analysis of Chapter 10 is that this plea's existence in the early manuscripts draws attention to the pedagogical element, to the fact that one of Wertheimer's goals in *Productive Thinking* was to demonstrate the uses of the Gestalt theory of thinking for "teaching."

Wertheimer's "drama" is set forth in Chapter 10 in ten "acts." Of these, Act I is entitled "The Beginning of the Problem." There Wertheimer discusses the thought experiment that Einstein conceived of in 1895 at Aarau—namely, the consequences of being able to catch up with a light wave and run alongside of it. According to Wertheimer the "problem" offered itself at that time to Einstein when he began to ask himself questions such as, "What *is* 'the velocity of light'?" (italics in original). If it is a relative quantity then one should be able to use it to determine the earth's velocity through the ether. Wertheimer writes that Einstein "was intensely concerned with it [the problem whether the velocity of light is a relative quantity] for seven years," at which point he realized that the solution of the problem involved a redefinition of the customary concept of time and hence of simultaneity.

We now come to Act II in Wertheimer's account: "Act II. Light Determines a State of Absolute Rest?" Wertheimer notes that before relativity theory, one could consider from the viewpoint of mechanics there was no absolute rest, but from the viewpoint of electromagnetism and optics, that is, "from the point of view of light," there was (since only with respect to the light ether is the speed of light exactly $c = 3 \times 10^8$ m/sec). However, "young Einstein had reached some kind of conviction" that one could not detect absolute motion, that is, that the notion of absolute motion was in conflict with the contemporaneous conception of light which did assume the existence of a state of absolute rest. Wertheimer reports that: "Back of all this there had to be something not yet grasped, not yet understood. Uneasiness about this characterized young Einstein's state of mind at this time." Wertheimer continues by writing that Einstein did not "in this period" (presumably before 1902) yet have "some idea of the constancy of light velocity." However, Einstein had begun to doubt that the velocity of light was a relative quantity. Einstein found questions concerning light "interesting, exciting. Light was to Einstein something very fundamental."

Wertheimer's third act is: "Act III. Work on One Alternative." "Serious work started," reports Wertheimer. Einstein's first attempts towards understanding the velocity of light consisted in assuming that this velocity is a relative quantity and then trying to modify Maxwell's equations. Furthermore, he "tried hard" to reconcile the laws of light propagation with those of mechanics. The conviction began to grow in Einstein's mind that the velocity of light is a constant quantity independent of the motion of its source. But then, Wertheimer continues, "Einstein was puzzled as to how one can arrive at a satisfactory theory of electromagnetic phenomena."

Let us pause here to comment upon the contents of these three "acts." Overlooking a mixup in dates, the historical and scientific content of the first three acts of the Gestalt scenario appear to bear a fair resemblance to actual events, as they emerge from Einstein's "Autobiographical Notes" and Einstein's remarks in interviews with Shankland. Thus, Einstein in his "Autobiographical Notes" did indeed accord a central position to the early thought experiment of the Aarau period, referring to it as a paradox in which "the germ of the special relativity is already contained." (In that essay, we may note, Einstein nowhere mentioned the Michelson-Morley experiment.) However, there exists no evidence to substantiate that it took Einstein seven years after conceiving of the thought experiment, that is, until 1902, to begin to question the customary concept of time. On the contrary, Einstein writes in the "Autobiographical Notes" that he reflected for "ten years" (i.e., 1895 to 1905) on this experiment.

Einstein, in a meeting with Shankland on 4 February 1950, is reported to have recalled that before 1905 he had worked on an emission theory of light similar to Ritz's, work which involved trying to modify Maxwell's equations (Shankland, 1963). Thus, Wertheimer's assertions on Einstein's attempted modifications of Maxwell's equations may be consistent with Einstein's statements made before 1950.

Wertheimer in "Act IV. Michelson's Result and Einstein"[15] presents a vivid discussion of the Michelson-Morley experiment, and notes that this experiment presented physicists with a "disconcerting result." He continues:

> For Einstein, Michelson's result was not a fact for itself. It had its place within his thoughts as they had thus far developed. Therefore, when Einstein read about these crucial experiments made by physicists, and the finest ones made by Michelson, their results were no surprise to him, although very important and decisive. They seemed to confirm rather than to undermine his ideas. But the matter was not yet entirely cleared up. Precisely how does this result come about? The problem was an obsession with Einstein although he saw no way to a positive solution.

Here we come to the heart of Wertheimer's account, as used by those anxious to ground the genesis of the special relativity theory in a crucial experiment. According to Wertheimer the Michelson-Morley experiment "had its place within Einstein's thoughts as they had thus far developed . . . ." The results of all ether drift experiments "were no surprise to him. . . . They seemed to confirm rather than to undermine his ideas." The veracity of these portions of the passage quoted in full above can be argued for on the basis of their consistency with Einstein's own statements made from 1907 to 1955.[16] However, Wertheimer does not stop with these assertions. Rather he writes of the Michelson experiments as being for Einstein "crucial experiments . . . very important and decisive. . . . The problem was an obsession with Einstein although he saw no way to a positive solution." Historical scholarship—in particular Holton's detailed study of (1973e)—does not support these assertions, for example, that to Einstein the result of the Michelson-Morley experiment became an "obsession" or that the experiment was the "crucial" experiment that led Einstein specifically to the special relativity theory.[17] Wertheimer's need to make such assertions, however, will soon become evident.

Wertheimer next discusses H.A. Lorentz's explanation of the Michelson-Morley result: "Act V. The Lorentz Solution." Here Wertheimer describes briefly Lorentz's proposal of the contraction hypothesis which helped to explain Michelson's null result, and Wertheimer concludes with the statement: "There was now a fine positive formula, determining the Michelson results mathematically, and an auxiliary hypothesis [Lorentz's], the contraction."[18]

In the next act it becomes clear why the Michelson-Morley experiment is so important, indeed "crucial," to Wertheimer's dramatic scenario: "Act VI. Re-examination of the Theoretical Situation." According to Wertheimer:

> Einstein said to himself: "Except for that result [Lorentz's contraction hypothesis] the whole situation in the Michelson experiment seems absolutely clear; all the factors involved and their interplay seem clear. But *are* they really clear? Do I really understand the structure of the whole situation, especially in relation to the crucial result?"

Wertheimer then describes the state of Einstein's mind while trying to understand "the structure of the whole situation, especially in relation to the crucial result [Michelson's]": "During this time he was often depressed, sometimes in despair, but driven by the strongest vectors." Wertheimer continues, "In his passionate desire to understand or, better, to see whether the situation was really clear to him, he faced the essentials in the Michelson situation again and again, especially the central point: the measurement of the speed of light under conditions of movement of the whole set in the crucial direction." According to Wertheimer, Einstein "felt a gap somewhere without being able to clarify it, or even to formulate it. . . . He felt that a certain region in the whole situation was in reality not as clear to him as it should be. . . ." Wertheimer writes that Einstein began to reason thus: "There is a time measurement while the crucial movement is taking place." Einstein then is said to have asked himself the following questions, about which Wertheimer places quotation marks: "Do I see clearly the relation, the inner connection between the two, between the measurement of time and that of movement? Is it clear to me how the measurement of time works in such a situation?"

Here in Act VI the parallelism becomes striking between Wertheimer's scenario of Einstein's thought and his explanation of productive thinking which I outlined previously: the problem situation is realized; Einstein was depressed, driven by the strongest vectors; he had a "passionate desire to understand"; an experiment with a "crucial result" focused his attention on a region of the field wherein he begins to perceive a gap.

Einstein (according to Wertheimer's dating and sequencing of his acts) even appears to have in 1902 asked himself questions couched in the terminology of Gestalt theory—a theory not yet formulated until 1912. Thus, Einstein asked himself whether he "really understood the structure of the whole situation especially in relation to the crucial result"; whether he saw clearly "the inner connection . . . between the measurement of time and that of movement. . . ." Phrasing these questions in this way allows Wertheimer to set up the drama in the next acts which will culminate in the field being recentered about the region defined by Michelson's "crucial result." Einstein will then discover in the new Gestalt "the inner connection . . . between the measurement of time and that of movement."

In summary, explicit evidence begins to appear that this chapter is a reconstruction of Einstein's thought according to the Gestalt theory of thinking rather than a report of a historical study.

The next act unfolds: "Act VII. Positive Steps Towards Clarification." Wertheimer reports Einstein's realization that "time measurement involves simultaneity." Thus, the concept of simultaneity must be re-examined. Wertheimer then discusses Einstein's arguments for the relativity of simultaneity that Einstein presented in his popular (1917a) exposition of relativity theory.

This book is one in which, Einstein says in his Preface, he discussed the main ideas of the relativity theory "on the whole, in the sequence and connection in which they actually originated." But it is therefore striking that he discusses there the Michelson-Morley experiment only *after* describing his realization that a clear understanding of the notion of distant simultaneity was necessary to formulate a consistent theory of mechanics and electromagnetism. This, however, has the effect of interchanging Wertheimer's Act VII and Act IV—an unfortunate, indeed an impossible scenario for Gestalt theory.

That Einstein did not in his popular book (1917a) use the invariance of the velocity of light to discover the relativity of simultaneity is important for interpreting the next act of the Gestalt scenario: "Act VIII. Invariants and Transformation." The reason is that Einstein's closing of the gap in the Gestalt reconstruction involved supposedly injecting the concept of an invariant quantity. Wertheimer begins Act VIII: "What followed was determined by two vectors which simultaneously tended toward the same question." The Gestalt character of Wertheimer's reconstruction is becoming clearer; thus, in consonance with the general explanation of Gestalt theory that I outlined previously, "the stresses and strains" in Einstein's thought now "yield vectors in the direction of improvement of the situation. . . ." The first of the two vectors is indicated in this supposed thought of Einstein: "The system of reference may vary; it can be chosen arbitrarily. But in order to reach physical realities, I have to get rid of such arbitrariness. The basic laws must be independent of arbitrarily chosen co-ordinates. If one wants to get a description of physical events, the basic laws of physics must be invariant with regard to such changes."

The second vector concerns the insufficiency of having only "insight into the interdependence of time measurement and movement." It is necessary to have "a transformation formula . . . to relate place and time values" of events in different moving systems. "The transformation," Wertheimer continues, would have to be based "on an assumption with regard to some physical realities which could be used as invariants."

Wertheimer next makes an interesting comparison between Einstein's thought at this juncture in his work towards a relativity theory on one hand, and a possible Gestalt psychological explanation for the discovery of the principles of thermodynamics on the other hand. After many unsuccessful attempts "by physicists to construct a *perpetuum mobile* . . . the question suddenly arose: how would physics look if nature were basically such as to make a *perpetuum mobile* impossible? This involved an enormous change, which recentered the whole field." Wertheimer continues: "similarly there arose in Einstein the following question, which was inspired by his early ideas mentioned in Acts II and III. How would physics look if, by nature, measurements of the velocity of light would under all conditions have to lead to the identical value? Here is the needed invariant! (Thesis of the basic constancy of the velocity of light.)." This quotation contains the "question" towards which the two vectors "simultaneously tended," and its answer. Einstein then discovered the "concrete and definite transformation formulas for distances in time and space" that left unchanged the velocity of light.

Let us comment on Act VIII as it has unfolded thus far. The vague manner in which the first vector is described reveals further Wertheimer's lack of expertise of the subject matter of relativity theory. The reason is that nowhere does Wertheimer mention the term inertial system. In fact Wertheimer, in Chapter 10, never states explicitly that Einstein's special relativity theory of 1905 refers only to inertial systems. Further lack of knowledge as to the meaning of the principle of relativity is Wertheimer's statement "that one might adequately call Einstein's theory of relativity just the opposite, an absolute theory." It is true that the special theory of relativity contains an absolute quantity—the velocity of light *in vacuo*. However, in order that the "basic laws of physics . . . be invariant with regard" to changes of coordinate system, that is, remain unchanged in form, necessitates that the quantities in these laws transform in ways that are specified by the requirements of form-invariance. For example, Einstein in his relativity paper of 1905 applied the requirement of form-invariance to Maxwell's equations and deduced the relative nature of the electromagnetic field quantities. To summarize, the invariance of the velocity of light is basically different from the invariance of laws of nature written in their mathematical form, and it is thus misleading to refer to Einstein's special theory of relativity as an "absolute theory"—though it might well be called an "Invariance theory," as Einstein himself referred to it initially. The requirement of the form-invarince of physical laws is a statement of the principle of relativity, a principle never mentioned explicitly by Wertheimer in Chapter 10.

Wertheimer's paralleling of considerations by physicists based upon the impossibility of constructing a *perpetuum mobile* and Einstein's thought towards the special theory of relativity can be substantiated from Einstein's "Autobiographical Notes." There Einstein wrote that applying considerations analogous to those which evolved from the apparent impossibility of constructing a "perpetuum mobile (of the first and second kind)" to the thought experiment from the Aarau period, was instrumental to his discovery of a "universal principle"—the principle of relativity, *not* the principle of the invariance of the velocity of light. Furthermore, the Michelson-Morley experiment is, in Einstein's account, given no role in the discovery of either principle. Wertheimer's reason for the omission of any explicit mention of the principle of relativity will become clear in Part II, as will Wertheimer's emphasis on Einstein's setting of the axiom of the invariance of the velocity of light *after* his discovery of the gap.

Wertheimer continues in Act VIII by discussing Einstein's reply to Wertheimer's inquiry of whether his choice of "the velocity of light as a constant was arbitrary." According to Wertheimer, Einstein replied that "we are entirely free in choosing axioms" and thus he could have chosen "the velocity of sound instead of light"; however, Einstein continued, "light is an 'outstanding' process." Wertheimer says that "questions like the following had occurred to Einstein" concerning the velocity of light: Is light the fastest velocity? Can bodies be accelerated beyond the velocity of light and would this require an infinite amount of force? Wertheimer asserts that such questions had been "unthought of before."

Einstein's statement on the freedom of choice of the system of axioms is consistent with his writings on his philosophy of science. However, Wertheimer's linking of Einstein's choice of the velocity of light as an invariant with the "questions" that occurred to Einstein seriously understates how important indeed were considerations on the nature of light to Einstein's thought during the period 1900 to 1905. The "questions" which, according to Wertheimer, Einstein asked himself relate only to the electrodynamics and mechanics of moving bodies. Historical studies by Holton (1973a), Klein (1967) and myself (1981b), have shown that fundamental to Einstein's thought during this period were problems concerning the fluctuations in the pressure of black body radiation.

Furthermore, in the early manuscripts of this "act" (manuscripts 1 and 2), there occurs a rather revealing erroneous statement by Wertheimer, which was ultimately crossed out. It referred to a question that, according to Wertheimer, Einstein was *supposed* to have asked himself while considering the nature of the velocity of light: "And perhaps mass is not a constant but decreases with increasing velocity and approaches zero as the velocity approaches the velocity of light?" I doubt whether Einstein ever pondered such a question, for it was well known that in any theory of the electron, the electron's mass becomes not zero but infinite as its velocity approaches the velocity of light.[19] One therefore is led by another route to wonder if Wertheimer did fully understand the underlying relativistic or prerelativistic physical concepts.

Wertheimer's "Act IX. On Movement, on Space, a Thought Experiment" discusses Einstein's presumed work towards a general theory of relativity and need not concern us here.

The final act is "Act X. Questions for Observation and Experiment." It is a short act and the message is: "Einstein was at heart a physicist. Thus, all developments aimed at real, concrete experimental problems." Einstein began work with the desire to clarify experimental problems, and "as soon as he reached clarification" he thought of "crucial experiments" to test his "new theses." Here Wertheimer reiterates his position, one still held by certain commentators, and one which makes him a useful ally to them—that the role of experiments, including "crucial" ones, in formulating a scientific theory cannot be overemphasized.

At this point in Einstein's thought, Wertheimer says, Einstein proposed "crucial experiments" to test the new theory. One may doubt that Wertheimer ever studied carefully Einstein's relativity paper of 1905; for if he had he would have noted that only at the very end did Einstein give three predictions to compare his theory with experiment, and that there is no evidence whatever that he considered them as crucial. To the best of my knowledge none of them was ever attempted circa 1905. In fact, the data that were considered by the majority of the physics community to be of crucial importance towards deciding between the predictions of the velocity dependence of the mass of a moving electron of the so-called Lorentz-Einstein theory and those of Abraham and Langevin-Bucherer were the data of the experimentalist W. Kaufmann. Yet Einstein did not number his prediction for the mass of the electron among the "crucial" tests he proposed at the end of the relativity paper, nor did he even make reference to the well known data of Kaufmann. Indeed, in the relativity paper of 1905 Einstein displayed little interest in questions concerning testability.[20]

A CASE STUDY IN THE GESTALT THEORY
OF THINKING

WERTHEIMER'S EXPLICIT GESTALT ANALYSIS OF EINSTEIN'S THOUGHT

We have now reached Wertheimer's Part II of Chapter 10, containing Wertheimer's explicit Gestalt analysis of the previous ten acts. Einstein's task, says Wertheimer, was a very difficult one, for "every step had to be taken against a very strong Gestalt—the traditional Gestalt of physics." Wertheimer refers to the first five acts as the "foreperiod": Act I: A Gedanken experiment causes Einstein to become "puzzled" over questions concerning the velocity of light as measured by moving observers, and over the notions of absolute motion and absolute rest: the problem situation arises. Act II: Einstein begins to doubt that the velocity of light is a relative quantity. Act III: As one alternative Einstein tries to modify Maxwell's equations to a form compatible with a variable velocity for light, but to no avail. The "conviction" begins to grow that the velocity of light is a constant quantity, independent of the motion of its source, and that in electromagnetism as in mechanics there is no absolute motion or absolute rest. Act IV: Michelson's "crucial" experiment convinces Einstein of "the other alternative." Act V: Although the Lorentz contraction hypothesis explains Michelson's result, and moreover can be understood "in terms of traditional physics—structure I," for Einstein it "did not seem to go to the root of the trouble." This completes the foreperiod. To Einstein, Michelson's result seemed to be "contradictory" to structure I.

Act VI: A portion of the field of structure I "became crucial and was subjected to a radical examination." Recall that Wertheimer described Einstein as being "often depressed . . . driven by the strongest vectors." "Under this scrutiny a gap was discovered (in the classical treatment of time)." Act VII: This was the "revolutionary moment": "The structural role of simultaneity in its relation to movement, underwent a radical change." The reason is that in structure I, no distinction is made between distant and local simultaneity; furthermore, in structure I, simultaneity is an absolute quantity. Since simultaneity depends upon relative motion, then so do other concepts in structure I, for example, length.

Act VIII: Here occurs the transition to structure II, relativity physics. The reason is that Einstein "in a passionate desire for clearness" at last discovers the relationship "between the velocity of light and the movement of a system." It is supposed to have come about as follows: Einstein's introduction of the "observer and his system of coordinates," that is, the measurer of simultaneity, "seemed" to introduce an "arbitrary or subjective factor." Wertheimer quotes Einstein as saying: "But the reality cannot be so arbitrary and subjective." To obtain a "concrete transformation formula between various systems" that is free of any arbitrary elements, "a basic invariant" is needed. According to Wertheimer, "This led to the decisive step—the introduction of the velocity of light as the invariant" and the recentering of physics about this point. "Bold consequences followed," reports Wertheimer, such as the *derivation* of the Lorentz transformations, and a new understanding of the space-time quantities in this transformation, as well as a new understanding of the crucial experiment of Michelson.

In summary, Wertheimer's analysis claims that a crucial experiment, that of Michelson, focused Einstein's thought upon a region of the field of structure I. Driven by the strongest vectors Einstein discovers here a gap; namely, the meaning of distant simultaneity. The gap is closed, that is, the problem is solved, by introducing as an invariant the velocity of light—a revolutionary development has occurred. The field is restructured about the new invariant; relationships between heretofore unrelated quantities such as space and time, have been established—structure II is a better Gestalt than was structure I.

A few comments must be made on the Gestalt analysis presented thus far by Wertheimer. The second axiom of relativity is given a prominent role; indeed, the principle of relativity is not even mentioned. Needless to say, this does not square with Einstein's theory. The principle of relativity is what, in fact, accounts for the result of Michelson; and it does so by definition—by setting up a physics in which no other result could have been expected.

Almost certainly, Wertheimer's reason for emphasizing the second axiom and neglecting the first is the central role played by invariant quantities in Gestalt psychology. Consider the illustration due to Koffka (1935) from the Gestalt theory of perception. A person riding uphill in a railroad car sees a building as being tilted relative to his frame of reference. On the other hand, a person in the building sees the railroad car as tilted relative to his (the building's) frame of reference. To paraphrase Wertheimer, both observers seem to introduce a degree of subjectivity or arbitrariness; however, both sets of measurements can be related by the one factor that is invariant to both observers—namely, the angle formed by the intersection of the railroad car and the building. The situation can be better understood if the field of perception is centered about this angle, thereby attaining a good Gestalt.

Let us now return to Wertheimer's explicit Gestalt analysis. In Act VIII the axioms are set after the realization of where the gap lies, and how to close it: "The axioms were only a matter of later formulation—after the real thing, the main discovery had happened." Act X: "Einstein could proceed to the question of experimental verification."

It is of crucial importance to Wertheimer's Gestalt analysis that Einstein did indeed set up the axioms after discovering the relativity of simultaneity, and then supposedly used only traditional logic to deduce the rest of special relativity theory. In support of this assertion Wertheimer, in footnote seven of p. 228 of the enlarged edition, recounts conversations with Einstein:

> . . . I wish to report some characteristic remarks of Einstein himself. Before the discovery that the crucial point, the solution lay in the concept of time, more particularly in that of simultaneity, axioms played no role in the thought process—of this Einstein is sure. (The very moment he saw the gap, and realized the relevance of simulaneity, he knew this to be the crucial point for the solution.) But even afterward, in the final five weeks, it was not the axioms that came first. "No really productive man thinks in such a paper fashion," said Einstein . . . . He added, "these thoughts did not come in any verbal formulation. I very rarely think in words at all . . . ."

Furthermore, Wertheimer reports, there was always a direction to his thought, a feeling of, as Einstein supposedly said, "going straight toward something concrete." These comments by Einstein are in accord with the general approach of Gestalt theory, and in fact are once again, for the most part, couched in the terminology of Gestalt theory.

A LETTER FROM WERTHEIMER TO EINSTEIN

Wertheimer in the "Acknowledgments" to *Productive Thinking* expressed his debt "to the distinguished men of science—Einstein above all—who made it possible for me to study intimately, in many conversations, how some of their greatest achievements in thinking developed." Then, in Chapter 10, Wertheimer reconstructs the "hours and hours" of conversations with Einstein, claiming to "tell the story of his thinking." Wertheimer asserts that this story is not contained in "Einstein's original papers."

However, on 9 August 1943 Wertheimer sent a letter to Einstein, the contents of which reveal a different picture of the genesis of Wertheimer's Chapter 10. The letter reads as follows:[21]

Dear Albert Einstein:

When long ago I read to you the chapter on relativity theory (from the planned book on productive thinking) I was very glad that you found it on the whole good. At that time it became however apparent that with reference to the time of your youth it is necessary to make clear (clearer than I had it in those days in the text) how the thought situation was prior to the Michelson experiment. To write this was not easy; the sketch should also be accessible to the layman.

In the presentation I have followed your exposition from that time as much as possible verbatim; these are now the first three "Acts" [i.e., up to, but not including "Act IV. Michelson's Result and Einstein"].

I would hope that it goes thus. If it should be erroneous at any point (or if some absolutely essential point is missing), I would hope that this can be put in order through some correction or short addition.

Would you be so kind as to look through it?—essentially the first three acts—and where relevant to make corrections in the margin.

In the remaining parts I have made no essential alterations with the exception of a few changes which became necessary owing to the first three acts being placed first (pages 23, 26), and some small additions from those conversations (page [illegible]).

Oxford Press is interested in the book and would like to have the whole manuscript in a few weeks; but I would not like to lay this chapter in final form before them until I have laid the changed version before you.

[The final paragraph contains comments on Wertheimer's declining health, *etc.*]

Signed
MW

(In a postscript Wertheimer writes that he is sending Einstein only the chapter on relativity theory and the table of contents, not wanting to impose on Einstein's time.)

The message is clear: Wertheimer is asking Einstein's opinion of his reconstruction of Einstein's thought processes in his early work on the special relativity theory. Furthermore, that the only portion of Chapter 10 which pertains "to your [Einstein's] exposition from that time as much as possible verbatim [presumably Wertheimer makes reference here to the talks in the period 1916] is the first three 'Acts' ," in which the Michelson-Morley experiment is not mentioned. Thus, it may be no accident that we have found the historical and scientific content of these three acts to be better and from the point of view of the physics less confused than the subsequent ones. The manuscript pages 23 and 26 to which Wertheimer refers almost certainly belong to a carbon copy of manuscript 2, which in my opinion was the one he sent to Einstein. The material on manuscript pages 23 and 26, with the exception of stylistic changes, appear in the printed version.

It has so far not been possible to locate the copy of the manuscript which Einstein received from Wertheimer, if indeed it was kept at all. The manuscript does not appear in the listing of manuscripts in the Einstein Archives. Mr. V. Wertheimer informs me that the two men spoke frequently by phone; thus, Einstein could have relayed his comments in this manner. My examination of the five manuscripts of this chapter at the New York Public Library shows in them no substantial changes in Chapter 10 nor any marginal notes to the effect that Einstein had made comments, nor any reference to the version that Wertheimer read to Einstein "long ago"—once again presumably in the 1916 period.

Einstein may have made no critical comments to Wertheimer at all, but might have discussed with Wertheimer the role of axioms in his thought. The result of such a conversation may have been the addition to manuscript 2, a carbon copy of which he may very well have sent to Einstein, of the long footnote on p. 228 of the enlarged edition which I discussed above.[22] Here Wertheimer mentions the Einstein-Infeld book *The Evolution of Physics* published in 1938. If this is the case, then one may conjecture that Einstein's only comment to Wertheimer on Chapter 10 was his agreement with Wertheimer on the role of axioms in scientific discovery. In addition, owing to Wertheimer's rapidly declining health—he died on 12 October 1943—Einstein may have wanted to respond quickly and positively to Wertheimer's manuscript.[23]

A Remark in a Letter from Einstein to Hadamard

Einstein apparently at some point did take the time to read Wertheimer's manuscript. We know this from the postscript to a letter that Einstein sent to Jacques Hadamard on 17 June 1944, in which he referred to it.

This letter contains Einstein's response to Hadamard's questionnaire on the working habits of mathematicians. Einstein's description here of his essentially visual manner of thought is similar to the one that Wertheimer quoted in footnote seven, Chapter 10, in his book *Productive Thinking*. At this time Einstein's postscript is the only critical printed or letter comment by Einstein to third parties regarding Wertheimer's work that I have been able to locate. The postscript to Hadamard reads thus:[24]

> Remark: Professor Max Wertheimer has tried to investigate the distinction between mere associating or combining of reproducible elements and between understanding [*organisches Begreifen*]. You may be interested in the manuscript (not yet published). I cannot judge how far this psychological analysis catches the essential point.

The "manuscript (not yet published)" can only have been the one that Wertheimer sent to Einstein with the letter of 9 August 1943. Einstein's "remark" can be interpreted in one of two ways: as a negative criticism, or as Einstein preferring not to comment upon a subject about which he felt inadequately informed. I opt for the former conjecture because Einstein was keenly interested in the subject of creative thinking. In fact, as is well known Einstein began his "Autobiographical Notes" (written two years later, in 1946) not by discussing physics, but with a section entitled "What, precisely, is 'thinking'?" Whichever way one interprets Einstein's "remark" on Wertheimer, it is clear that Einstein was not vouching for Wertheimer's contribution.

CONCLUSION

In summary, Wertheimer's discussion of Einstein's discovery of the relativity of simultaneity has been shown to be, and to be intended as, a reconstruction according to the Gestalt theory of thinking. Wertheimer realized that Einstein's discovery was for Einstein the major breakthrough towards a special theory of relativity, and that it implied a heretofore unsuspected connection between space and time, which in turn led to a restructuring of physical concepts. To Wertheimer, all the elements for a Gestalt scenario seemed to be present. In order to fit this episode into the framework of the then developing Gestalt theory of thinking, Wertheimer had to reshape the historical development; in particular, he had to overemphasize the role of the Michelson-Morley experiment. Wertheimer, in his letter to Einstein of 9 August 1943 explained that he had followed Einstein's "exposition from ['long ago'] as much as possible verbatim" only in the first three "Acts." In the concluding acts Wertheimer unduly emphasized the second axiom in order to conform to the importance of invariants in the Gestalt theory. There is so far no evidence that Einstein was impressed with Wertheimer's Chapter 10, or that he even agreed with anything more than the first three "Acts" and Wertheimer's assertion that "the axioms came later."

What is striking in Wertheimer's Gestalt analysis is the importance of crucial events and of discontinuities. Thus, in the case of Einstein Wertheimer writes of a *crucial* experiment studying *crucial* movement in a *crucial* direction; the *crucial* experiment serves to focus Einstein's attention on a region of the field that becomes *crucial*; a *gap* is realized; the gap is filled—a *revolutionary* movement. The progress toward and transition from one Gestalt to another, that is, from one structure to another, is not continuous. I have shown that such a scenario is not sufficient to explain Einstein's discovery of the special theory of relativity if one looks at the available historical evidence.

During the period 1900 to 1905 there were hard data which Einstein considered as being of importance to his thought towards the relativity theory: ". . . observations of stellar aberration and Fizeau's measurements on the speed of light in moving water. 'They were enough' ," Shankland (1963) reported Einstein as saying.[25] In addition, there was the Gedanken experiment, conceived of at Aarau. It was a period during which Einstein studied the works of Hume, Kant, Mach and Poincaré, among other philosophers. It was a period during which Einstein worked on problems concerning the nature of radiation. It must also have been a period during which his concepts of space and time underwent transformation. How did this happen?

Einstein wrote that "scientific thought is a development of prescientific thought." Here, in my opinion, is a valuable clue to Einstein's thought and thus also an indicator as to how we can best use theories of the cognitive processes to study his work not only towards the special theory of relativity, but his later preference for field theories as well. As I have argued in Chapter 1, we can with a great deal of certainty interpret Einstein's statement quoted above as follows: To Einstein scientific theories provide us with a more sophisticated and deeper understanding of the world about us than the world-picture that we constructed since childhood. Good scientific theories should reflect prescientific knowledge. Thus, for example, "action-at-a-distance . . . is unsuited to the ideas which one forms on the basis of the raw experience of daily life" (1946). The reason is that we obtain our elementary knowledge of the external world by means of pushing, pulling and turning solid objects, all effects appear to be caused by actions that are contiguous. This very likely accounts for Einstein's preference for a field-theoretical description of nature.

From our prescientific interactions with the world about us, we form such concepts as solid object, space, and time. The most natural concept of time is that of the absolute time of Newtonian physics. Yet Einstein by 1905 concluded that time was not absolute. Might there be some structure in Einstein's thought towards this discovery that has a parallelism with his prescientific thought? Among other questions that suggest themselves, and are not unrelated to the one I just posed, are: How did Einstein's research on the nature of radiation contribute to the transformation of the concepts of space and time? Can psychological analysis shed any further light on the question of why it was Einstein and not Poincaré (to whom the Michelson-Morley experiment was crucial) who discovered the relativity of simultaneity? Wertheimer's Gestalt analysis did not provide us with any insight into these questions. In addition to being too often ahistorical, there are other factors that render Gestalt psychology as unsatisfactory by itself to discuss the questions posed above, and more generally the problem of what is the nature of scientific discovery— for example, Wertheimer's Gestalt analysis is too narrowly directed, too dependent upon the actual existence in Einstein's thought of crucial events and discontinuities, and there is too big a jump in Wertheimer's analysis between Einstein's thought according to structure I and according to structure II.

The important contribution of Gestalt psychology toward understanding the nature of scientific discovery is Wertheimer's *Prägnanz* principle, that is, the tendency towards an organization that possesses simplicity, symmetry, and regularity.[26] Gestalt psychology considers the *Prägnanz* principle as an a priori organizing principle. Indeed, the tendency for biological systems and for human thought towards such an organization is stunning and did not go unnoticed by Einstein.[27] For example, Einstein began his papers of 1905 on light quanta and on relativity with arguments based upon symmetry.

Let us pursue this point further for the relativity paper of 1905. In contrast to the usual style of doing physics circa 1905 (or ca 1984), Einstein did not begin with a survey of existing experimental data and then discuss current tensions between theory and experiment; rather he began with the discussion of a quasi-aesthetic criticism of the commonly-held interpretation of Maxwell's electrodynamics:

> That Maxwell's electrodynamics—the way in which it is usually understood—when applied to moving bodies, leads to asymmetries which do not appear to be inherent in the phenomena is well known. Consider, for example, the reciprocal electrodynamic interaction of a magnet and a conductor.

We do not know whether Einstein and Wertheimer discussed Einstein's treatment of 1905 of the case of magnet and conductor; but if they did, then it is indeed unfortunate that Wertheimer did not develop a Gestalt scenario around this case. The reason is that Einstein, in an unpublished manuscript written about 1919, to which attention was first drawn by Holton (1973g), explains how important to his thought towards discovering the relativity of simultaneity was the case of magnet and conductor:[28]

> *The fundamental idea of general relativity theory in its original form.* In the construction of special relativity theory, the following, [in the earlier part of this manuscript] not-yet-mentioned thought concerning the Faraday [experiment] on electromagnetic induction played for me a leading role.
>
> According to Faraday, during the relative motion of a magnet with respect to a conducting circuit, an electric current is induced in the latter. It is all the same whether the magnet is moved or the conductor; only the relative motion counts, according to the Maxwell-Lorentz theory. However, the theoretical interpretation of the phenomenon in these two cases is quite different. . . .
>
> The thought that one is dealing here with two fundamentally different cases was for me unbearable [*war mir unerträglich*]. The difference between these two cases could be not a real difference but rather, in my conviction, only a difference in the choice of the reference point. Judged from the magnet, there were certainly no electric fields, [whereas] judged from the conducting circuit there certainly was one. The existence of an electric field was therefore a relative one, depending on the state of motion of the coordinate system being used, and a kind of objective reality could be granted only to the *electric and magnetic field together,* quite apart from the state of relative motion of the observer or the coordinate system. The phenomenon of the electromagnetic induction forced me to postulate the (special) relativity principle. [Footnote:] The difficulty that had to be overcome was in the constancy of the velocity of light in vacuum which I had first thought I would have to give up. Only after groping for years did I notice that the difficulty rests on the arbitrariness of the kinematical fundamental concepts [presumably such concepts as simultaneity].

Thus Faraday's experiment played a "leading role" in the "construction of special relativity theory," that is, it was not the only factor involved, and so it is included in the Gestalt psychological scenario in Chapter 7. This scenario addresses the questions posed previously concerning the transformation of Einstein's concepts of space and time, and to the criticisms of a purely Gestalt-psychological approach to Einstein's invention of the relativity of simultaneity. So it explores how Einstein's work on the nature of radiation and his readings in physics and philosophy affected his "groping for years," until he realized "that the difficulty [in the problem of magnet and conductor] rests on the arbitrariness of the kinematical fundamental concepts [presumably such concepts as simultaneity]."

It is very likely that further insights into the nature of scientific discovery could be obtained by supplementing Gestalt analysis with other structuralist theories of the cognitive processes—for example, Jean Piaget's genetic epistemology which asserts a "parallelism between the progress made in the logical and rational organization of knowledge and the corresponding formative psychological process." A premise basic to genetic epistemology is that there is a direct relationship between biological organization and cognitive processes. The thrust of the view of genetic epistemology is consistent with Einstein's view that "scientific thought is a development of pre-scientific thought." The genetic epistemological analysis in Chapter 7 explores this point.

Other types of analysis of the cognitive processes will no doubt also be necessary—for example, the concept of *themata* as put forth by Holton has yielded insights into Einstein's thought and into the nature of scientific discovery, or a psychoanalytic study in the style of Erik Erikson that has been applied by Manuel (1968) to Newton. The use of theories of the cognitive processes to investigate the nature of scientific discovery is a field that promises new riches. Wertheimer, too, was a pioneer; but it is good to remember that his first aim was establishing and illustrating a field of psychology and not that of the history of science.

EPILOGUE

What, then, *was* the genesis of the special theory of relativity? Thorough historical studies, taking account of Einstein's published papers and archival material, have only begun to plumb the depth of this problem. A complex picture emerges—as we would expect of a man such as Einstein—which does not allow of a simple reconstruction; particularly not one that is intended chiefly to help serve the purpose of establishing a new school of psychological studies rather than to serve the purpose of establishing the historical facts of the case.

A goal of Chapters 6 and 7 is to widen the net of historical research into Einstein's invention of the special theory of relativity by application of cognitive psychology in a manner consistent with what we know of the actual historical development.

NOTES

(A version of this chapter is in *History of Science, 13,* 75–103 (1975). I acknowledge the permission of this journal to reprint portions of the article.)

1. Private communication from Wertheimer's elder son Mr. Valentin Wertheimer. Wertheimer discovered that sound travels to the brain at different speeds from the left and right ears. The left-right localization of sound is a function of the difference in time with which the sound wave strikes the two ears. Wertheimer designed a helmet with two protruding tubes at the location of the ears, which can be adjusted to ensure that both ears are stimulated simultaneously. For further discussion of Wertheimer's post-war research on this problem see Koffka (1935).

2. The kinship that the Gestalt psychologists felt for the physicists surfaced many times in their writings: Köhler in his lectures delivered in 1966 at Princeton University took passages from Clerk Maxwell's *Treatise on Electricity and Magnetism,* published in 1873, and from a set of lectures given by Max Planck in 1909 to demonstrate that they were unknowingly working and expressing their thought along the guidelines of Gestalt psychology—a subject not formulated until 1912. See Köhler (1969).

3. Professor Michael Wertheimer is Max Wertheimer's younger son. I thank Professor Rand Evans for alerting me to this essay, and Mr. V. Wertheimer for a copy of Einstein's original foreword in German. The volume was never published and hence this is, as far as I know, the only place where an English translation can be found. I have slightly corrected the translation.

4. The Gestalt theory of thinking circa 1943 was not as well developed as the Gestalt theory of perception. Koffka (1935) expressed his hope "that Wertheimer will before long publish the results of the work which he did in this field [the Gestalt theory of thinking] for many years."

5. Wertheimer presents in Chap. 9, "A discovery of Galileo," a Gestalt reconstruction of Galileo's thought that supposedly led to his deduction of a law of inertia using a perfect sphere and an inclined plane, as presented in *Dialogue concerning the two chief world systems,* translated by S. Drake (Berkeley: University of California Press, 1967). Wertheimer conjectures that Galileo was led by "the structural principle of the whole," a Gestalt law that borders on the quasi-aesthetic, to devise a crucial experiment. Yet another difference between Galileo's supposed crucial exper-

iment and Einstein's supposed crucial experiment (*i.e.,* the Michelson-Morley experiment), is that Galileo's is possible only in principle—one cannot observe forever the motion of a body rolling on a plane of infinite extent.

6. In analogy with vectors in physics, vectors in Gestalt psychology possess "direction, quality and intensity."

7. Shankland (1973) writes: "This writer finds the account of Wertheimer entirely consistent with his own notes and recollections of Einstein's attitude in 1952 toward the Michelson-Morley experiment." However, they are entirely antithetical to Shankland's publication of (1963) of his discussions with Einstein in 1952, and in particular with every quotation of Einstein's words in 1950, 1952 and 1954 given in Shankland's 1963 article. The 1973 paper is consistent with Shankland's view of the matter which he had published in 1950 prior to his interviews with Einstein.

8. Grünbaum (1973) writes that his "philosophical analysis" of the genesis of special relativity theory is "attested by the historical account that Einstein gave to Wertheimer as well as by his 'Autobiographical Notes'."

9. Gutting writes (1972): "By far the most thorough and explicit account of the discovery of special relativity is that given by Max Wertheimer, based on his own extensive conversations with Einstein in 1916."

10. Schaffner (1974) writes: "In the Einstein literature there seems to be only one source which explicitly touches on the crucial stage of Einstein's reasoning when he conjectured that the classical conceptions of time and simultaneity might require reanalysis. The source is Wertheimer's account." Schaffner continues: ". . . it may not be unimportant that, though the above [Wertheimer's] scenario was developed prior to the availability of Shankland's very recent and significant 'Conversations with Albert Einstein. II' [1973] it would appear that this new evidence is in complete accord with the above [Schaffner's] reconstruction. Shankland finds the Wertheimer account completely reliable . . . ."

11. Thus, Gutting's (1972) assessment of Wertheimer's account of Einstein's thought is incorrect: "There is no evidence that STR as a whole came to him [Einstein] as a flash of insight or a lucky guess. Rather, as Wertheimer points out, each step Einstein took was guided by the internal logic of his inquiry."

12. Wertheimer (1959) quotes Einstein as saying: " 'I am not sure,' Einstein said once in this context, 'whether there can be a way of really understanding the miracle of thinking. Certainly you are right in trying to get a deeper understanding of what really goes on in a thinking process'." This is consistent with Einstein's other comments on thinking. *Cf.* Einstein's "Autobiographical Notes." The Gestaltists from the beginning disagreed with the positivistic viewpoint, as did Einstein explicitly in his published writings from 1933 on—see Koffka (1935) and Holton (1973d). For a positivist's critique of Gestalt psychology see von Mises (1968). Von Mises reminds the reader that Gestalt psychology is an outgrowth of Mach's research:

> *Gestalt psychology studies specific aspects of phenomena connected with sense impressions. It forms a special chapter, also discussed at length by Ernst Mach, of empirical natural science, of which elementary and structural psychology* [E. Mach, *Popular Scientific Lectures,* translated by T. J. McCormack (La Salle: (Open Court 1943), 214–235], *are other chapters. It is completely unjustified to contrast with Gestalt psychology every other psychological approach as "physicalistic" or "empiricist."*

Von Mises continues: (italics in original):

> 5. *The Whole and the Sum.* The favorite formula by which the followers of the holistic doctrine like to assert their superiority over the lower level of empirical science is this: *The whole is more than the sum of its parts.* This can, indeed, be called the model of a pseudo sentence.

It is indeed interesting that positivistically inclined philosophers circa 1984 have come full circle and use, when convenient, Gestalt psychology in support of their viewpoint.

13. With the generous assistance of Mr. V. Wertheimer, I was able to arrange the manuscripts in the order in which they were written and edited (they are undated). All of the manuscripts are very similar. There are two originals—which I shall designate as 1 and 5—with carbons on which were made editorial changes. The changes on the carbons 2, 3 and 4 were then transferred back to the original copy 5. In the final stages (manuscripts 3 through 5) there appears almost exclusively the writing of Mr. V. Wertheimer and Max Wertheimer's colleagues, Professors S.E. Asch, W. Köhler and C.W. Mayer. It is very likely that Wertheimer had already died by the time the final stylistic changes were transferred to manuscript 5, which was then sent to press.

14. Wertheimer (1959) continues: "Besides the printed material would not be enough for the psychologist, who is interested in characteristics of the growing process of thought which is not usually put down in writing." In Einstein's case Wertheimer did not claim to be basing his work on "printed" material, but on conversations with Einstein.

   Similarly, Frank E. Manuel (1968) in his adventurous foray into the psychology of discovery alerts his readers thus: "It will quickly become plain to the reader that on more than one occasion I steered my way between a Scylla of historians of science and a Charybdis of psychoanalysts."

15. Wertheimer refers to the Michelson-Morley experiment as "Michelson's experiment."

16. For example, the letter of Einstein to F.G. Davenport written 9 February 1954 and quoted in Holton (1973e). The part of the letter relevant to the discussion here reads thus:

   > In my own development Michelson's result has not had a considerable influence. I even do not remember if I knew of it at all when I wrote my first paper on the subject (1905). The explanation is that I was, for general reasons firmly convinced how this could be reconciled with our knowledge of electro-dynamics. One can therefore understand why in my personal struggle Michelson's experiment played no role or at least no decisive role.

   Indeed, in this essay Holton warned that Wertheimer's "work has to be used cautiously."

17. Holton (1973e) in a detailed historical study of the relationship of the Michelson-Morley experiment to Einstein's thought toward formulating the special relativity theory found that one is forced to agree with the evidence for the veracity of Einstein's own comments, that is, those written by Einstein himself or his words quoted or reported by physicists who interviewed him. See *Thematic Origins,* the letter of Einstein to Davenport, in Holton (1973e), and Shankland (1963), pp. 48, 55. Holton is not denigrating the actual role of experiment in Einstein's thought. In fact, he concludes his essay by taking great care to explain that "experiments are essential for the progress of science"; however, the role of experiment in the scientific method and in the writing of the history of science must be carefully weighed, "the experimenticist fallacy of imposing a logical sequence must be resisted." To elevate the role of the Michelson-Morley experiment to a "crucial" one, as Einstein never did in any of his writings, is

simply and demonstrably wrong, though the urge of peda-
gogues or experimenticist historians or philosophers to do so can
be explained. Indeed, the majority of physics teachers circa 1983,
and many even today, would probably present Einstein's relativity theory as explicitly motivated by the Michelson-Morley experiment.

18. The first sentence of this passage is perplexing; for is it not the
case that the formula for the Lorentz contraction *is* the mathematical statement of the contraction hypothesis? A discussion of
Einstein's reported comments made to Wertheimer on the 'ad
hocness' of the Lorentz contraction are of interest but go beyond
the theme of this chapter. For further discussion and references
on the status of Lorentz's contraction see my (1974, 1981b, and
1984a).

19. Einstein was aware of Max Abraham's (1902–1903) theory of
the electron. Moreover, such a question as whether the velocity
of light is the ultimate velocity had not been "unthought of
before" Einstein. See Miller (1981b).

20. Kaufmann's data were referenced in Abraham's papers of 1902
and 1903. It is true also of the other two papers which Einstein
published in vol. 17 of the *Annalen der Physik* that in them he
proposed tests at the very end, briefly and almost casually. I am
not intimating that Einstein did not care about testability. That is
not true. See his "Autobiographical Notes" for further discussion.

21. Unpublished letter. I am grateful to the Estate of Max Wertheimer for permitting the use of this letter. The translation is
mine.

22. I am grateful to Mr. V. Wertheimer for pointing this out to me.

23. Such, for example, was clearly the intent of Einstein's letter to
Freud of 21 April 1936, quoted in Jones (1963), p. 493, and
Freud's reply is on p. 494.

24. A copy of this letter is in the Einstein Archives at the Institute for
Advanced Study. Except for the sentence, "You may be interested in this manuscript (not yet published)," the letter appears
in Hadamard (1954).

25. Noteworthy is that Shankland reports Einstein as prefacing this
statement with the words "*only after* 1905 had it [the Michelson-
Morley experiment] come to my attention" (italics in original).

26. Gestalt psychology plays an important role in certain recent

studies in the history and philosophy of science, particularly in Hanson (1958) and Kuhn (1962). This is discussed in Chapter 7.

27. For example, see Arnheim (1971) for discussions of Gestalt psychology toward a study of "visual perception as a cognitive activity"; and Gombrich (1956) who uses examples from art to illustrate the *Prägnanz* principle in action—Gombrich refers to the *Prägnanz* principle as the "simplicity hypothesis."

28. Holton in (1973e) once again warns that "Wertheimer's own remarks [in *Productive Thinking*] on psychology are useful, but those on physics are far less soundly based."

CHAPTER 6

On the Limits
of the IMAGination

Logic . . . remains barren unless it is fertilized by intuition.

H. Poincaré (1908a)

The very fact that the totality of our sense experience is such that by means of
thinking (operations with concepts, and the creation and use of definite functional
relations between them, and the coordination of sense experiences to these concepts) it can be put in order, this fact is one which leaves us in awe, but which
we shall never understand. One may say "the eternal mystery of the world is its
comprehensibility."

A. Einstein (1936)

[Cognitive science] shows how one can construct a normative theory—a "logic" if
you will—of discovery processes [and] shows how one can treat operationally
and formally phenomena that have usually been dismissed with fuzzy labels like
"intuition" and "creativity."

H.A. Simon (1973)

Mental imagery in auditory, sensual, and visual modes has played a central role in creative thought. Wolfgang Amadeus Mozart's auditory imagery permitted him to hear a new symphony "*tout ensemble.*" The great French mathematician and philosopher Henri Poincaré's "sensual imagery" led him to sense a mathematical proof in its entirety "at a glance." Albert Einstein's creative thinking occurred in visual imagery, and words were "sought after laboriously only in a secondary stage" (Hadamard, 1954).

Einstein's visual imagery was fundamental to his framing of the special (1905) and general (1915) theories of relativity. The unexpected changes they brought to such basic notions as space and time prepared the way for the next great upheaval in physics wrought in 1925 by Werner Heisenberg's physics of the atom (quantum mechanics). Heisenberg accomplished this formulation without a mental image of the atom because by 1925 all attempts to impose imagery on atomic theory had failed. Heisenberg went on to propose yet another mode of imagery, one that was tempered by mathematics instead of perceptions and that went beyond the restricted metaphors offered in 1927 by his mentor Niels Bohr.

Through analyses of case studies in the creative thinking of Poincaré, Einstein, Bohr, and Heisenberg, this chapter traces the transformation of imagery that turned out to have been necessary for progress in twentieth-century physics. The transformation is from a sensual imagery (Poincaré), to a visual imagery based on objects that had actually been perceived (Einstein), to restrictions on Einstein's sort of imagery in order to apply it to the atomic domain (Bohr), and then to a mode of imagery tempered by the mathematics of subatomic elementary particles (Heisenberg). Tracing this fascinating transformation requires exploring the origins and dynamics of mental imagery, its role in the construction of scientific concepts, and then the limits of the imagination that led these scientists to new vistas of imagery. Poincaré, Einstein, Bohr, and Heisenberg are ideal subjects for this analysis because the depth of their scientific research required them to investigate the nature of thinking itself.

In summary, one goal of this chapter is to show that an ingredient essential to scientific research of the highest creativity is what Poincaré described as "our need of thinking in images" (1904b).

In the course of investigating the changing role of imagery in atomic physics I noticed that the historical narrative exhibited traces of notions from cognitive science—for example, the interplay between mathematical symbols and mental imagery. So, I considered it worthwhile to see what light cognitive science could shed on all three case studies. Thus, a second goal of this chapter is to use history of science as a source and laboratory for models of cognitive processes proposed by cognitive scientists. The first part surveys issues in cognitive science that are pertinent to the case studies of scientific creativity discusssed in the three sections that follow.

This chapter focuses on four exceptional individuals out of a sizable scientific community, and yet the psychological analysis uses notions obtained from laboratory subjects of average intelligence. In my opinion, there is nothing amiss with this methodology because as Jacques Hadamard wrote in what may now appear as the antediluvian days of cognitive psychology (1954):

> It is the exceptional phenomenon which is likely to explain the usual one; and, consequently, whatever we can observe that has to do with invention or even, as in this study, this or that kind of invention, is capable of throwing light on psychology in general.

Application of cognitive science to these thinkers permits the testing of notions obtained from laboratory experiments on minds that set the milieu for twentieth-century science, philosophy, and to some degree psychology as well.

SOME NOTIONS FROM COGNITIVE SCIENCE

AN OVERVIEW

What is Cognitive Science? Why is it needed? We lack a science of the mind, of intelligence, of thought, a science concerned with knowledge and its uses.

D.A. Norman (1981)

Some aspects of the historical analyses in the sections that follow this depend upon the individual scientists' introspective accounts of their thinking.

Material in previous chapters supports introspections by Einstein, and Poincaré, as well as most of the recollections of Bohr and Heisenberg. These scientists emphasized the role of mental imagery in their work. Their testimonies against productive thinking in modes that are purely syllogistic or verbal is reasonable because thinking in these modes can proceed only linearly or stepwise. Moreover, words achieve their meanings in context, and this often entails imagery or at minimum intuitive aspects that go beyond the verbal material. Productive thinking is to a high degree intuitive, moving freely over a multidimensional field of data that were either heretofore unconnected or connected inappropriately. As Rudolf Arnheim writes (1971): "What makes language so valuable for thinking, then, cannot be thinking in words. It must be the help that words lend to thinking while it operates in a more appropriate medium such as visual imagery"—where, to give the imagination even freer rein, replace the word visual with mental. In the sections to follow evidence from cognitive psychology will be presented that further supports the essential role of mental imagery in productive thinking.

Not suprisingly, only the cognitive scientists who grant a fundamental role to mental images mention the introspective accounts of scientists who relied on visual thinking. Being somewhat unaware of recent historical research they consider these introspections to be "merely anecdotal" (Shepard, 1978), thereby missing a golden opportunity to support their case.[1] On the other hand, one of the leading computer science spokesmen, Herbert A. Simon, takes seriously introspections by scientists whose nonvisual thinking seems to lend itself to reduction to trial-and-error search procedures, for example, Poincaré. In fact, Simon goes on to make the bold assertion that cognitive science can produce a logic of discovery that can "treat operationally and formally phenomena that have usually been dismissed with fuzzy labels like 'intuition' and 'creativity' " (Simon,

1973; see epigraph to this chapter). Simon's claim is explored further in this chapter.

I turn next to developing these subdisciplines of cognitive science.

THREE VIEWS

The guiding inspiration of cognitive science is that, at a suitable level of abstraction, a theory of "natural" intelligence should have the same form as the theories that explain sophisticated computer systems.

J. Haugeland (1981)

A principal tenet of cognitive science is the information-processing paradigm, which states that the mind is a symbol-manipulating machine that is a computational system. This key point crystallized in the work of Herbert Simon and Allen Newell in 1972 and is based on results that Alan Turing obtained before the invention of digital computers (see Lachman, 1979). Cognitive scientists differ on the degree to which thinking is reduced to symbol manipulations and to the consequences on thinking of such symbol manipulations as mental images. One faction asserts that mental images are merely a particular aspect of a more general and abstract processing system. To another faction, mental images play a fundamental role in thinking. The first view considers mental images to be akin to lights set flashing on the exterior of a computer by the computer's program, where smashing these lights would not affect the computer's internal workings. Thus mental images have no causal role in thought processes; they merely exist and so are referred to as epiphenomena. The second view contends otherwise, and claims support from a theory of mental imagery. For lack of a better term, those who consider mental images to be merely epiphenomena are here called anti-imagists, and those who deem otherwise are pro-imagists. In order to develop the differences within and between these factions it is apropos to discuss their uses of computer technology.

Computer users rarely deal with the computer's hardware. The programmer can just as well consider the levels between the hardware and the program to be the machine itself; this machine is referred to as the "virtual machine" (Pylyshyn, 1980). In the terminology of Zenon Pylyshyn, who is a principal spokesman for the anti-imagists, there is a "cognitive virtual machine" about whose structure no assumptions need be made; thus this machine can be the brain with all of its neurophysiological complexities for the moment ignored. The cognitive virtual machine provides the "cognitive functional architecture" or computer language that is fixed and "hopefully universal to the species" (Pylyshyn, 1980, 1981b). Pylyshyn defines those mental processes that are explainable biologically—rather than by the rules and representations of abstract symbols—as "cognitively impenetrable" and hence part of the functional architecture. All other processes are defined as cognitive processes that can be simulated with mental algorithms. The algorithms are executed on the functional architecture, just as in ordinary computer programming. Consequently, the execution is accomplished in a computer language in which meaningless symbols are manipulated according to formal rules (syntax). The abstract symbols are of a particular sort called propositions and possess three properties: They are abstract because their meaning is not connected to any words or pictures; they possess truth value which is carried by the rules of mathematical logic; and they have rules of formation or syntax. The symbols are then given interpretation or are represented (semantics)—for example, the symbols can be interpreted as numbers. Even in mathematical logic the truth of the interpretation is not always straightforward since symbol systems can be interpreted in many different ways, but only certain of them make sense—and this judgment is often made intuitively.[2]

The computer science subset of the anti-imagists claim that the biological substratum is composed of levels of technologies that can *in principle* be encompassed within ever-more sophisticated theories of syntax. They boldly claim that the mind is a "semantic engine"—that is, that the syntax completely determines the semantics.[3]

The pro-imagists agree that images are rooted in propositional encodings and that there is a functional architecture. They go on to assert that not only do mental images play a causal role in thinking, but once formed they remain encoded in a "literal encoding" (Kosslyn, 1981) that may be composed of, for example, polar coordinates. The literal encodings are stored for future use. Data from experiments on how subjects recall the verbal description of spatial relations among commonplace objects indicate the following points (Johnson-Laird, 1981; Kosslyn, 1981): (1) mental models, for example, mental images, are better remembered than propositional representations of a phenomenon; (2) we can, and often do, reason by developing mental models rather than strictly by syllogisms; and (3) inferences drawn from mental models lead to a more profound understanding of the problem situation. In other words (and here is a point of interest for historical studies), there seems to be a complex interplay between logical syllogistic reasoning and mental images.

In summary: All factions agree that there are mental images. The fundamental issue is the image's mental representation in the functional architecture of the mind. For our purposes there are two main aspects to this issue, namely, the mental image's content and its format (Kosslyn, 1977; Farah and Kosslyn 1982). The content of a representation is what is being represented. The disagreement between anti-imagists and pro-imagists arises over the issue of format. The format of an internal representation is its encoding, which for imagery could be propositional or possibly literal as well as propositional (as the pro-imagists maintain). In discussing Poincaré I will comment on the problem of format. For Einstein I discuss content only. In the investigation of Bohr and Heisenberg I analyze both content and format.

The views of the three factions concerning mental imagery can be summarized thus far as follows. According to Haugeland (1981), by eliminating the biological substratum, computer science does "without any messy physiology or ethics [and sends] philosophy packing" because by assuming that "people *just are* computers" the mind-body problem, for example, becomes a "red herring" (italics in original). According to Johnson-Laird and Kosslyn, among others, the mental images most useful in problem solving are the product of propositional encodings, which they can surpass in powers of inference. To the pro-imagery group, mental images need not always be propositionally encoded, an assertion for which they claim empirical proof. It is to the "truly spectacular" data on mental imagery that I now turn, along with the ways in which these data are explained by the different factions.

SOME DATA

After fifty years of neglect during the heyday of behaviorism, mental imagery is once again a topic of research in psychology. Indeed, with the emergence of a truly spectacular body of experiments, imagery is one of the hottest topics in cognitive research.

N. Block (1981)

Roger N. Shepard's experiments (Shepard, 1971; Brown, 1981) are taken to be among the principal reasons for the emergence of research in mental imagery from the great eclipse of almost half a century caused by behaviorism's emphasis on sensory-response data.[4] Shepard's experiments provide an externalization of mental events—that is, his data could be interpreted as a glimpse into the mind's eye of the subject. For example, Shepard would project the letter *R* in its standard upright orientation on a screen and then replace it by a rotated version. He found that the time required for a subject to decide whether a geometrical figure is rotated from the orientation of a standard one is linearly related to the angle of rotation. These data led him to propose a hypothesis on the internal representation of the rotation process, namely, that subjects mentally rotate objects just as they would if the objects were actually perceived, and, thus, that mental images have spatial extent and are not epiphenomena. Furthermore, continued Shepard, the "very same mechanisms are operative in imagery and perception" (1978b). Thus he was able to explain why rotations are performed gradually with the rotated figures passing through every intermediate state instead of flipping instantaneously into congruence with the standard, and, therefore, why figures at increasing angular disparities from the standard take longer to rotate into congruence.

Assuming, as seems to be the case from Shepard's experiments, that mental images have spatial extent, then images can be scanned and more time should be required to scan longer distances across images. In a series of experiments initiated in 1973, Stephen M. Kosslyn, a principal spokesman for the pro-imagery view, obtained data that supported the linear relation between scanning time and distances scanned.[5]

Data of these sorts, in conjunction with indications of the interplay between descriptive (nonpictorial) and depictive (pictorial) information, are the empirical foundations for the computer-simulated model of the theory of mental imagery that Kosslyn proposed in detail in 1981.[6]

A THEORY OF IMAGERY

However, if a coherent theory that treats imagery as a distinct "mental organ," a theory having explanatory power and predictive utility, *can* be developed, this alone should make us hesitant to abandon the construct.

S.M. Kosslyn (1979; italics in original)

In agreement with the reductionist goal of cognitive science, Kosslyn seeks an abstract theory whose "general model" does not simply account for behavioral data, but provides an account that is "mechanistic, knowing a given initial state should allow one to specify factors that will determine the next state."

In Kosslyn's theory mental images have spatial content and are functional in thought processes. In the interest of cutting through several long-standing philosophical and psychological objections to the notion of an image-in-my-head, he sharpened the notion of a mental image:[7] the mental image is a quasipictorial representation that results from the processing of depictive information that is encoded propositionally and literally. The quasipictorial representation is displayed in a spatial medium, or "visual buffer," that is innate and hence part of the functional architecture.[8] The theory is composed of the visual buffer and the underlying deep representations that are encoded. The visual buffer is simulated by a two-dimensional surface matrix in which cells are filled in, or is accomplished in practice by dots made by a printer connected to the computer. The deep representation is twofold: (1) The perceptual memory containing information about the remembered literal appearance of an object that is encoded in polar coordinates (r, θ); and (2) facts about objects that are encoded in lists of propositions. Consequently, for Kosslyn the deep representation contains memories of visual information witnessed previously that are in literal encodings and also are in propositional encodings. Images are generated through mapping deep representations into surface representations on the visual buffer. These images are tested by other deep representations regarding the image's clarity, size, and orientation.[9] Besides accounting for extant data, Kosslyn's model also has predictive power.[10]

In summary, Kosslyn postulates that although the representation of mental imagery is at first propositional, certain mental images of objects that have actually been perceived can be stored (i.e., remembered) in a literal format. The deep representation is mapped onto a visual buffer that is a specified part of the functional architecture and so is assumed to be cognitively impenetrable. (The anti-imagists are having difficulty identifying processes other than primitive reflexes that are cognitively impenetrable—see Pylyshyn, 1980.)

Since 1973 cycles of exchanges have occurred principally between Pylyshyn and Kosslyn, with each countering the other's opinions of imagery and then offering revised positions. A review of these exchanges is revealing of the deep problems that confront understanding not only the dynamics of mental imagery, but also the limitations on metaphorical thinking.

Thesis and Antithesis

Pylyshyn (1973) has recently summarized and developed [his view], and Kosslyn and Pomerantz (1977) have provided counterarguments. Not surprisingly, neither the arguments nor the counterarguments have been definitive . . . .

S. Kosslyn (1979)

The following points of the exchange bear directly on the historical case studies; in fact, they will emerge from them.

Pylyshyn, a spokesman for the anti-imagists, has taken the following positions:

(1) That people imagine processes in the same way as if they were actually occurring is referred to as the perceptual metaphor, or image-perception link. The anti-imagists claim that the perceptual metaphor for imagery has neither explanatory nor predictive power. Their principal objection, according to Pylyshyn (1981a), is that metaphorical explanation leaves the relationship between the "primary and secondary objects of the metaphor" too open; consequently, interpretation of rotation data is trivial because the picture metaphor is based on language from the world of perceptions. Pylyshyn prefers a propositional encoding of the rotation process.[11]

The pro-imagery (see Kosslyn, 1977, 1979, 1981) replies to these criticisms are as follows:

(1) Kosslyn, too, is critical of Shepard's extreme position on the image-perception link, owing to its simple analogies and its tendency to pass the theoretical buck to workers in perception. Nevertheless he supports the assumption that imagery resembles perception because of its usefulness as a working hypothesis and because it provides a step toward a unified theory of imagery and perception. Thus Kosslyn's model allows to imagery certain structures that are used in perception, for example, the visual buffer. Kosslyn proposes that imagery construction and transformation may not be shared equally by imagery and perception: "After all surely one of the functions of imagery is the ability to visualize the outcome of some operation that cannot for whatever reason be performed in the physical world" (1979). In support he mentions the role

of the thought experiment in creative scientific thinking.

(2) The image-perception link was tacitly assumed by Kosslyn and Shepard because they had implicitly instructed their subjects to perform mental operations as if the objects were actually present. Pylyshyn (1981a) offered data showing that the slope of the curve of relative orientations of two figures and the time it takes to compare them "depends on various cognitive factors such as figural complexity." Thus the visual buffer is not innate since it can be explained by rules or representations that are not biologically based— that is, the visual buffer is cognitively penetrable.

(2) As for Pylyshyn's data, Kosslyn considers that figural complexity complicates only the mental task of rotation, thereby cognitively penetrating the comparison phase but not the rotation phase.

(3) Imagery is more constrained than description because we cannot imagine every object that we can describe, for example, four-dimensional space. Pylyshyn discriminates between (imagine)$_S$, which is to imagine an object that we have actually seen or to imagine constructing an object from parts we have actually seen (i.e., an image-perception link); and (imagine)$_T$, which is to think of something in the abstract, like four-dimensional space.

(3) Kosslyn cites data that support the contention that not all imagery simply mirrors the analogous perceptual phenomena. For example [and this applies to (1) as well], there are laboratory tasks involving prism adaptations that are so novel that "'naive' subjects" could hardly have the tacit knowledge of the analogous perceptual effects (Finke, 1980)—that is, of the laws of optics that would permit them to unravel (i.e., cognitively penetrate) the rotation and scanning phenomena they have viewed through the array of prisms. Furthermore, to

counter charges of biased data, experiments have been performed in which the experimenter deliberately creates expectations for outcomes that are contrary to those predicted (Finke, 1980). The results support the pro-imagery faction.

(4) As to the abstractness of propositions, Pylyshyn (1973) claimed that since people can go from mental words to mental images, then the necessity is unavoidable to use a propositional encoding that also plays the role of an interlingua. Furthermore, a single code is parsimonious.

(4) The argument for a single code is dangerous because it could lead to an infinite regress (Anderson, 1978).[12]

(5) As to the truth content of propositions, Pylyshyn (1973) considers mental images inadequate for representing the world since they are neither true nor false, but merely exist.

(5) "Propositions need not be the only way to model mental processes" (Kosslyn, 1977).[13]

As Kosslyn wrote (see epigraph), neither these "arguments nor the counterarguments have been definitive . . . . " A possible reason for this stalemate, as well as an attempt at a sort of synthesis between the anti- and pro-imagery views, was proposed by John R. Anderson (1978).

SYNTHESIS

No sooner does one side produce what they take to be a damaging result than the other side finds a way to compensate for this by adjusting the process that accesses the representation while leaving fixed the assumed properties of the representation itself.

Z. Pylyshyn (1981b)

In response to this impasse Anderson proposed the so-called "agnostic" view that behavioral data cannot discriminate between the imaginal and propositional points of view. Anderson's reason is that since these data concern a subject's response to sensory input, they cannot decide uniquely issues concerning the internal representations. For example, data such as Shepard's describe only the stimulus and response. In support of this conjecture, Anderson sketched a proof that an imagery theory can be transformed into an equivalent propositional theory and then applied it to Shepard's rotation data.[14] Although the proof's details have been criticized (see Anderson, 1979), it has not been disposed of and both sides have had to respond to the spectre it raises.

Kosslyn, who has a stake in the empirical data, dismisses Anderson with almost short shrift. As a "statement of faith . . . . it will be possible" to distinguish empirically between two views (Kosslyn, 1981).

Pylyshyn was more receptive to Anderson's result, perhaps because the anti-imagists' response to the empirical tide has often taken a form that Kosslyn has (1981) called "Rube Goldberg" models. Pylyshyn believes that "one should not appeal solely to data . . . since no finite amount of data alone could ever uniquely determine a theory . . . but one should also consider the explanatory power of the model—that is, how well it captures important generalizations, how constrained it is . . . how general it is, and so on." Thus contrary to the way in which certain philosophers of science have retrodicted the course of science, we have here a case in which one of two competing researchers declares a stalemate to be a step toward victory. The anti-imagery opinion is based on deemphasizing empirical data in favor of such time-honored notions as parsimony, completeness, and, I shall assume, aesthetics (which is what Pylyshyn includes in his "and so on").

Other anti-imagists take the interesting view that it is still too early "to know in advance what kind of experimental evidence will ultimately appear decisive in supporting (or rejecting) ontological claims for our theoretical entities," that is, internal representations (Lachman, 1979). In order to support this tack they reach out to a historical episode in physics in which the ontological status of atoms was settled by Einstein's 1905 theory of Brownian motion, although empirical confirmation of Einstein's theory had to wait until 1908. Until then it was unclear what sort of experimental evidence would suffice to support the claims of atomism that had been vigorously advocated since the late nineteenth century.[15]

I now turn to case studies in creative thinking, which may serve as a laboratory for cognitive science, as well as perhaps provide a means for testing the agnostic view. In each case a historical scenario is developed first that is based on the detailed analyses in chapters 1–4, and the analysis from cognitive science follows. For continuity in development I have omitted historical reference to material discussed in previous chapters.

HENRI POINCARÉ

POINCARÉ'S INTROSPECTION ON INVENTION

It is by logic that we prove, but by intuition that we invent.

H. Poincaré (1904b)

What, in fact, is mathematical invention?

H. Poincaré (1908b)

On 8 May 1908 Henri Poincaré, at fifty-four years old France's premier scientist, addressed l'Institut Général Psychologique in Paris with a lecture entitled "Mathematical Invention." Straightaway, in this often-cited lecture, he posed the question, "What, in fact, is mathematical invention?"[16] Elsewhere Poincaré had described the creative mathematician as one who thinks in images that can be visual or "sensual," where the latter sort could lead toward mathematical invention (1900b, 1904b).[17] In the 1908 lecture he focused on sensual imagery, defining it as the "ability to perceive the whole of the argument at a glance." Like Mozart, whose auditory image of a new and as-yet unwritten piece was not of the "parts successively" but "all at once" (Mozart, in Ghiselin, 1952; see also Gardner, 1982),[18] Poincaré's sensual imagery of a mathematical proof was not of a "simple juxtaposition of syllogisms," but of their "order . . . the feeling, so to speak, the intuition of this order." For both men the form of the new creation was of the essence, and details followed.

As an example of creative thinking, in (1908c) Poincaré described his invention in 1881 of certain mathematical quantities (Fuchsian functions) that almost overnight singled him out as an outstanding mathematician. Here I focus on his introspective analysis into the dynamics of the unconscious. According to Poincaré creative thinking occurs in cycles of conscious work—unconscious work—conscious work—verification.[19] Conscious work is followed by a period of rest which actually involves "unconscious work": "One is at once struck by these appearances of sudden illumination, obvious indications of a long period of previous unconscious work."[20]

Poincaré described unconscious work as the period during which "everything happens as if the inventor were a secondary examiner"—that is, as if the inventor could step out of his mental framework to observe his mind's unconscious workings. Since to Poincaré, "invention is selection," then how does the unconscious select and assemble the appropriate combination of mathematical facts? Poincaré's reply mirrors his intuitionism: "The rules that guide choices are extremely subtle and delicate, and it is practically impossible to state them in precise language; they must be felt rather than formulated. Under these conditions how can we imagine a sieve capable of applying them mechanically?" Nor, according to Poincaré, could even the choice of the appropriate combination of facts be articulated: "This too is most mysterious [because the] useful combinations are the most beautiful, I mean those that can charm that special sensibility that all mathematicians know."[21] Thus the subliminal ego, or unconscious, tests all possible combinations of mathematical facts. The mathematician's "special aesthetic sensibility plays the part of the delicate sieve" that filters out all but the few combinations that are "harmonious" and "beautiful."[22] These combinations find their way into conscious thought, and "this, too, is most mysterious."[23] But this process must be prepared for, because conscious work causes unconscious work that may trigger creative thinking. Whereas conscious work requires "discipline, attention . . . in the subliminal ego . . . there reigns what I would call liberty." The liberation of thought in the unconscious, permits "absence of discipline [and] this very disorder permits unexpected combinations . . . . In a word, is not the subliminal ego superior to the conscious ego?"

Toward the end of the nineteenth century scientists interested in the link between prescientific and scientific thinking began to publish introspections. But unique to Poincaré's introspection was his elaboration on the workings of the unconscious. This interest may have resulted from his familiarity with the psychologists Jean-Martin Charcot and Pierre Janet, and with the work of the philosopher Emil Boutroux (his brother-in-law), whom Poincaré cited in the 1908 essay and who was a friend of William James. Poincaré's interest in the psychology of concept formation is amply demonstrated in his classic writings on the foundations of geometry. It is an understatement to say that he was aware of every current of research in Paris. In fact, from correspondence in the Poincaré archives we know that Poincaré often attended the soirées of a future fervent supporter of Freudian psychoanalysis and an intimate of Freud himself, Marie Bonaparte.

A Psychologist's Profile of Poincaré

Il paraît même tout à fait négliger l'image visuelle.

E. Toulouse (1910)

Poincaré's introspection is supported by a fascinating document that seems to have been overlooked by modern cognitive scientists, namely, the profile of Poincaré by the psychologist E. Toulouse, entitled *Henri Poincaré*. So far as I know, this is the only systematic psychological profile done face-to-face with a major scientist. In 1895 Toulouse embarked on the ambitious project of examining with the "methods of clinical medicine and of the psychological laboratories" men who demonstrated "superior minds" through their work. His conversations with Poincaré occurred in 1897 and the book was published in 1910 with Poincaré's imprimatur.

According to Toulouse, Poincaré was of medium height (5'4"), portly (154 lb.), slightly stooped in posture, and had an air of distraction about him that was legendary. Even while attending meetings at l'Académie des Sciences, l'Académie Française, or at faculty meetings at the University of Paris, he was either sketching figures (usually symmetrical designs) or scribbling mathematics on memos or even on the back of envelopes (Fig. 6.1). I have found that Poincaré's notebooks from l'Ecole Polytechnique reveal that he was an inveterate doodler, once again of symmetric figures (Fig. 6.2). Toulouse reported that Poincaré worked from 10 A.M. to noon, and then from 5 P.M. to 7 P.M.. The evening was reserved for journal reading.[24] Poincaré suffered from a mild insomnia that he linked to chronic indigestion; of the amount of time he spent in bed, usually from 10 P.M. to 7 A.M., he slept effectively for only an average of seven hours, and even the slightest noise could awaken him. Poincaré reported that just before falling asleep he experienced intensely vivid hypnagogic visual images that he was unable to remember clearly but that usually concerned silent people moving about. In "Mathematical Invention," Poincaré wrote that the ideas that came to him during his "semi-hypnagogic state," either at night or in the morning, were not always trustworthy.[25] On passing into a deeper sleep these unpleasant images were replaced by dreams that he could not remember at all. Poincaré was adamant that he never solved mathematical problems while asleep and dreaming.

Toulouse compared Poincaré's responses to various tests, for example, memorizing of numerical sequences, alphabetical sequences, and poetry, with those of Emile Zola and the sculptor, Jules Dalou. For

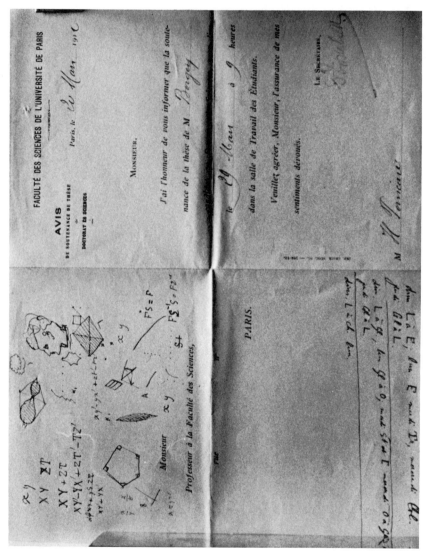

Fig. 6.1(a). Marginalia on a memo indicate how Poincaré sometimes occupied himself during a thesis defense (*soutenance de la thèse*).

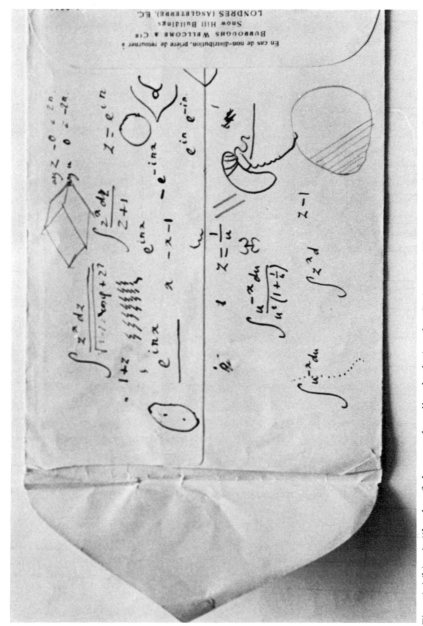

Fig. 6.1(b). A "back of the envelope" calculation by Poincaré. (Courtesy of Estate of Henri Poincaré)

Fig. 6.2. A sample of Poincaré's doodlings in a notebook from his student days at l'Ecole Polytechnique. (Courtesy of Estate of Henri Poincaré)

example, Toulouse would read a passage to each of the three subjects and then ask them to reproduce it in writing. Zola performed best, reproducing the dictated passage almost verbatim. Dalou omitted many details, while freely rearranging and simplifying the text. Poincaré omitted some details but reproduced the logical order of events—that is, he retained the form of the text. Toulouse concluded that Poincaré was poor at rote learning. He was not, however, asserting that Poincaré had a poor memory, but that instead of attempting to memorize a collection of numbers or words by brute force, Poincaré would try to fit them into patterns. Similarly, continued Toulouse, Poincaré did not memorize the individual functions of laboratory apparatus or mathematical formulae, but would try to figure them out from first principles. These results square with Poincaré's ability to take in at a glance the trend and form of a mathematical proof. On the basis of memory tests, such as the one described above, Toulouse declared Poincaré's imagery to be "*auditif.*" This does not capture Poincaré's creative thinking because, for example, words or sounds were unnecessary for Poincaré's invention of Fuchsian functions. Poincaré's depiction of his imagery with the term "sensual," although less precise than auditory, is more meaningful. In fact, Toulouse himself was puzzled by his conclusion that Poincaré was an "*auditif*" and "usually neglected visual images altogether." Did not Poincaré consider himself to be an intuitionist? Here, as is too often the case when psychologists examine scientists, Toulouse's unfamiliarity with scientists' methods misled him because Poincaré's sensual intuition required no visual images.

Toulouse's conclusion on "*le problème du génie*" was that there is a tendency for the "primary play of associations" among elements in the field of knowledge to achieve the proper synthesis that is an "illumination." Poincaré, continued Toulouse, referred to this inarticulable process as "*le travail de l'inconscient.*"

POINCARÉ AND COGNITIVE SCIENCE

Keeping in mind Poincaré's introspection and Toulouse's psychological profile, I now draw on aspects of cognitive science from the earlier section of this chapter, "Some Notions from Cognitive Science," in order to investigate whether they shed any new light on Poincaré's creative thinking.

Is Poincaré's method of thinking amenable to the computer science approach? After all, the search for correct combinations seems ripe for computer simulation—that is, if the "fuzzy" notion of aesthetics is excluded. In fact, Poincaré's invention of Fuchsian functions seems a likely candidate for the "logic of discovery" that Simon alluded to in the epigraph to this chapter. Simon's logic of discovery is embodied in computer programs that can detect patterns in data: "Law discovery means only finding patterns in the data that have been observed; whether the pattern will continue to hold for new data that are observed subsequently will be decided in the course of testing the new law, not discovering it" (1966). Thus Simon's logic of discovery is a computerization of Norwood R. Hanson's (1958) scenario of scientific discovery, which is closely connected to theories of perception. But in the case of Poincaré's invention of Fuchsian functions in contrast with an invention in the physical sciences, it is unclear what the mathematical "facts" were that Poincaré and his rival Felix Klein both "saw" in 1881, but in which only Poincaré could "see" the pattern statement that made the facts intelligible. Indeed, neither could Poincaré articulate a method for identifying useful facts beyond observing that they are "simple," "oft-recurring," and that it is the "quest for this especial beauty, the sense of the harmony of the cosmos, which makes us choose the facts most fitting to contribute to this harmony" (Poincaré, 1907).

Simon (1966) treats creative thinkers such as Poincaré by postulating that their "discovery is a form of problem solving and there are no qualitative differences between . . . work of high creativity and journeyman work." Creative thinkers just have better methods of search for solutions, that is, better heuristics. Thus Simon redefined the problem of creativity into a form amenable to laboratory investigations with subjects of average mental capabilities. Applying this "minimalist" view (i.e., economical à la Mach) to Poincaré, Simon further postulated that unconscious thought is the same as conscious thought. Simon then conjectured that Poincaré's thinking is composed of two mechanisms that he calls "familiarization" and "selective forgetting." Familiarization is conscious work that involves using data to move toward a goal along the nodes and branches of a decision tree. In selective forgetting, or incubation, the decision tree disappears from memory, leaving behind the successful attempts that are the nodes for the next phase of familiarization.[26] While this search procedure is the basis for several interesting problem-solving computer programs, it leaves something to be desired concerning Poincaré's invention of Fuchsian functions.[27] For example, can aesthetic sensibility be defined operationally post hoc as a heuristic or should it be considered a mental algorithm? It seems here, as in every logical empiricistic or "minimalist" reconstruction, there are terms and concepts that slip through the net of an operational view so extreme as to be disavowed by its founder, P. W. Bridgman.

Poincaré's sensual imagery enabled him to achieve astonishing results in mathematics, where he left to others the task of providing rigor. In physics, on the other hand, his forte was criticism, elaboration, and providing rigor. Thus in 1905 he clothed a current theory of matter in a mathematics that was destined to find its deepest meaning within another theory for whose invention sensual imagery was insufficient—Albert Einstein's special theory of relativity.

ALBERT EINSTEIN

EINSTEIN'S INTROSPECTION ON THINKING

What, precisely, is 'thinking'?

A. Einstein (1946)

Although he disagreed with Poincaré on scientific and philosophical matters, Albert Einstein had a deep regard for Poincaré's views on the origin of scientific knowledge. Like Poincaré, Einstein began his most widely read piece of introspection, the "Autobiographical Notes" of 1946, with a question, "What, precisely, is 'thinking'?" He replied that from sense impressions "memory pictures" emerge. A certain "picture" serves as an "ordering element" for the potpourri of memory-pictures and that picture is a "concept." Thinking is the free "play" with concepts, and words follow. In order to explore the origins first of Einstein's creative thinking, and then of Heisenberg's, it is useful to digress for a moment into Kantian philosophy in order to present the sort of imagery that I find to have been influential in developments of science and engineering in German-speaking countries (see Chapters 2, 3 and 4).

KANT AND LIGHT

An *Anschauung* of interest to this chapter is the promotion of the primitive picture of a water wave into the *Anschauung* of light as a wave phenomenon. From the late nineteenth into the twentieth centuries, physicists have thought so much about light as a wave that they can "see" light waves and magnetic lines of force as well. As the physicist Ludwig Hopf, who had been Einstein's assistant from 1909 to 1911, wrote (1931), "Nature does not direct itself to our *Anschauung*, but our capacity for *Anschauung* must direct itself to nature. Our scientific work is a larger process of accommodation to nature." That is, in Einstein's words, "scientific thought is a development of pre-scientific thought" (Einstein, 1934).

A defining characteristic of the quantity we call light is interference. For example, when light is directed at an opaque screen scored through with two slits there results on the other side a pattern made up of regions of lightness and darkness that can be observed in the variations on the dial of a photographic exposure meter. This double-slit interference pattern can be explained with a wave theory of light as follows. We *imagine* that the light waves that strike the screen are like the familiar spherical water waves that issue from the point where a stone has struck the surface of a pond. The result of each incoming spherical wave striking the two slits is that each slit becomes the source of spherical waves, and these two new sets of waves meet and intermingle. If their crests meet, the two waves reinforce each other and the result of this reinforcement is constructive interference, which is a region of brightness. The other extreme case occurs when a crest and trough meet, which results in destructive interference and a region of darkness.

We turn next to the effect of *Anschauung* on Einstein's thoughts on the nature of light.

EINSTEIN AND *ANSCHAUUNG*

In his pioneering study of Einstein's use of visual imagery, Gerald Holton (1973g) pointed out the importance of Einstein's sojourn of 1895–1896 at the Kanton Schule, Aarau, Switzerland, where "everything changed for him." Here I explore this connection further.

The school was founded by followers of the Swiss educational reformer, Johann Heinrich Pestalozzi, who emphasized the innate "power of *Anschauung*." Knowledge, he wrote in his 1801 book, *How Gertrude Teaches Her Children*, arises from sense impressions that are "irregular, confused." Through education begun with observing the world about oneself "clear ideas" emerge, until finally one develops intuitions according to a notion of *Anschauung* analogous to Kant's. Pestalozzi realized the fecundity of visual thinking based first on a close link to objects that had actually been perceived (a close image-perception link), and then through abstraction to *Anschauungen*.[28] Einstein's introduction to Kant's notion of *Anschauung* came through reading the *Critique of Pure Reason* at age thirteen. This notion was emphasized further at the Kanton Schule, and then again at the Zurich Polytechnic Institute, where the curriculum focused on applied electricity at a time when the *Anschauung* of magnetic lines of force were puzzled over throughout German-language engineering journals.

Needless to say, I am not defending the innateness of *Anschauung*. What I am stressing is that *Anschauung* as a philosophical-scientific notion was used in German-speaking countries, and Einstein was a product of that environment.

ANSCHAUUNG, PICTURES, AND THOUGHT EXPERIMENTS

Elsewhere I have discussed the role of the *Anschauung* of magnetic lines of force in Einstein's invention of special relativity theory (Chapter 3 and Miller, 1981b). Here I cite two examples in which Einstein wielded with stunning results his view that "scientific thought is a development of pre-scientific thought"—namely, the two thought experiments that led to the special and general theories of relativity.[29] These experiments concerned a mixture of *Anschauungen* with mental imagery based on a strong image-perception link. The thought experiment of 1895 that led to the special relativity theory of 1905 concerned the experiences of a moving observer who tries to catch up with a point on a light wave whose source is at rest.★ What is involved in this thought experiment is the picture of a moving observer, the intuition of catching up with something that at first was moving faster than you, and the *Anschauung* or customary intuition of light as a wave phenomenon. By the customary intuition of light I mean depicting light as an entity that has properties that are abstracted from certain properties of water waves. Throughout the next decade Einstein pondered this thought experiment.

By 1905 Einstein realized that not only could the thought experimenter never catch up with the point on the light wave, but that the point always moved at the same velocity relative to his laboratory—namely, the velocity of light. Imagine what a mind boggling result this is, because it runs counter to our intuition of catching up with objects from the world of perceptions. Imagine, for example, that you pull onto a highway and ahead of you is another car that was already on the highway and is traveling faster than you. You step on the gas in order to pass this car, but no matter how fast you go the other car travels at the same velocity relative to you. It turned out that the notion of time is at the root of this phenomenon. Time is a

---

★N.B. This experiment is analyzed in some detail in the Appendix in Chapter 3.

relative quantity, and so it depends on the motion of an observer. This is contrary to our customary intuition. As Einstein recalled in 1946 the high velocity of light compared with the other velocities that we encounter daily, had prevented our appreciating that the "absolute character of time . . . unrecognizedly was anchored in the unconscious." Indeed, Einstein's own appreciation of this point did not come easily (see Chapter 3). Thus, in 1905 Einstein went beyond customary intuition because he realized that the conclusions drawn from the mental imagery of customary intuition cannot always be trusted.

The thought experiment of 1907 that led to the 1915 general relativity theory concerned an observer in free fall and the consequences of there being "for him during this fall no gravitational field—at least in his immediate vicinity" (pictures only). But these pictures were not merely image-perceptual representations, they were theory laden. For example, in the 1907 thought experiment Einstein "saw" objects that dropped with him, and thus fell at relative rest, in the context of Newton's law of motion applied to the falling observer and to an observer on the ground. Although scientists have many times "seen" objects falling side by side, Einstein "saw" what cognitive scientists refer to as the "deep structure" in this scene; that is, he "saw" the relation between gravity and acceleration so long as he also assumed the exact equality of gravitational and inertial masses.

In order to explore further the links that Einstein emphasized among prescientific and scientific thought, his visual mode of thinking, and *Anschauung,* it is essential to recall his definition of a "concept" as a "memory-picture" that serves as an ordering element for perceptions. This is because like the concepts in the thought experiments of 1895 and 1907, the basic concepts in the special relativity theory—measuring rods and clocks—have unambiguous correlates in the world of perceptions. This was probably intentional since in 1923 he introduced the "stipulation of meaning," according to which "concepts and distinctions are only admissible to the extent that observable facts can be assigned to them without ambiguity." This stipulation had been important to his research toward the relativity theories. It was almost certainly of great concern to him in 1909 when he reflected in print on the light quantum, which he had introduced in another masterpiece of his annus mirabilis of 1905. In (1905b) Einstein proposed that to understand certain processes in which light is converted into matter it would be useful to consider that light propagated through space like a hail of shot, or light quanta, instead of as a wave.

ANSCHAUUNG AND LIGHT

Possibly because in the 1905 light quantum paper Einstein focused on the particulate property of light and its usefulness for eliminating an asymmetry in radiation theory, he did not use the notion of *Anschauung*. To the best of my knowledge Einstein reserved the term *Anschauung*, in its Kantian sense, for speculations on how the nature of light affects our notion of physical reality. For example, in an important paper of 1907, in which he first recalled his view of physics in 1905, Einstein called attention to his suprising finding that the state of light in a small volume is determined by a finite number of light quanta, in contrast to the classical wave theory of light which specified that it resulted from adding together infinitely many waves (1907a). Deftly playing on an often-quoted phrase from Heinrich Hertz's 1894 *Principles of Mechanics,* Einstein wrote (1907a) that "the diversity of the possible processes is in reality less than the diversity of the possible processes in the meaning of our customary *Anschauung*." Whereas Hertz had written: "We become convinced that the manifold of the actual universe must be greater than the manifold of the universe which is directly revealed to us by our senses," Einstein had realized otherwise.

The first time that Einstein used the term *Anschauung* extensively was in his (1909b) paper, "On the Development of Our Intuition [*Anschauung*] of the Existence and Constitution of Radiation." As the title indicates Einstein explored the clash between the *Anschauungen* of the time-honored wave mode of light and the particle mode that he had proposed in 1905. Physicists long puzzled over and resisted the light quantum, preferring instead the wave picture of light. For example, they considered it unimaginable that light quanta could explain a double-slit interference pattern. Since light quanta were assumed to have a well-defined boundary and were indivisible, how could they pair off so as to produce an interference pattern with periodically recurring regions of lightness and darkness? In addition to the possibility that light was a wave *or* particle, physicists were also perplexed over the possibility that light was a wave *and* a particle—for how could something be spread out in space like a wave and at the same time be localized like a particle? In the sections "Bohr's

Approach" and "Heisenberg's Approach," we discuss how Bohr and Heisenberg resolved these problems in 1927.

In 1909 Einstein introduced the notion of *Anschauung* thus: "The relativity theory altered our *Anschauung* of the nature of light, since it interpreted light not as a consequence of the state of a hypothetical medium but as existing independently of matter." By declaring the "hypothetical medium," or ether, to be "superfluous," the special relativity theory had freed physicists of the anthropomorphic *Anschauung* of light as something requiring support in transit like waves in water or sound in air. Since Einstein considered the principles and concepts of special relativity theory to be extensions of those in Newtonian mechanics, despite special relativity's new notions of space and time, he persisted in later years to emphasize the continuities in these two theories. But, as Einstein wrote in a letter to his friend Conrad Habicht in spring 1905, he considered light quanta to be "*sehr revolutionär*" (Seelig, 1954), and we can now suggest why— namely, that light quanta were what Einstein defined in 1946 to be a "wonder" since their characteristics were in conflict with already formed concepts, that is, with image-perceptual figures. Einstein never came to grips with the riddle of light quanta. He preferred instead to seek a unified field-theoretical description of matter based on such themes as continuity and action by direct contact, themes that are among the mainstays of prescientific thought.

In summary, the contrast is startling between the simplicity of the mental images in Einstein's visual thinking and the far-reaching theories that he drew from them. By their very nature the relativity theories are axiomatically based and represent levels in a hierarchy of theories, each one more complete than the preceding one—for example, the general relativity theory encompasses more phenomena than the lower-level special theory of relativity. As the axiom system becomes richer, the theory's basic concepts become less satisfactory from the standpoint of the stipulation of meaning. In 1936 Einstein referred to this price to be paid for increased unification as the "poverty of concepts." Could it have been that Einstein never achieved his lifelong goal of the highest level in this hierarchy—the unified field theory—because of limitations on his mode of visual thinking? Such a highly-placed theory would be impoverished in concepts because its concepts would be at best indirectly connected with objects from the world of perceptions. Perhaps that was the reason why there were no more thought experiments like the ones of 1895 and 1907.

EINSTEIN AND COGNITIVE SCIENCE

With the historical analysis of Einstein's visual thinking in hand, let us investigate whether cognitive science offers new insights. We recall Pylyshyn's notion of cognitive penetrability (see "Thesis and Antithesis"). It applies to the images of measuring rods, clocks, and stones, that is, "concepts," in Einstein's thought experiments of 1895 and 1907. But is not cognitive penetrability the assertion that these concepts have meaning within the theoretical context of the thought experiments, that is, that they are theory laden? Whether a concept or fact is theory laden has been an important issue in science and philosophy. Only the most trivial facts are not theory laden. Thus, it is perhaps not unexpected that the anti-imagists have yet to identify a nontrivial reflex or image that is cognitively impenetrable. The images in scientific thought experiments are always theory laden and so are cognitively penetrable.

Putting aside for the moment the questions of the origins of the images, we find the images incontestably functional in Einstein's thinking. The images in Einstein's thought experiments support the data of Kosslyn and Shepard because they crystalized problems to the point where solution ensued. Consequently they played a causal role in Einstein's thinking. Einstein's 1895 thought experiment supports Shepard's interpretation of the "spatial" character of the transformation of his mental images. For as the thought experimenter conceives of himself as accelerating, he measures—in accordance with prerelativity physics—a decreasing velocity of light relative to him, for example, by noting the steady displacement of fringes in an interferometer. Then as he catches up with a point on the light wave he observes the gradual transformation from the mode of a linearly translating vibratory state to a standing vibratory wave state—for example, a state in which one end of a rope is tied to a wall while the other end is driven by hand to form a wave pattern.

Yet it is unclear how the anti- and pro-imagists can deal with the dynamics of how primitive images are abstracted to become *Anschauungen* except by postulating a feedback mechanism by means of which literal representations are somehow enriched by the more general propositional representations. But unless the propositional representations are better defined, and the relation between syntax and semantics is further delineated, then a feedback hypothesis is, as Poincaré wrote in another context, the sort of hypothesis that explains everything and therefore explains nothing. We recall that an *Anschauung* is an abstraction from primitive sense perceptions, which is appended to a theory as a visualization for its mathematics. In certain cases, such as a magnet's lines of force, an *Anschauung* is promoted to be an integral part of the theory itself. The construction and dynamics of an *Anschauung* is beyond the present state of any existing theory of imagery. In fact, on a more basic but no less profound level, Kosslyn's theory does not yet provide a mechanism for how mental imagery is transformed from childhood through adulthood, a shortcoming that Kosslyn himself recognizes (Kosslyn, 1981). Moreover, we have found that *Anschauung* is a philosophical-cultural component of thinking. But does this not run counter to the assumption that there is a universal basic encoding? This is part of a fundamental problem in the psychology of syntax (see Lachman, 1979).

Because of its plasticity, a variant of Kosslyn's theory would provide an attractive approach to the study of *Anschauung*. Yet, in the end, the contrast between the simplicity of Einstein's mental images, their theoretical context and the dazzling theories that they spawned is so startling as to undercut available cognitive scientific models.

From creative thinking based on a theory-laden image-perception *Anschauung* mode, Einstein spun two theories whose consequences further liberated the next generation of scientists from the world of perceptions—for example, Niels Bohr and particularly Werner Heisenberg.

NIELS BOHR AND WERNER HEISENBERG

Indeed, we find ourselves here on the very path taken by Einstein of adapting our modes of perception borrowed from the sensations to the gradually deepening knowledge of the laws of Nature. The hindrances met on this path originate above all in the fact that, so to say, every word in the language refers to our ordinary perception.

N. Bohr (1928)

What quite frequently happens in physics is that, from seeing some part of the experimental situation, you get a feeling of how the general experimental situation is. That is, you get some kind of picture. Well, there should be quotation marks around the word "picture." This "picture" allows you to guess how other experiments might come out. And, of course, then you try to give this picture some definite form in words or in mathematical formula. Then what frequently happens later on is that the mathematical formulation of the "picture" or the formulation of the "picture" in words, turns out to be rather wrong. Still, the experimental guesses are rather right. That is, the actual "picture" which you had in mind was much better than the rationalization which you tried to put down in the publication. That is, of course, a quite normal situation, because the rationalization, as everybody knows, is always a later stage and not the first stage. So first one has what one may call an impression of how things are connected, and from this impression you may guess, and you have a good chance to guess the correct things. But then you say, "Well why do you guess this, and not that?" Then you try to give rationalizations, to use words and say, "Well, because I described such and such." The picture changes over and over again and it's so nice to see how such pictures change.

W. Heisenberg (*AHQP*: 11 February 1963)

The roles of Niels Bohr and Werner Heisenberg in the development of quantum theory provide a most fascinating case study of mental imagery because they demonstrate forcefully the scientist's need for some sort of anchor to the world of perceptions. In fact, without the imposition of any notions from cognitive science, there emerges naturally from a case study of these two scientists some basic criticisms of the anti-imagists and replies similar to those of Kosslyn's.

For example, as Bohr wrote in the epigraph to this section, our customary imagery is abstracted from objects we have actually perceived. But this sort of imagery fails for atoms because it is not only inappropriate, it is nonfunctional as well. While Bohr's 1927 principle of complementarity permitted a restricted use of *Anschauungen*, Heisenberg's subsequent research led him to conclude that the construction of images need not be shared by perception.

Customary Intuition

In 1913 the 29-year-old Danish physicist Niels Bohr proposed a new and daring theory of the atom. He asserted that although his atomic theory violated classical physics, it was still possible to give the theory a "simple interpretation . . . with the help of symbols taken from ordinary mechanics"—that is, prerelativity classical physics based on what Heisenberg would later call our "customary intuition." The theory violated classical mechanics by permitting only certain electron orbits, and it violated classical electromagnetism by restricting the orbital electron's emission of radiation to its transit between orbits. But the system of syntax and semantics of "ordinary mechanics" was the only one available. For example, the symbols x and v, which obey the rules of formation or the grammar of the differential calculus, are assigned the perception-laden meanings of position and velocity. Since in 1913 atomic models were formulated in the language of "ordinary mechanics," then any assumptions concerning the distribution of the electrons in Rutherford's "sphere of electrification" surrounding the nucleus necessarily took the guise of orbital electrons. (Actually, Rutherford's initial experiments were inadequate for specifying the sign of the charges on the nucleus and its surrounding electronic cloud.)

Most physicists thought Bohr's atomic theory a considerable achievement, since it could account for a wide range of empirical data, for example, spectral lines of hydrogen, and it provided a basis for the periodic table of elements.

In 1923 another pioneer of atomic physics, Max Born, wrote of the high hopes that rode on Bohr's theory in which the "laws of the macrocosmos in the small reflect the terrestrial world." This was most reassuring in an era when the relativity theories had so deeply affected notions of physical reality. However, by 1925 empirical data had eroded the functional imagery of the solar system atom.

In a review paper of 1926 entitled "Quantum Mechanics," the 25-year-old wunderkind of German science, Werner Heisenberg, discussed the state of physics that had led him in 1925 to formulate the quantum mechanics. He emphasized the failure of our "customary intuition [*Anschauung*]" when extended to atomic dimensions, and recalled the necessity to liberate oneself from "intuitive pictures." He said in effect that although the old Bohr theory had the "benefit of direct visualizability [*Anschaulichkeit*]," it was fraught with internal contradictions.

IMAGERY LOST

But in 1925 when visualization of the atom itself had been lost, mathematics was to be the guide. This situation suited Heisenberg and in 1925 he followed a promising line of research to its fruition. That is, he based the new quantum mechanics on properties of the atom that are measurable experimentally, for example, spectral lines that serve as the atom's signature, instead of the unobservable electron orbits. Without a mental image of the atom most physicists were adrift in the atomic domain. This situation provoked a direct response from the 43-year-old Austrian physicist, Erwin Schrödinger. In 1926 Schrödinger formulated a wave mechanics in which the *Anschauung* of waves was associated with the orbital electrons. Thus, for example, he envisaged the lone orbital electron in the hydrogen atom as an electrically charged wave completely surrounding the nucleus.[30] In print Schrödinger had written that his original impetus to formulate the wave mechanics was that he was "repelled" by the quantum mechanics' "lack of visualizability [*Anschaulichkeit*]" and by its mathematics. In correspondence Heisenberg referred to Schrödinger's wave pictures as "trash," and to the wave mechanics as useful only for calculational purposes. The atmosphere grew tense. Heisenberg, still uncomfortable with Schrödinger's original earlier proof of the equivalence of the wave and quantum mechanics, proceeded to demonstrate the untenability of Schrödinger's conception of the electron as a charged wave.

At this point in the development of quantum theory problems were approached with a mathematical scheme that lacked a complete set of correspondence rules, or, in other words, with an incomplete semantics. Moreover, mental imagery was still constrained to images of objects from the world of perceptions. It was during this fall season of 1926 that Bohr and Heisenberg met at Bohr's Institute in Copenhagen to begin their intense struggle toward a physical interpretation of the wave and quantum mechanics. Nothing less was at stake than the interpretation of physical reality itself.

At first they were unable to extend visual thinking based on the customary *Anschauung*-image-perception link to the behavior of electrons and light quanta in the thought experiments they devised. Heisenberg recalled the "despair" he and Bohr both felt at their inability to understand the meaning of the perception-laden terms wave and particle. After all, what other way is there to describe entities in the physical world than as waves and particles, since intermediate realities, whatever they may be, are not open to our perceptions? Yet the wave and particle modes resulted in conundrums when used in thought experiments—for example, how can particulate electrons that are shot at a screen with two slits produce the same interference pattern as light waves except perhaps by splitting in two and then recombining in some complex manner on the other side? But is not the electron—the basic unit of electric charge—supposed to be indivisible? Here the arguments against a particle mode of light were applied to the mind-boggling situation of objects that had been assumed to be material particles, but could also behave like waves. Just as the particle nature of light seemed inconceivable, so did the wave nature of the electron.

Bohr's Approach: Restricted Metaphors

In early 1927 Bohr and Heisenberg arrived at apparently separate resolutions to such conundrums. Bohr's insight, as he recalled in a lecture delivered on 16 September 1927, involved his coming full circle back to a realization of 1913 that the restrictions imposed by language on our capacity to form images for scientific theories originated "above all in the fact that, so to say, every word in the language refers to our ordinary perceptions" (see epigraph to the section, "Niels Bohr and Werner Heisenberg"). He compared the dilemma in the atomic domain to the impasse faced by special relativity in which the very high velocity of light produced effects unperceivable by the senses but nevertheless recognized by Einstein in 1905 as the central point around which relativity would have to be formulated—for example, the nonabsoluteness of simultaneity. Just as the scale of phenomena in the world of perceptions is set by the very large velocity of light (3×10^8 m/sec), in the atomic domain the scale is set by the very small quantity called Planck's constant (6.63×10^{-34} joule/sec). The perhaps unfamiliar units are unimportant here; it is the relative sizes of these fundamental constants of nature that is of the essence. For example, as Einstein had realized in 1905, the largeness of the velocity of light tricks us into assuming the absoluteness of simultaneity. On the other end of the scale of nature, as Bohr realized in 1927, the very minuteness of Planck's constant places in bold relief the restrictions that our language imposes on imagery. In the atomic domain Planck's constant connects the wave and particle modes into a wave-particle entity that is hidden from us by constraints on our sense perceptions; for example, Planck's constant links an electron's wave and particle attributes into a single equation.[31]

Bohr's dilemma was that although images (either from *Anschauungen* or from the image-perception link) are necessarily distinct from the laws of physics, we are forced to phrase these laws in a language tempered by sense perceptions because it is the only language we have. Bohr encompassed both horns of the dilemma with the principle of complementarity that he introduced in a lecture of 16 September 1927. This far-reaching principle has two main parts: (1) In the atomic domain an essential difference lies between pictures (or *Anschauungen*) and the actual development of atomic systems. For in this domain physical laws require a "departure from visualization in the usual sense" (1928). (2) The perceived mode of an atomic entity depends on the experimental arrangement in use. For example, light and electrons display their wave mode, that is, behave as waves, when the double-slit interference apparatus discussed earlier is used. But wave and particle modes cannot be exhibited in a single experiment because they are mutually exclusive. Yet, both modes are required to characterize fully an atomic entity.

HEISENBERG'S APPROACH: IMAGERY TRANSFORMED

There are presently between Bohr and myself differences of opinion on the word *'anschaulich.'*

W. Heisenberg (letter to W. Pauli of 16 May 1927, in Pauli, 1979)

Quantum mechanics: syntax and semantics

Unlike Bohr, Heisenberg focused initially on his own quantum mechanics, whose mathematical formulation lent itself to an interpretation of matter as composed of unvisualizable subatomic particles. This approach enabled Heisenberg to deduce such far-reaching results as the intrinsic restriction on the accuracy of simultaneously measuring an elementary particle's position and velocity. In quantum mechanics the product of the accuracies of position and velocity measurements turned out to be directly proportional to Planck's constant. Although Planck's constant is extremely small, it is not zero, and so the more accurate the determination of the electron's position is, the less accurately can its velocity be ascertained; this is one of Heisenberg's famous uncertainty relations. Because of the minuteness of Planck's constant, these relations are alien to the physics of macroscopic systems, where Planck's constant can be taken to be zero. Consequently, according to classical physics in measurements on macroscopic systems either one or both of the uncertainties in velocity and position can be in principle reduced to zero.

In the 1927 paper where Heisenberg presented these results he resisted any customary imagery of atoms and permitted the theory to determine the limits on the meaning of symbols like v for velocity and x for position coordinate. These symbols are among the necessary holdovers from classical physics and so are linked to the world of perceptions; they are necessary for talking about the theory and for use in experimental observations. But at Bohr's insistence Heisenberg accepted the necessity of both wave and particle representations in order to achieve a more consistent and general interpretation of the quantum theory than only particles would provide. On closer examination by Bohr, it turned out that Heisenberg's uncertainty relations were a particular case of complementarity. Although Heisenberg agreed with the complementarity principle's limits on metaphors from the world of perceptions, he remained wary of them, owing to their previous disservices.

This opinion is traceable to the fall of 1926 when Heisenberg began to think deeply about the nature of physical reality on the submicroscopic level. In his first paper devoted to discussing the state of quantum mechanics, "Quantum Mechanics," he wrote that the "electron and the atom possess not any degree of direct physical reality as the objects of daily experience." Thus, from this time on Heisenberg reserved the term *Anschauung* for intuitions constructed from objects actually seen, and *Anschaulichkeit* for perception-laden terms like position and velocity, whose interpretation is dictated by the mathematics of quantum mechanics. The mode of visualizability (*Anschaulichkeit*) in Heisenberg's uncertainty principle paper of 1927 is descriptive—that is, Heisenberg's inversion of the Kantian notions of *Anschauung* and *Anschaulichkeit* led him to conclude that in atomic physics visualizability need not have a visual component. His subsequent research in nuclear physics and the interaction between light and electrons in the 1930s led him to a depictive mode of visualizability that was more far-reaching than the restricted metaphors of the complementarity principle.

Heisenberg's transformation of concepts has been traced in some detail in Chapter 4. The point that I want to develop here is that Heisenberg's new meaning for *Anschaulichkeit* can be either depictive or descriptive, depending on the research problem he had at hand.

Nuclear physics: metaphor becomes physical reality

Heisenberg introduced the visual mode of *Anschaulichkeit* as a result of his 1932 work on nuclear physics. The key problem in Heisenberg's view of nuclear physics was twofold: First, how to explain the stability of a nucleus that contained charged protons and neutral neutrons—that is, how to represent an attractive force between a charged particle and a neutral particle. The customary notion from electricity is of attraction between oppositely charged particles (gravitational attraction is too small to hold a nucleus together). Second, how to explain the emergence of electrons from nuclei that are transformed into lighter nuclei by emitting electrons. According to quantum mechanics, nuclei would have incorrect properties if they contained electrons. Heisenberg's resolution of the problem that he set for himself took him back to another of his dazzling inventions in physics—namely, the exchange force that had permitted him in 1926 to explain several basic characteristics of the helium atom. Since, according to quantum mechanics, the two electrons in a helium atom are indistinguishable, they can be considered only metaphorically to trade their positions periodically, thereby producing an exchange force. Heisenberg's exchange force played an important role in the successful application of quantum mechanics to molecules in 1927, whereas the old Bohr theory had not been able to deal adequately with even the simple H_2^+-ion, in which the exchange force operates through the metaphorical sharing of the single electron between the two protons (Figs. 6.3 (a) and (f)). Since the exchange force also ensures the molecule's stability, Heisenberg ingeniously applied the notion of exchange to the interior of the nucleus. Although every published discussion of exchange forces before Heisenberg's 1932 paper on nuclear physics used the word "*Austausch*" for "exchange," in 1932 Heisenberg specifically suggested substituting the word "*Platzwechsel*" or "migration." Thus Heisenberg's attractive exchange force between a neutron and proton operates through an electron that "migrates" from a composite neutron (i.e., a neutron composed of an electron and proton) to an elementary proton, which captures the electron and becomes a neutron. Although Heisenberg insisted that the allusion to a real "migration" of an electron was unintentional, the metaphor of motion was closer to the actual phenomenon in the nucleus than was the exchange phenomenon in molecules (Figs. 6.3 (b) and (g)).

In order to render the "picture" [*Bild*] of the migration more "intuitive" [*anschaulich*], Heisenberg elaborated on the actual migration of an electron, some of whose characteristics were inconsistent with electrons found outside the nucleus. But, after all, was it not Heisenberg's tenet that the mathematics of a theoretical construct determines what is visualizable? Thus the mathematical formulation of the nucleus, based on the unvisualizable mathematical description of the exchange force of molecular physics, led Heisenberg to the *Anschaulichkeit* of a migrating electron. In fact, at first physicists were somewhat disturbed over the nonvisualizability of the exchange force in molecular physics (see the section in Chapter 4, "Elementary-Particle Physics"). Heisenberg's support for the notion of a migratory electron with incorrect properties was that quantum mechanics probably did not apply to electrons within the nucleus.

Although other sorts of exchange forces were proposed that agreed with available data better than Heisenberg's, forces that did not require inappropriate migratory electrons, Heisenberg's *Anschaulichkeit* of a nuclear exchange force with something exchanged was considered fruitful. It was along this line that Enrico Fermi in 1934 further elaborated on Heisenberg's theme of interpreting mathemat-

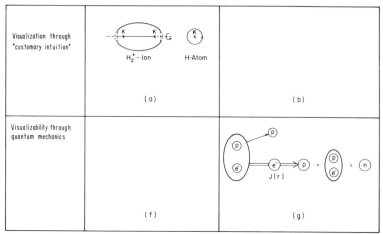

| Visualization through "customary intuition" | (a) | (b) |
| Visualizability through quantum mechanics | (f) | (g) |

Fig. 6.3. The two rows of frames show how quantum theory distinguishes between visualization and visualizability. Frames (a) through (e) contain visualizations according to pictures constructed from objects and phenomena actually perceived. Frames (f) through (j) depict visualizability according to quantum mechanics. Thus, frame (f) is empty, as are frames (b), (c) and (d). The changing notions of physical reality that began with Heisen-

ics through abstracted mental imagery. Fermi's "*anschaulich*" thinking was based on the analogy of an atomic transition between energy levels (Fig. 4.4) and the transition of a neutron into a proton and vice versa. Although Fermi's nuclear theory solved the vexatious problem of the origin of the electrons that are emitted from heavy nuclei, it was inadequate for dealing with the neutron–proton force.

Almost immediately Tamm and Iwanenko extended Fermi's *anschaulich* thinking to the neutron and proton interacting through an exchange of particles. But diagrams abstracted from atomic transitions persisted (Fig. 6.3 (h)). Still another change in visualizability (*Anschaulichkeit*) was required.

The means for this change resided in Hideki Yukawa's 1935 meson theory of nuclear forces. Yukawa reverted from the usual terminology of nuclear "exchange" force back to Heisenberg's original 1932 switch of terminology from "exchange" force to "migration" force. It is noteworthy that in his seminal paper on nuclear forces, Yukawa used the German word for migration rather than exchange, thereby transforming metaphor into physical reality. But it was not until Yukawa's migrating particle was supposedly discovered in 1937 that physicists began to vigorously explore his theory. In 1943 a depiction of Yukawa's nuclear force appeared in Wentzel's book as a didactic device (Fig. 6.3 (i)).

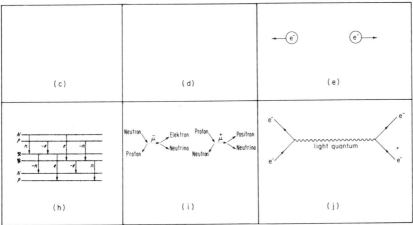

berg's remarkable extension of the exchange force from molecular physics to nuclear physics are in frames (g) through (j). Comparison of frames (e) and (j) illustrates the startling contrast between physical reality according to the world of perceptions and the atomic domain. We note that frames (a), (e), (g), (h), (i), and (j) are from Figs. 4.2, 4.8(a), 4.3, 4.6, 4.7, and 4.8(b), respectively.

The diagrams representing Yukawa's mathematics were further developed into the far-reaching diagrammatic methods of Richard P. Feynman (compare Figs. 6.3 (e) and (j)). The lines in Feynman diagrams (see also Fig. 4.9 (b)) are trajectories in space and time of elementary particles, with which they have become synonymous. In fact, Heisenberg referred to Feynman diagrams as an "*anschaulich*" representation—that is, as a new mode of visualizability.

The impact of Wentzel's book on Feynman remains to be ascertained owing to the complexity of the route that led Feynman to his diagrammatic methods. Nevertheless, it is possible that just as in many cases of productive thinking, the problem situation attains a higher plateau of clarification once the proper diagram is drawn. Conversely, the theories of Schwinger and Tomonaga are the fruits of by-and-large creative nonimaginal thinking. Yet, not surprisingly, more physicists are comfortable with Feynman's diagrams.

Elementary-particle physics: the merging of imaginal and nonimaginal thinking

The historical evidence from Chapter 4 led to the conjecture that Heisenberg's descriptive mode of *Anschaulichkeit* was the other path to Feynman's diagrammatic methods. In order to deal with certain fundamental problems in the interaction between electrons and light,

Heisenberg in 1934 jettisoned mental imagery for this sort of problem and took as *Anschaulichkeit* mathematical formulas that he believed to be trustworthy because they treated noninteracting electrons; interactions could be included by degrees of approximations.
This was essentially his tack in 1925 after he had abandoned images
of the atom in order to begin anew with a theory based on trustworthy formulas and measurable quantities. In 1934 the quantities he
defined to be unobservable were the infinitely large contributions
that emerged from current theory for certain intrinsic properties of
the electron such as its mass. Heisenberg eliminated these infinitely
large quantities through a calculational method that allowed them to
be subtracted. But this so-called "subtraction physics" was unwieldy
to use and could not remove all the infinities that plagued the quantum-theoretical treatment of the interactions between light and electrons. Nor could any of Heisenberg's other attempts from 1934 to
1943 at a theory of elementary particles deal with these problems,
among others. In fact, by 1943 he conceived of the state of fundamental theory to be so serious as to require recourse to his strategy of
1925. So, Heisenberg proposed another descriptive theory of elementary particles that dealt with only experimentally measurable
quantities and which could serve as a probe toward the new theory
that was valid at interaction distances less than 10^{-13} cm. (The contemporaneous theory failed at energies that are comparable to this
distance; in quantum mechanics there is an inverse relation between
energy and distance.) The quantity basic to the new theory was the
S-matrix. Heisenberg was prepared once again, as he had been in
1927, to seek the "intuitive content" of the future theory. It turned
out that a fundamental length of interaction was unnecessary in the
new theory of the interactions between electrons and light that was
formulated in 1948 by Schwinger, Tomonaga, and Feynman.
Nevertheless, Heisenberg's S-matrix formalism retained its validity.
In 1949 Freeman J. Dyson proved that Feynman's diagrammatic
methods are essentially a set of rules for calculating the S-matrix.
Heisenberg's descriptive mode of thinking in the subtraction physics
and then in the S-matrix theory became depictive. This could only
have been the reason for Heisenberg's referring to Dyson's 1949
work as providing an *anschaulich* representation.

In summary, the complexity of Heisenberg's switch of internal
representations resulting from his scientific research led him beyond
merely inverting the Kantian view of perception, no mean feat in
itself. He carried the wave-particle duality beyond Bohr's original
formulation of complementarity: Planck's constant led not only to

the wave-particle duality and to the mutual exclusiveness of these two modes of existence in experimental setups, but also to a distinction between visualization and visualizability. Visualization is the result of our customary intuition (*Anschauung*) and leads to the restricted metaphors of the complementarity principle. Visualizability is the property of *Anschaulichkeit* which is the new form of imagery that is constructed without resort to objects actually perceived. There can be no visualization of atomic entities; on that, Bohr's analysis of perceptions was abundantly clear. But atomic entities are visualizable because their intrinsic characteristics are revealed through the mathematics of quantum mechanics—to that, Heisenberg had been driven.

In 1927 Heisenberg had permitted the mathematics of the quantum theory to decide the meaning of and restrictions on the theory's symbols, which are tied to metaphors of a perception-laden language. By 1935 he permitted the theory to decide the representation of the entities that the theory purports to describe. Consequently, the course of Heisenberg's research had led to permitting the syntax of the quantum theory to determine its semantics, that is, the meaning of the symbols and the representation of the objects that the theory characterizes are determined by the grammar or in this case the rules of mathematics. In the world of perceptions such notions as visualization and visualizability are interchangeable. But in the atomic domain they become mutually exclusive, owing to the fusion in this domain of the wave and particle modes of light and matter.

Having begun his scientific career as a good positivist, Heisenberg came to embrace the philosophical view of the first book that he had read on atomic physics in 1919. As he wrote in 1976: "Our elementary particles can be compared with the regular solids of Plato's *Timaeus*. They are the Archetypes of the Ideas of matter." Heisenberg's new mode of imagery is constructed from the mathematics of atomic physics and is comparable to the imagery sought by Plato's cave dwellers.[32]

Heisenberg's new image, or *Anschaulichkeit,* has influenced developments in the atomic and subatomic realms. On the course of his research Heisenberg recalled in 1963 that the "picture changes over and over again and it's so nice to see how such pictures change" (*AHQP*; 11 February 1963).

BOHR, HEISENBERG, AND COGNITIVE SCIENCE

The struggle for some sort of imagery by Bohr and Heisenberg seems to square with Pylyshyn's criticism that mental imagery is constrained to copies of objects from the world of perceptions (see the section "Thesis and Antithesis"). In the course of their deliberations in late 1926 to interpret the wave and quantum mechanics physically, Bohr and Heisenberg could not (imagine)$_S$ the behavior of electrons and light quanta in thought experiments, although they could describe them mathematically, that is, (imagine)$_T$. Bohr's dilemma in 1927, and his original assessment in 1913 of the interpretation of the symbols in the new atomic theory, presaged what would surface in later years as criticisms by anti-imagists of the image-perception link in visual thinking. The complementarity principle can be taken as a step toward replying to these criticisms through its restrictions on perceptual metaphors. The restrictions on imagery in the atomic domain also result in restrictions on the pro-imagery hypotheses because ab initio, according to complementarity, mental images are connected only loosely to the theory's mathematical, that is, logical, apparatus and are not completely functional.

At first blush Heisenberg's 1927 strategy appears to provide an anti-imagery approach in its extreme because he threw out (imagine)$_S$. Heisenberg permitted the theory to interpret only its symbols and not the objects it discussed. But that tack turned out not to be correct and to require the restricted metaphors of complementarity—that is, some sort of imagery. Dissatisfied with metaphors based on objects actually perceived, Heisenberg subsequently returned to scientific research that permitted the quantum theory to become a semantic engine. The fertility of this approach provides support for Kosslyn's reply to Pylyshyn that we can imagine things that we have not seen; in fact, that belief underlaid Schrödinger's philosophical need to propose the wave mechanics in the first place (see the section "Visualizability Regained" in Chapter 4). But the new image of atoms cannot be constructed from parts of images of the *Anschauung* genre. Indeed, according to Kosslyn, certain facets of image construction and transformation may not be shared between imagery and perception, as Heisenberg had shown some four decades previously.

Yet Heisenberg's imagery could not have been realized without first following the path suggested by Einstein in 1905. In 1927 Bohr described this path as "adapting our modes of perception borrowed from the sensations to the gradually deepening knowledge of the laws of Nature" (1928). For without the perception-laden language of classical physics as a guide, a beachhead could never have been established in the atomic domain; this was the task of Bohr's theory during 1913 to 1925.

In summary, the new imagery of modern elementary-particle physics had to be achieved slowly by degrees, starting from the image-perception link and going on to redefine visualizability (*Anschaulichkeit*). All this progress was accompanied by a complex interplay among syntax, semantics, and mental imagery.

CONCLUSION

Thus twentieth-century physicists were forced to liberate their thinking from the world of perceptions. Coincidentally so was imagery liberated from the world of perceptions, to be shaped and reshaped by a mathematical formalism that was appropriate for probing the characteristics of nature in a realm that is beyond mental imagery, as this concept had been interpreted before 1927.

The three views of cognitive science have not led to unequivocal answers in these case studies. In fact, more questions have been generated than answered. While the format of Poincaré's sensual imagery was too sophisticated to be dealt with adequately by, for example, Simon's model, the simplicity of the content of Einstein's imagery escaped the pro-imagery view. Nevertheless, cognitive science possesses traces of what may have occurred at the nascent moment, and can aid our study of this elusive phenomenon that both Poincaré and Einstein expected to remain forever mysterious.

For example, cognitive scientific analysis has served to further clarify creative thinking, particularly that of Heisenberg. There the complex transformations of a philosophic-scientific kind were triggered more by the need for some sort of imagery than by empirical data.

Conversely, the history of science can be brought to bear on problems in the psychological foundations of cognitive science such as the agnostic view. For example, the case study of Heisenberg bears on problems of content and format and reveals changes in internal mental representations that concerned cognitive transformations between imageless and imaginal thinking. Thus although behavioral data may not determine a victor in the imagery debate, further studies in the history of science may provide a promising route to understanding creative thinking.

This chapter focused on transformations of mental imagery using cognitive theories that for the most part could not address the dynamics of creative scientific thinking, which is the theme of the next chapter.

NOTES

1. Of all the pro-imagists I have studied, only Roger N. Shepard utilized vignettes from the history of science to illustrate the

importance of visual thinking—for example, Friedrich Kekulé's work on a theory of molecular structures, Nikola Tesla's design of complex electric motors, James Watson's molecular models to describe DNA, and Albert Einstein's thought experiment that led to the special theory of relativity. But to the historian these creative episodes possess essential differences: Kekulé's illumination was a visual image that occurred in a dream; Tesla's illuminations appeared to him suddenly while he was consciously thinking of something entirely different from electromagnetism (in one case a Goethe poem); Watson's mental gymnastics with complex models for DNA was prefaced by his having spent long periods manipulating molecular models constructed from apparatus that resemble a child's tinker toy; and Einstein's thought experiment did not solve any problem, but posed a paradox whose resolution required 10 years of work and, as far as we know, no more exceptional visual thinking. Thus, Einstein's thought experiment differed from Kekulé's (which resulted in a sudden problem solution), Tesla's (which by degrees led to a solution), and Watson's (which was the conscious mental play prepared for by hands-on play with structural models). Tesla, and others who designed machinery, were of particular interest to Shepard because they recalled having disassembled and rotated their mental images in order to inspect every facet of the machinery. The wonderfully eccentric Tesla delighted in recounting that once having designed a motor mentally, three weeks later he could return to the still running motor and mentally check the parts for wear (Shepard, 1978a). A characteristic common to all these episodes is the conscious preparation that preceded the mental imagery, although in Einstein's case there was no "Aha Erlebnis" phenomenon. See Gruber (1981) for a study of Aha experiences.

2. Here there is a deep problem facing computer science that had been realized by Poincaré in the context of mathematical logic; namely, that "pure logic cannot give us [a] view of the whole; it is to intuition that we must look for it" (Poincaré, 1904b).

3. Among the sharpest critics of computer science is John R. Searle (1981, 1982) who has accused them of "simply play acting at science." According to Searle computers are merely useful tools for cognitive psychology, a view that has been dismissed by computer science (see Newell and Simon, 1981).

4. Mental imagery, or more generally mental representation, has

been of central concern throughout much of the history of philosophy and its offshoot, psychology. For example, in the *Thaetatus*, Plato compared mental images to impressions made on a wax tablet that are stored away for later use. Aristotle emphasized the importance of visual thinking when he wrote that "thought is impossible without an image."

Among the major philosophical movements of the seventeenth- and eighteenth centuries the notion of mental imagery was central to British empiricists such as Thomas Hobbes, John Locke, and David Hume. Take, for example, Hume's differentiation between ideas and impressions. Ideas, Hume wrote in *A Treatise of Human Nature,* are the "faint images" of those perceptions that struck the mind with the highest degree of "force and liveliness," that is, impressions. Hume likened mental images to be surrogate percepts that evoke responses similar to those that occurred when the object was present physically. The association of ideas with images by Hume, among others, is fraught with problems, some of which Aristotle himself had realized—for example, that images need not be identical with all ideas, but can serve as their vehicle. Bishop George Berkeley wrote at length on the problem of imaging something that cannot be perceived as, for example, the abstract idea of a man compared to an anatomical definition of a man.

In the *Critique of Pure Reason*, Immanuel Kant replied to Berkeley's criticism of abstract ideas and images, and to Hume's too tight and direct connection between percepts and ideas. Kant gave freer rein to the imagination, while simultaneously taking the notions of space and time to be synthetic a priori intuitions that serve to begin the construction of knowledge from sense impressions. However, according to Kant, the construction of knowledge and of mental images is unquantifiable because the mental image is never isomorphic to the object of experience. The manner in which a mental image is generated, Kant continued, without its referent being physically present "is an art concealed in the depths of the human soul, whose real modes of activity nature is hardly likely ever to allow us to discover, and to have open to our gaze." Despite this rather explicit statement concerning the nonquantifiability of thought processes, Kant's imposition of underlying articulable regulative principles, such as causality, has led cognitive scientists to quote this passage as support for their research programs that are based on the assumption that the mind is a computational system (e.g., Kosslyn, 1981).

Early psychologists focused on the notion of the "idea" which they believed could be represented ultimately by images. On the basis chiefly of introspective accounts Wilhelm Wundt at Leipzig, the father of modern psychology, took images to accompany all thought processes. But in the last decade of the nineteenth century Oswald Külpe and others of the Wurzburg School found that certain simple association tasks require no images. As Kurt Koffka (1935) has written, this result "began to shock the psychological world" and launched the "imageless thought controversy." The Wundtians at Leipzig were unable to respond convincingly because, as Kosslyn writes (1981), "the main problem here was not the fact of disagreement, but the fact that until very recently there were no good ways of empirically studying claims about any kind of mental representations." Nor was the case of the imageless faction presented satisfactorily. The impasse was taken to be resolved in about 1915 by a positivistic approach to psychology that considered the study of imagery to be an illegitimate problem since mental imagery was not open to direct measurement. Rather, the new school of psychology would study human behavior. John B. Watson, who coined the term behaviorism, wrote in 1915 of the "rubbish called consciousness"—"prove to me . . . that you have auditory images, visual images . . . ."

By the 1960s behaviorism had run its course. Owing both to pressing practical problems, and to developments in linguistics and artificial intelligence, the great eclipse of imagery research ended. Among the discussions of the course of imagery research from time immemorial to the present see Merz (1965, Vol. 2), Kosslyn (1981), and Richardson (1969). Richardson describes such practical problems requiring research into mental imagery as the emergence into consciousness of vivid imagery in truck drivers making long night runs over turnpikes, radar operators, and jet pilots flying straight and level over long distances.

5. For example (see Kosslyn, 1981), subjects were asked to memorize a set of drawings and later to visualize them one at a time. Then they were requested to focus on one end of an object, for example, the rear of a speed boat. Next, they were asked to locate a particular property of the object that may or may not have been in the drawings, for example, a porthole. The time to

scan for the named property was found to be linearly proportional to its distance from the starting point. Another group of
subjects asked to image the entire picture retrieved properties
located at different positions on the image in the same times.
Kosslyn interpreted these data as a step-by-step shift of the image across a visual field. On a propositional account distances
between pairs of objects are listed in the memory and response
time increases with distance since it is a function of the values
associated by each entry. According to Kosslyn the "imagery
account is somewhat more plausible, while the propositional
account seems entirely ad hoc. A propositional framework based
on discrete units and links is inherently ill-suited for representation of processes that occur continuously" (Kosslyn, 1977).

6. For example, the interplay between depictive and long-term descriptive information was illustrated by an experiment in which
subjects viewed a 3×6 array of letters. After the array was removed, the subjects were told that the array would be referred to
either as 6 columns of 3 or 3 rows of 6. Subjects took longer to
image the array when it was described in the first way. For other
experiments see Kosslyn (1979, 1981).

7. The notion of "in-my-head" has been severely criticized by Ludwig Wittgenstein and then raked over the coals by Gilbert Ryle
in his widely-read book *Concept of Mind*. Wittgenstein wrote in
the *Tractatus*: "All philosophy is a critique of language . . . A
proposition is a picture of reality." See also Wittgenstein's *Blue
and Brown Books,* especially pp. 171, 182–184. Ryle wrote in
Concept of Mind: "It is 'in my head' that I go over the Kings
of England, solve anagrams and compose limericks. Why is
this felt to be an appropriate and expressive metaphor? For a
metaphor it certainly is. No one thinks that when a tune is
running in my head, a surgeon could unearth a little orchestra
buried inside my skull or that a doctor by applying a stethoscope
to my cranium could hear a muffled tune, in the way in which I
hear a muffled whistling of my neighbour when I put my ear to
the wall between our rooms . . . . Indeed, if we are asked
whether imagining is a cognitive or a noncognitive activity, our
proper policy is to ignore the question. 'Cognitive' belongs to
the vocabulary of examination papers." The notion of mental
image as a picture-in-my-head is also referred to as the picture
metaphor. According to the picture metaphor mental images are
like photographs that are stored away for future retrieval, like

the situation in Plato's wax tablet notion. For after all, is not the language of mental imagery appropriate for describing pictures and the process of perceiving pictures? The picture metaphor brings with it the notion of a mind's eye to see images internally, which in turn requires a mind's eye brain to elucidate what the mind's eye sees, and so on. It was at this juncture that Kosslyn realized the importance of operationalizing the picture metaphor.

8. The visual buffer can be compared to what philosophers such as Poincaré have called representative space that is constructed by our tactile, visual, and motor systems. Thus, it is neither isotropic, nor homogeneous, nor infinite in extent (see Chapter 1).

9. Kosslyn's computer simulation model contains skeletal images stored in files listed by name and represented by literal encoding in polar coordinates (r, θ). The image is generated by printing out points that depict first the skeletal or bare outline image of the object. This skeletal image is filled in by propositional encodings that contain lists of facts about the image whose skeletal outline is generated from a literal encoding. The four processing components in the deep representation are IMAGE, PICTURE, PUT, and FIND. Kosslyn writes (1979): "We claim that people have operations that accomplish the same ends as these procedures, although obviously the human operations are not identical to their counterparts in the simulation." PICTURE takes information encoded in an underlying literal encoding and maps it into the visual buffer through specification of (r, θ). IMAGE is an executive routine that interfaces between the surface and deep representations. In the computer simulation IMAGE interfaces with the user who uses it to indicate which image in size and orientation to generate. For example, suppose we want to generate the skeletal image of a car. The IMAGE file for a car is called that contains the (r, θ) coordinates of the car's properties. Then PICTURE is called to print on the surface matrix the points that form the skeletal image. The elaborated image can then be generated as follows. IMAGE checks the propositional file for other parts. For example, it finds that the car has a rear tire. PUT is called to seek the propositional file for rear tire in order to seek its location on the car; the rear tire belongs under the rear wheelbase, where "under" is the relation to the foundation part "wheelbase." If either the relation or foundation part is missing, IMAGE will switch to search for other parts listed in the car's

propositional file. If the relation and foundation parts can be located, then PUT looks up the propositional file for their description which is given in terms of spatial relations. FIND is then called to scan the skeletal image for the foundation part, which is then passed back to PUT. Then PICTURE is called to map the relation on to the surface image. Thus, for example, on smaller images fewer foundation parts can be found and hence less time is spent in generation. Just as mental images do not remain transfixed, a part begins to fade as soon as it is processed. Another routine REGENERATE can be called to refresh the image. Among other processing routines whose names are self-explanatory are PAN, SCAN, and ZOOM.

10. For example, relative to claims of how ordered sets of properties are searched, the association strength between an object and part should affect times similarly in image inspection and image generation. And that up to certain scanning distances the time required to transform an image by shifting is greater than that required to permit the image to fade and then to regenerate at the correct size and orientation (i.e., a blink transformation).

11. For example, the propositional account of Shepard's data is that the image of an object can be encoded with a network of propositions describing its shape and orientation, and containing a rotation operator. Kosslyn's reply to the propositional account is that the imagery account seems "somewhat plausible and relatively straightforward" (Kosslyn, 1979). The reason is, continues Kosslyn, that the propositional account requires the imposition of ad hoc constraints to account for why images are rotated continuously; for, after all, a rotation of 180° should be accomplished quicker than one of 45° because it can be propositionalized as a spatial reflection.

12. For example, in order to translate from a code 1 to a code 2, it would be necessary to translate code 1 into a code 3 and then code 3 to code 2. But then a new code is required to translate code 1 into code 3, and so on (Anderson, 1978). Paivio has proposed that there is a dual-code theory for image and verbal modes, asserting that long-term memory representaion for images is separate from that of verbal behavior (see Kosslyn (1981) for criticism).

13. Kosslyn (1979, 1981) writes that theories relying heavily on computer encodings are so powerful that they are not easily operationalizable, and so a host of theories can be formulated

post hoc for each experiment; these theories usually lack predictive power and are "primarily metatheoretical commitments to the form that a theory and model will ultimately take" (Kosslyn, 1981). Furthermore, emphasizes Kosslyn using an argument due to Putnam (1973), it is unnecessary to assert that a phenomenon is explained only if it is reduced completely to a formal system— for example, to explain why a round peg does not fit into a square hole requires only notions of form and not elementary-particle physics.

14. Anderson (1978) continues by emphasizing that at issue is also the important point that one cannot probe a representation in the abstract, but the "representation in combination with certain assumptions about the processes that use the representation." For example, how the letter *R* is internally represented (imaginal as a display on a two-dimensional grid or propositional with a list of propositions describing its shape), and how the letter is processed (rotated by computer subroutines for rotating a matrix representation in small steps or through the propositional calculus).

15. Their other tack has been to appeal to Thomas S. Kuhn's scenario for the advance of science. So, for example, we find that after 532 pages of almost nothing but the computer science approach, the concluding sentence in the Lachman (1979) tome is: "It is our final hope that our treatment of the information processing paradigm has been persuasive."

16. This lecture was reprinted as "L'Invention mathématique" in Poincaré's *Science and Method*. Although he had written previously on intuitionism versus logicism, Poincaré felt moved to make this lecture owing to the lack of response by most major mathematicians and scientists to a questionnaire entitled, "An Inquiry into the Working Methods of Mathematicians," that had appeared in the journal *L'Enseignement mathématique* during 1902–1904. Poincaré's essay is often included in collections of introspective accounts of the creative process and is widely cited in books on cognitive psychology, but with the incorrect title "Mathematical Discovery" or "Mathematical Creation." These mistranslations do an injustice to Poincaré's philosophic viewpoint. Poincaré considered that the sole objective reality are the relations among perceptions. Consequently, he meant mathematical "invention" to be a new construction from mathematical facts. This interpretation emerges from "L'invention math-

ématique." When necessary I have retranslated parts of Poincaré's essay.

17. Poincaré illustrated creative thought aided by geometrical images by the research of the eminent German mathematician Felix Klein. Klein's work on mathematical functions that describe certain properties of curved surfaces was aided by studying the distribution of electricity over curved metallic surfaces; that is, he had constructed models. Then there is Poincaré's invention of algebraic topology based on models of two and three dimensional surfaces (i.e., analysis situs).

18. The authenticity is disputed of the Mozart letter from which these passages were taken. Nevertheless, the contents square with other of Mozart's descriptions of his creative style. For example, in a letter of 30 December 1780 to his father he wrote of work on *Idomeneo* that "everything is composed, just not copied out yet" (Hildesheimer, 1983).

19. Toward following Poincaré's introspective analysis it is unimportant to understand the details of Fuchsian functions: "Ce qui est interessant pour le psychologie, ce n'est pas la théoreme, ce sont les circonstances" (1908b). (Fuchsian functions are automorphic functions of a single complex variable that are invariant under linear fractional transformations.)

In detail Poincaré's scenario can be broken into seven acts:

Act I. Poincaré spent fifteen days, for an hour or two each day, attempting to prove that no such functions existed by "trying a great number of combinations" of mathematical facts.

Act II. One night, contrary to habit, Poincaré drank black coffee and was unable to sleep: "A host of ideas kept surging in my head; I could almost feel them jostling one another, until two of them coalesced, so to speak, to form a stable combination." Upon awakening Poincaré realized that he could establish the existence of one class of Fuchsian functions. Written verification of the results followed quickly, as did further generalizations.

Act III. Poincaré journeyed from Caen (where he was a faculty member at the University) to a geological conference at nearby Coutances. The "incidents of the journey made me forget my mathematical work." About to embark on a sightseeing drive at Coutances he stepped up into the carriage when the "idea came to me, though nothing in my former thoughts seemed to have prepared me for it." The idea was that the defining transformations for Fuchsian functions were identical with those of non-

Euclidean geometry. No written verification was necessary. Poincaré felt "absolute certainty" and immediately resumed conversing with the other passengers. Back at Caen he verified the result "à tête reposée."

Act IV. Poincaré turned next to certain "arithmetical questions" with no immediate success and "without suspecting that they could have the connection with my previous researches." Disgusted with this unsuccessful work he went on vacation to the seashore and "thought of entirely different things." While walking on a cliff beside the sea he experienced a flash of illumination like the one at Coutances "with the same characteristics of conciseness, suddenness, and immediate certainty"; namely, that the application of arithmetical transformations to his work on Fuchsian functions permitted further generalizations. Verification followed back at Caen.

Act V. He deduced further classes of Fuchsian functions but "one still held out, whose fall would carry with it that of the central fortress." Although Poincaré's efforts toward constructing this function failed, they enabled him to "understand better the difficulty, which was already something. All this work was perfectly conscious."

Act VI. Poincaré traveled to Mount Valerien outside Paris to fulfill his military obligations, and so his "mind was preoccupied with other matters." During this period, while crossing a street the "solution of the difficulty . . . came to me all at once."

Act VII. On returning to Caen Poincaré "composed [his] definitive treatise at a sitting and without any difficulty." See Gruber (1981) for further analysis of these Acts. Arieti's (1976) brief attempt at analysis merely paraphrases Poincaré. Hadamard's (1954) discussion, although intrinsically interesting, is discursive.

20. Poincaré's notion of stages of thinking was not unique—for example, the German philosopher-scientist Hermann von Helmholtz's description in (1891) of his creative episodes was similar. Von Hemholtz recalled a period of hard work culminating in intellectual fatigue: "Then after the fatigue of the work had passed away, an hour of perfect bodily repose and rest was necessary before the fruitful ideas came. Often they came in the morning upon awakening . . . . " On the phenomenon of illuminations occurring on awakening in the morning, von Helmholtz quoted Gauss: "The law of induction was discovered January

1835, at 7 a.m., before rising." On the workings of the unconscious, von Helmholtz recalled Goethe's words:

What man does not know
Or has not thought of
Wanders in the night
Through the labyrinth of mind.

Whereas the type of problems on which Poincaré did creative work often yielded to solution in relatively short periods of time, those pursued by von Helmholtz sometimes took "weeks or months" resulting in a "sharp attack of migraine." Besides the basic difference in the areas on which Poincaré and von Helmholtz pursued their basic research, so differed their approaches or styles. Whereas Poincaré almost invariably found the royal road to invention, von Helmholtz described himself as a mountain climber who "ascends slowly and toilsomely [and] who, finally, when he has reached his goal, discovers to his annoyance a royal road on which he might have ridden up if he had been clever enough." The royal road, continued von Helmholtz, is the one displayed in scientific papers. The analysis of creative thinking into stages is common to von Helmholtz's and Poincaré's introspections and was the impetus for Graham Wallas' (1926). For critical comments on this book see, for example, Hadamard (1954), grounded in particular in Wallas not being a scientist.

21. This deep sentiment on thinking was meant no doubt to be read by logicians who were attempting to reduce mathematics to an axiomatic basis from which in turn could be generated a cookbook procedure to construct new theorems. As Poincaré wrote elsewhere in 1908, logicians claim "to have shown that mathematics is entirely reducible to logic, and that intuition plays no part in it whatever." And in reply to one logician, Bertrand Russell, in 1909 Poincaré wrote that "there is no logic and epistemology independent of psychology." No doubt he would have written similarly of computer science.

Poincaré's antireductionist stance in mathematics carried over into the life sciences. For example, in essays of (1900) and (1904c) he wrote of the impossibility to reconstruct the "unity of the individual" having analyzed the atomic structure of its cells: "Would a naturalist imagine that he had an adequate knowledge of the elephant if he had never studied the animal except through a microscope?" (Precisely this antireductionist argument has been rediscovered by others, e.g., Polanyi (1962) who used a frog.) Poincaré's argument that the content of living organisms is greater than the sum of its parts is applicable also to the interpretation problem in mathematical logic, where the interesting interpretations are the sensible ones, and we make this judgment from our ability to recognize them by examining the completed structure.

Haugeland (1981) suggests avoiding the problem of complete reduction by redefining reductionism in cognitive science to be "systematic reductions." This is the process of explaining a technological layer by means of those in which it is instantiated, and so on. The very bottom layer is explained in the traditional sense, that is, by the laws of physics.

22. For example, the mathematician G.H. Hardy wrote: "The mathematical patterns, like the painter's or the poet's, must be *beautiful*; the ideas, like the colours or the words, must fit together in a harmonious way. Beauty is the first test: there is no permanent place in the world for ugly mathematics . . . . It may be very hard to *define* mathematical beauty, but that is just as true of beauty of any kind" (Hardy (1940) italics in original).

23. Poincaré's view of the creative process contains what we may construe to be his attempt to exclude the *homunculus* from thinking. Neisser (1967) has succinctly described this agent: "Who does the turning, the trying and the erring? Is there a little man in the head, a *homunculus,* who acts the part of paleontologist vis-à-vis dinosaur?" Poincaré assumed that the mathematician's "special aesthetic sensibility" served as the agent of selection. By interaction with the subliminal ego whose "automatic actions . . . blindly forms" a large number of combinations of facts this agent selects out only the most beautiful ones which somehow find their way into the conscious. In computer simulation of thinking, be it the purely descriptive encodings, or the mixed descriptive-depictive encodings, the *homunculus* problem is declared to be resolved through the definition of thinking or perception as information processing. Thus the computer functions that permit access to various subroutines are operationally the *homunculi* or mind's eye; these guiding routines are referred to as "executive routines," and they are "in no sense a *programmulus* or miniature of the entire program" (Neisser, 1967). For example, an executive routine can divert the flow of a calculation according to whether a certain variable is greater or less than zero.

24. Similarly, the mathematician G.H. Hardy worked only from 9 a.m. to 1 p.m. whereupon he retired to the cricket field. There is an interesting difference in working habits between Poincaré and von Helmholtz that seems to hold for many mathematicians and physicists. Whereas mathematicians often keep to a strict regimen of conscious work, this is not always the case for physicists.

25. Poincaré's hypnagogic imagery was the typical sort. For further discussion of hypnagogic imagery that is pertinent to this essay see Richardson (1969) and Shepard (1978a).

26. Hadamard (1954) had proposed an analogous sequence in which Simon's selective forgetting is referred to as the "forgetting hypothesis"; needless to say, there was no attempt on Hadamard's part to suggest automation of this process.

27. Needless to say, computer science proponents are optimistic over cracking the problem of creativity. Consider Patrick Langley's computer program whose name is an adequate description of its method for processing data—BACON. Using input data for the relative distances of planets from the sun, BACON rediscovered Johannes Kepler's third law. But Langley, himself, has emphasized BACON's limitations; among them are that BACON operates with far better rules of research than Kepler, who was confronted with a maze of imprecise data, and BACON may contain the result to be discovered. But BACON also omits Kepler's preoccupation with number mysticism and neo–Platonism, not to mention his idiosynchratic personality. Can a computer be programmed with enough mathematics to invent Fuchsian functions? At present it is unlikely unless, that is, the program were named POINCARE.

    McDermott's (1976) is aimed at self-criticism—indeed, "self-ridicule"—of computer science, a field that "has always been on the border of respectability." In the section entitled "Wishful Mnemonics" we find: "Remember GPS? By now, 'GPS' is a colorless term denoting a particularly stupid program to solve puzzles. But it originally meant 'General Problem Solver,' which caused everybody a lot of needless excitement and distraction. It should have been called LFGNS—'Local-Feature-Guided Network Searcher.'"

28. Pestalozzi (1801) wrote (italics in original): "What have I especially done for the very being of education? I find I have fixed the highest, supreme principle of instruction in recognition of *Anschauung as the absolute foundation of all knowledge*." A recent

biographer of Pestalozzi has emphasized that whereas the ideas in Pestalozzi's most important book, *How Gertrude Teaches Her Children,* have been widely accepted, the book itself is "now hardly ever read": the prose is difficult because even as he wrote the book Pestalozzi was still shaping his educational system; and the central term in his system is virtually untranslatable into English, and even in German the word *Anschauung* possesses a multitude of philosophically weighted meanings through which Pestalozzi shifted with little warning (Silber, 1960). For example, we find: "Nothing is more difficult to grasp in Pestalozzi's doctrine than what exactly is meant by the untranslatable word *Anschauung*" (Green, 1913). In fact, a painstaking 1894 translation of *How Gertrude Teaches Her Children,* contains the word *Anschauung* in parenthesis whenever it is rendered other than "sense-impression." Detailed notes to this translation contain over two pages devoted to *Anschauung* and stress that Pestalozzi's own usage covers the spectrum from sense-impressions (in the infant's mind) to observation, perception, appreception, and intuition ("knowledge obtained by contemplation of ideas already in the mind, which have not necessarily been derived from the observation of external objects"). Sometimes Pestalozzi shifted through all of the above meanings in a single paragraph. Kantian shades of interpretation color all of Pestalozzi's uses for *Anschauung,* coinciding exactly with "intuition." Certain of Pestalozzi's and Kant's notions of the construction of knowledge are similar, but were conceived of independently. Pestalozzi first aired his thoughts in the book *Leonard and Gertrude* that was published in 1781, the same year as Kant's *Critique of Pure Reason.* That fervent representative of Kantian philosophy, Johann Gottlieb Fichte, found in Pestalozzi's 1781 book "many of the same results as in Kant." Pestalozzi's influence on Fichte can be seen in Fichte's addresses of 1807–1808 in Berlin, "Reden an die deutschen Nation." Fichte transferred Pestalozzi's notion of the fundamental power of the human mind, i.e., *Anschauung,* to an innate characteristic of the German people (Merz, 1965; Höffding, 1955). Friedrich Nietzsche would elaborate further on this hypothesis with unforeseen consequences. This darker side of *Anschauung* I do not discuss here, except to mention that the meaning of *anschaulich* in the relativity theory as compared to classical physics was a central point in Philipp Lenard's criticisms of Einstein (see Beyerchen, 1977).

Although Pestalozzi was aware of the works of such philosophers as Rousseau, Voltaire, Hume, and Kant, among others, he always claimed that recent developments in philosophy were "beyond his capacity" (Silber, 1960). Gratified to be compared with such "crystal-clear thinkers" as Kant, Pestalozzi preferred "feeling philosophers" such as Johann Gottfried Herder and Friedrich Heinrich Jacobi.

Although the "organization of American technology in the first half of the nineteenth century tended naturally to follow the pattern set by the world of art" (Ferguson, 1977), the introduction of Pestalozzi's teachings into mid-nineteenth century America no longer found fertile ground. For since the 1860s engineering education in the United States has stressed analytical over nonverbal thinking. Ferguson (1977) writes that "in engineering school a course in 'visual thinking' is regarded as an aberration rather than as a discipline that should be incorporated into an engineer's repertoire of skills . . . and the course in which it occurs is picked up by the New York Times." See Hindle (1981) and Ferguson (1977) for discussion of engineering education in the United States during the first half of the nineteenth century.

29. Although offering Gedanken experiments was not uncommon in the literature, by comparison with Einstein's those offered, for example, by Ernst Mach in his (1905b) "*Über Gedankenexperimente,*" pale. Mach defined a Gedanken experiment as an "idealization or abstraction" of existing physical conditions." Among the examples he gave were Gustav Kirchhoff's description of the perfect black body and the mathematical process of integration on a line as the limiting case of summing up many line elements. Clearly, Mach's Gedanken experiments were the usual sorts of abstractions from direct experience that throughout were linked with sense perceptions: "Experience produces a thought experiment which is then spun further to be compared with experience and modified."

30. In 1923 Louis de Broglie had proposed that just as light can be a wave or particle so can atomic entities. By late 1925 the physical reality of the light quantum had been established, and by 1926 supporting evidence had begun to accrue for the wave nature of atomic entites like electrons. But these new modes of existence for light and matter were not yet understood, as we shall see below.

31. The equation is $\lambda = h/p$, where h is Planck's constant, λ is the particle's wavelength, and p its momentum.

32. Here Heisenberg may well have used the term "Archetype" in a redefined Jungian sense. According to Jung, Archetypes did not exist prior to sense experiences; rather, they are primordial forms that the primitive mind constructed from perceptions. On the other hand, Platonic Ideas existed before all else. Thus, as Jung wrote in his "Über den Archetypus" of 1936, the Archetypes, as it were, "put Platonic Ideas on an empirical basis" (Jacobi, 1971). Our analysis of Heisenberg's notion of perception in subatomic physics leads to an interpretation of his meaning of Archetypes as primitive images to which we are led by the mathematics of the quantum mechanics. These images give us some further clues to the Platonic Ideas of matter.

    On the other hand, the elementary particles of the strict Copenhagen interpretation are more like those of Democritus than of Boltzmann. According to the Greek atomists, atoms differ from ordinary matter in ways that render them inaccessible to our sense perceptions; but atoms are not Ideal quantities in the Platonic meaning of Ideal.

CHAPTER 7

Scenarios in Gestalt Psychology and Genetic Epistemology

What occurs when, now and then, thinking really works productively? What happens when, now and then, thinking forges ahead? What is really going on in such a process? . . . Two directions are involved: getting a whole consistent picture, and seeing what the structure of the whole requires for the parts.

M. Wertheimer (1959)

Genetic epistemology attempts to explain knowledge, on the basis of its history, its sociogenesis, and especially the psychological origin of the notions and operations upon which it is based . . . . The fundamental hypothesis of genetic epistemology is that there is a parallelism between progress made in the logical and rational organization of knowledge and the corresponding formative psychological processes.

J. Piaget (1970a)

The photograph shows Jean Piaget sitting in his garden, talking with Bärbel Inhelder, his collaborator of some fifty years, and co-author with him of many books and papers. (Courtesy of Jacques Vauclair)

U SING AS RESEARCH TOOLS Gestalt psychology and Jean Piaget's genetic epistemology, and the results of previous chapters as data, we examine Albert Einstein's thinking that led to the relativity of simultaneity (i.e., creative thinking in the individual), and the genesis of quantum theory during the period 1913–1927 (i.e., the growth of a theory among several scientists). The scenarios of these two episodes in the history of science are developed here in conformity to the guidelines of Gestalt psychology and genetic epistemology, while striving for the historical accuracy in Chapters 1–4. Chapter 5 introduced the portions of Gestalt psychology that will be used here. Regarding genetic epistemology, the direction of the investigation in this chapter is best set by Piaget himself in the epigraph to this chapter.

The first section of this chapter surveys the portions of genetic epistemology that are required for the analyses which constitute the second section. This first section concludes by setting the extensions to Piaget's theory necessary for investigating his claim that a psychological theory of how mental structures are developed in the child's mind can be applied to how scientists probe nature in order to formulate scientific theories. The principal problems to be addressed are: Does Einstein's invention of the relativity of simultaneity, and the genesis of quantum theory during 1913–1927, exhibit structures that have parallels with those in the Gestalt or genetic epistemological views of the development of knowledge? If so, then what can be said concerning the genesis of scientific ideas? It turns out, not surprisingly, that neither theory of the cognitive processes is entirely adequate for describing the discovery and genesis of scientific theories or the nature of scientific discovery. Yet there emerge certain hallmarks in the growth of science that are indigenous to both cognitive theories—for example, the conservation or invariance either of substance (e.g., the atom or light quantum), or of a physical quantity such as the velocity of light in vacuum. Thus we can begin to discern the origins of the important notion of conservation that pervades physical theory, and which Gerald Holton refers to as a methodological thema (1973).[1] As Piaget emphatically stated (1965): "Our contention is merely that conservation is a necessary condition for all rational activity."[2]

To the best of my knowledge, applications of theories of psychology to case studies in the history of science have appeared in print only twice previously, both written by psychologists: Max Wertheimer's (1959) Gestalt-oriented examination of Einstein's supposed testimony of how he had discovered the relativity of simultaneity, and Howard E. Gruber's (1974) investigation of Charles Darwin's notebooks to discern themes analogous to those in Jean Piaget's genetic epistemology. From the standpoint of the history of science, Wertheimer's analysis was unsuccessful; Gruber extracted many interesting psychological insights into the nature of thinking, while demonstrating the necessity for extending the guidelines of genetic epistemology for this sort of research.[3]

A SURVEY OF GENETIC EPISTEMOLOGY

Jean Piaget's genetic epistemology is a structuralist theory of knowledge that resulted from his classic studies of the construction of knowledge in children from birth through their teens.[4] Like Gestalt psychology, genetic epistemology is biologically based, although Piaget emphasizes (1971) that the "Gestalt notion . . . is still perfectly viable once it is cut loose from its purely Gestalt chains." Examples of these chains are: there is a priori character in such Gestalt-psychological notions as the *Prägnanz* principle; and the Gestalt laws that govern the approach to equilibrium in the field of knowledge do not emphasize the effect of the interaction between the organism and its environment. The formation of the structures of Piaget's theory can, to some degree, be traced and are not dependent ab initio on notions concerning perception. Although Piaget is somewhat unclear on perception, he emphasizes that as intelligence develops, perception is increasingly regulated and controlled.

According to Piaget, there are three main periods in the development of representative activity: sensorimotor, egocentric representative activity, and operational. The first period is completed by the advent of deferred imitation. I emphasize here the second and third periods because special relativity and quantum theory developed from the completed structures of classical mechanics and electromagnetism, which contain mathematical structures indigenous to those of the highest order in Piaget's hierarchical view of the development of intelligence. It will turn out that assimilation to classical mechanics and electromagnetism by Einstein and Bohr revealed that these structures were not in equilibrium and consequently required further development. But, to set the stage, it is useful to begin with a survey of Piaget's view of the origin of intelligence in the sensorimotor period.

According to Piaget the infant begins to act on the world about him by means of innate groping reflexes that constitute the lowest-level schemes of action. Higher-level schemes are constructed by means of the child's actively incorporating perceptions into schemes, and then adjusting the schemes to fit the situation—that is, by what Piaget refers to as assimilation and accommodation.

Just as biological and physical systems tend toward states of equilibrium, Piaget assumes the mind to be an open system possessing a self-regulatory mechanism that he refers to as equilibration. Equilibration coordinates the interplay between assimilation and accommodation and results in structures or schemes in equilibrium (i.e., an understanding of a certain situation). Through the assimilation-accommodation process, by the age of two or three years the child has discovered the practical group of displacements and the group's invariant quantity, which is the permanent object. Consequently, children learn that an object does not cease to exist when it disappears from their visual field. The construction of object permanency completes the sensorimotor period, and the child enters the period of egocentric representative activity that Piaget divides into two stages—preconceptual thought and intuitive thought. Coincident with the discovery of object permanency is the first appearance of the semiotic function: the child can represent an object without the object's actually being present; this marks the onset of preconceptual thinking.

Piaget believes the perconceptual stage of representation to overlap with the part of the sensorimotor period in which the child can form fleeting pictures of objects in the world of perceptions. At the start of semiotic functions, the scheme of the permanent object is closely linked with actual perceptions of that object. Piaget refers to thinking with schemes whose functions are connected with perceptions as preconceptual thinking. A concept, on the other hand, is a scheme that is independent of imagery: "The concept is general and communicable, the image is singular and egocentric" (Piaget, 1962). Concepts are not constructed until the formal operational period.

Whereas during the sensorimotor period there are fleeting pictures of objects in the mind's eye, the construction of object permanency leads to deferred imitation and the image. The image is a signifier, or concrete symbol, and the object or scheme to which it is related is the signified. The signifier is the product of accommodation, while the signified is the product of assimilation. The signifier is useful in attempts to know the object, but it is not the object itself. At first the image is closely related to the internalized sensorimotor reproductions; in later periods it results from internalized reproductions. Coincidentally with the development of intelligence, the image becomes less indispensable because the subject's actions become increasingly internalized. Although the construction of object permanency renders the delicate interplay between assimilation and accommodation momentarily in equilibrium, the possibility of deferred imitation disequilibrates this situation; that is, the universe as set out by the child cannot be understood by his present schemes because of their dependence on the image and the resulting imbalance between assimilation and accommodation.

The approach to the next equilibrated structure is aided by the processes of symbolic play and imitation. Symbolic play is the primacy of assimilation over accommodation: data are assimilated to schemes to which they are only more or less related. Imitation is the dominance of accommodation over assimilation: data are assimilated to particular signifiers (i.e., models). Thus imitation is the process of scientific work that can best be likened to the explication of models.

But symbolic play is the route to scientific discovery. In fact, in the only direct statement on the creative process that I have found by Piaget, he writes that "creative imagination . . . is the assimilating activity in a state of spontaneity" (1962).

Piaget refers to the sort of reasoning at the onset of representative thinking as "transduction," which results from lack of equilibrium between distorting assimilation and partial accommodation. True generality of thought is not achieved because assimilation is centered on particular objects (i.e., schemes), as in play, and accommodation is centered on the typical sample or image instead of the complete set. This leads to an algebralike structure that possesses reversibility by inversion and tautology, which Piaget refers to as "grouping."[5] Consequently, this level of thinking is part-way between "symbolic or imitative coordinations and deductive reasoning" (1962).

Piaget refers to the remaining intermediary steps toward achieving full reversibility as "intuitive thinking." For example, a child is presented with two sets of sticks that have the same graduations in sizes and then is asked to match the two sets for size. Children show three typical stages for this process between the ages of four and seven. First, they arrange them in pairs. At a later stage they find the right order and serial correspondence by trial and error. But if the arrangement is destroyed, the child's intuition is not yet developed fully enough (i.e., is not yet sufficiently free of imagery) to be sure that correspondence is still preserved between the two sets. Finally, in the third stage, the correspondence is made and equivalence preserved even if the arrangement is modified. The reason for this is because the third stage coincides with the onset of the operational level of thinking, the level at which both sorts of reversibility are fully developed. Another example useful for our purposes is where a child of around age four becomes convinced that two horizontal line segments are of equal length. But if the same line segments are compared when they are vertical or at some angle to the horizontal, then he is no longer sure of their equality. The configurations of either the sticks or the pair of parallel lines are by definition structures or schemes. During the final stage of intuitive thinking the image is no longer the image of an individual object, but of a scheme (1962).

Another example of achieving equilibrium between assimilation and accommodation through construction of the scheme of full reversibility is in one of Piaget's classic experiments, in which a child comes to realize that a given amount of clay remains the same no matter what shape it is rolled into. Thus, reasoning is restricted no longer to static configurations of states. The states are subordinated to transformations that are independent of the subject's internalized actions. The image can now be the image of a scheme instead of the image of an object.

The final stage of representative activity is achieved during the operational level of thinking. The thinking structures of operational representative activity are defined to be operations because they possess both sorts of reversibility, are fully internalized, and are governed by laws that incorporate both transformations of states and conservation, for example, group-theoretical structures. The stage of operational representative activity is subdivided into the stages of concrete and formal operations. In the concrete operational stage thinking is not yet independent of images. Furthermore, the two sorts of reversibility cannot be combined. The construction of higher-level structures results from the subject's acting on his environment, that is, by further experimentation, and from the ongoing unquantifiable enriching process that Piaget calls "reflecting abstraction." In *Biology and Knowledge* Piaget writes that by reflecting abstraction he means a "rearrangement, by means of thought, of some matter previously presented to the subject in a rough or immediate form. The name I propose to give this process of reconstruction with new combinations, which allows for any operational structure at any previous stage or level to be integrated into a richer structure at a higher level, is 'reflecting abstraction'."

Piaget attempted to steer a course between what he deemed to be the Scylla of Kantian apriorism and the Charybdis of associationism. But his critical view of associationism, coupled with his French style of rationalism, led him to lean toward Kantian apriorism. Reflecting abstraction is an inherent quality of the mind that lies dormant until the advent of symbolic activity. Reflecting abstraction results in a rearrangement of material from lower levels that provides the grist to be acted on by higher-level schemata. The subject achieves complete independence from representing reality by "its deceptive figurative appearances" (1969). Accommodation becomes completely generalized and no longer needs to be translated into images. This structure, or operative level, is formal operations. Thinking is now in terms of operations on operations, that is, the internalized coordinations of two or more actions, for example, retaining a fulcrum's balance by simultaneously changing both the weight and the distance of the weight from the fulcrum. At the formal operational level the image is deemphasized for thinking, but when needed it can be invoked through operational structures, which in the cases to be discussed are scientific theories. The transition from picture to image to symbol is completed in the formal operational period.[6]

Needless to say, when Bohr, Einstein, and Heisenberg began their scientific work, they had years earlier entered the operational stage of thinking. It is with trepidation that I can even propose to apply to their research a recapitulationist argument based on how a child's thinking develops. Clearly, extensions of certain notions from genetic epistemology will be required, and I propose the following:

1. The construction of knowledge by children and physicists is not exactly the same. Through the assimilation-accommodation mechanism the child's schemes develop the mathematical and physical attributes of the levels of genetic epistemology that are the child's version of physical reality. Physicists, however, deal with schemes that possess ab initio the proper mathematical attributes to set them in the operational stage.
2. Thus a theory's physical interpretation of new data places the theory into a lower stage of genetic epistemology.
3. Because children's attendant notions of physical reality are in flux, they are not always committed to holding onto a particular scheme. But physicists tend to hold to a theory because it is a structure that permits a view of physical reality that they have achieved through hard-won empirical data and previous research. Consequently, I propose the following as a second definition of assimilation: The application of a scheme to problems involving empirical data, data from thought experiments, or aesthetic-philosophical commitments or predispositions.
4. With Piaget I define "structure . . . as the set of *possible* states and transformations of which the system that actually pertains is a special case" (italics in original, 1970b). Here I propose that the relevant structures are those of conservation, reversibility, and systematization.

GENETIC EPISTEMOLOGICAL SCENARIOS

Einstein's Invention of the Relativity of Simultaneity

As the states of the system to be studied, I take the velocities of light, which are measured in an ether-fixed reference system S, and in an inertial reference system, S_r; I designate these velocities as c and c_r, respectively. The transformations are those prescribed first by the structures of classical physics and then by special relativity theory.

Einstein attempted to assimilate the data of the 1895 thought experiment (see Appendix in Chapter 3) to the operatory structure of classical physics, with little accommodation. This symbolic play squares with Einstein's statement of 1946 that "thinking is of this

nature of the free play with concepts." As Holton (1973g) has written, the words "*Bild* and *Spiel* [occur] with surprising frequency in
Einstein's writings." Einstein's notion of "wondering" fits well into
Piaget's definition of symbolic play and creativity. For Einstein
wrote (1946) that owing to his predominantly visual mode of thinking, he could " 'wonder' quite spontaneously about some experience
[that] comes into conflict with a world of percepts which is already
sufficiently fixed within us" (e.g., classical physics in the 1895
thought experiment). The "intuition" of the thought experimenter
predicts that, as in classical mechanics, the laws of physics should be
the same in the ether-fixed system S as in the inertial system S_r. For
the purpose of describing the behavior of light, both classical mechanics and electromagnetism have been found insufficient: the former's velocity addition law led to results at variance with empirical
data, and the latter theory asserted that in S_r light could be observed
to be a standing wave.

A further anomaly in classical electromagnetic theory was revealed
to Einstein through his assimilating the structure of classical physics
to Max Planck's law of radiation; Einstein realized that Planck's law
of radiation violated both electromagnetic theory and mechanics.

From Einstein's own writings we know that by mid-1905 his
statistical researches revealed that classical mechanics, electromagnetism, and thermodynamics all were undependable when applied to
physical systems of the order of the electron's volume. According to
genetic epistemology these researches consist of symbolic play and
imitation; they led Einstein to a further examination of classical electromagnetism and to his derivation of a new velocity addition law
(Chapter 3). Although Einstein took this law to be valid only to
order (v/c), it led him to construct the scheme of a permanent or
conserved quantity, namely, the velocity of light in vacuum, which
is a determined constant c in every inertial reference system and is
independent of the motion of the emitter—that is, $c = c_r$. The equilibrium was temporary, however, because this scheme did not contain full reversibility. There is inversion because S_r could be taken
also to be the resting system, that is, the reference system at rest
relative to the physical system under examination. But there is no
reciprocity because Lorentz's modified transformations do not treat
S and S_r symmetrically. One cannot, therefore, relate the subclasses
of systems S and S_r to the total class of inertial reference systems.
Mathematically, the set of modified Galilean transformations possesses both sorts of reversibilities (see Fig. 7.1).

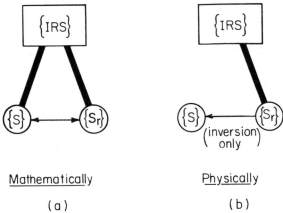

Fig. 7.1. $\{IRS\}$ are the total classes of inertial reference systems and $\{S\}$ and $\{S_r\}$ are subclasses of inertial reference systems. Mathematically (a), both $\{S\}$ and $\{S_r\}$ have the properties indigenous to inertial reference systems, and there is inversion and reciprocity among the $\{S\}$ and $\{S_r\}$. Physically (b), that is not the case in ether-based theories of electromagnetism because the $\{S\}$ are fixed in the ether. In (b), $\{IRS\}$ are composed of only $\{S_r\}$, and $\{S_r\}$ can be related to $\{S\}$ only by inversion (i.e., S_r could also be taken to be a body's resting system). In (b), whereas the $\{S_r\}$ subclasses are limited in number since S_r is an inertial reference system, there can be unlimited numbers of S; thus the class $\{S\}$ can be as big or bigger than $\{IRS\}$. In other words, $\{S\}$ are preferred subclasses of reference systems.

Transduction and imitation merge in Einstein's thinking as a result of the following thought experiment. Einstein began the 1905 relativity paper with a thought experiment concerning the current generated in a conducting loop that is in inertial motion relative to a magnet, that is, electromagnetic induction. Although the magnitude and direction of the induced current depend on only the relative velocity between magnet and conductor, according to the ether-based Lorentz electromagnetic theory the induced current can be explained in two different ways. From the standpoint of an observer on the magnet the induced current arises from the force exerted on charges in the loop as the loop moves through the magnet's magnetic field. From the standpoint of the conducting loop the current arises because an electric field that originated in the moving magnet exists at the loop's site. Hence, two different explanations are required for an effect that depends on only the relative velocity between loop and magnet. In the relativity paper Einstein interpreted this redundancy as an asymmetry that is "not inherent in the phenomena." In an unpublished 1919 essay Einstein described the situation in electromagnetic induction as "unbearable." He chose to focus on the relation between observers on the magnet and on the conductor, instead of attempting to explain how the current arose in the conductor, or to solve the complicated and, in Einstein's opinion of the physics of 1905, intractable problem of the source of the moving magnet's electric field.

Einstein invoked the principle of relativity from mechanics, which considers the inertial reference systems of magnet and conductor as equivalent, thereby constructing the scheme containing reversibility. According to Einstein's principle of relativity, the laws of mechanics, electromagnetism, and optics are the same in every inertial reference system. Since his thinking was still linked to images, both sorts of reversibility could not be used together—that is, the situation in Fig. 7.1(b) was still present in Einstein's mind because he had resolved in principle, but not yet mathematically, the situation of electromagnetic induction involving magnet and conductor. At this point Einstein's reflecting abstraction, further enhanced through readings of, among others, Ernst Mach, and David Hume convinced him that imposing a Newtonian unity on Lorentz's transformations required the relativity of time, and therefore that Lorentz's ether was "superfluous."

At the level of formal operations Einstein realized that the group of space-time transformations was rich enough to include conservation of the velocity of light, and that full reversibility resolved the "unbearable" situation in electromagnetic induction. As Einstein later wrote, an important and new fact of his special theory of relativity was that the relativistic transformations "transcended its connection with Maxwell's equations." Thus the level of formal operations was achieved when Einstein liberated himself from dependence on perceptions to locate what Piaget at this level of thinking refers to as, "reality in a group of transformations"—reality, that is, concerning the proper value for the velocity of light in vacuum. At this juncture, according to genetic epistemology, Einstein invented the relativity of simultaneity because the group structure implied full reversibility of S and S_r and, consequently, the physical reality of time's dependence on the motion of a reference system. Achieving full operativity through mathematical structures accords with Einstein's comment that his thinking was predominantly visual and that mathematical work was undertaken in the final stages of research. However, historical research provides material to suggest that Einstein never freed his thinking from images (see Chapter 6). Nor does genetic epistemology accent the importance of symmetry in Einstein's visual thinking.[7]

Nor does it seem reasonable that Einstein invented something as basic as the relativity of simultaneity through only mathematical considerations. Rather, as is discussed in Chapters 1 and 3, he realized after discovering the new addition law for velocities that the local time might be the "physical" time. Philosophical considerations followed. This break in the heretofore most reasonable historical sequence points up a basic weakness in genetic epistemology's emphasis on formal mathematical operations in the formal operational period, in which period considerations based on visual imagery are excluded.

DEVELOPMENT OF QUANTUM THEORY FROM 1913–1927

In contrast to special relativity theory, which was the work of a single person, the development of quantum theory during 1913–1927 was the product of a predominantly small scientific community. The states or systems are the wave and particle modes of matter, and the transformations are based on the mathematical notions in use during the particular period of theory development.

As with special relativity theory, we begin with the suitably redefined egocentric representative period. In quantum theory this period's hallmark is the permanence given to Rutherford's atom by Bohr's atomic theory. Owing to assimilation of the scheme of classical mechanics to his 1911 data from the scattering of α-particles off metal foils, Ernest Rutherford concluded that the atom was a miniature Copernican planetary system. He did not attempt any accommodation of the scheme of classical mechanics to these data. In fact, according to classical electrodynamics the orbiting electrons should radiate energy and eventually fall into the nucleus. On the other hand, for the most part atoms are stable. In 1913 Niels Bohr incorporated ab initio the stability of matter into a new theory of Rutherford's atom by postulating the existence of a lowest orbit. While an orbital electron did not radiate in one of the allowed orbits, its transition to a lower orbit was accompanied by the emission of light that was detected empirically as a spectral line. In transit the electron was unvisualizable; it disappeared and appeared again, like the smile of the Cheshire cat. The equilibration accomplished by Bohr's accommodation of the scheme of classical mechanics to the scheme's assimilation of Rutherford's data resulted in permanence of the Copernican atom. Rutherford's 1911 conception of the atom (the product of assimilation) is the signified, and Bohr's 1913 atom (the product of accommodation) is the signifier. To Bohr, discontinuities were essential to account for both the stability of matter and the discreteness of spectral lines. But these discontinuities violated classical mechanics and electrodynamics. According to classical mechanics the electron is allowed to occupy any orbit, and classical electrodynamics asserts that the orbiting electron is always radiating. In addition, both these disciplines assert the possibility in principle of tracing in space and time the continuous development of physical systems.

The signifier or image represents the atom, but neither is it the atom itself nor is the signifier the image of the scheme of Bohr's atomic theory. For, as Bohr emphasized in 1913, the images from classical mechanics entered the atomic theory through the interpretation (language) of mathematical notation taken from ordinary mechanics, that is, from the world of perceptions.

The period 1913 through late 1927 is the preconceptual period of representative activity. The decade 1913–1923 exhibits the predominance of accommodation over assimilation, which is the imitative activity that emphasizes analyzing problems with models. During 1913–1923 the Bohr theory of the Copernican atom became further elaborated owing to inclusion of the correspondence principle, with its mathematical prescription for applying the classical physics of the macrocosmos to the microcosmos. Central to the correspondence principle was a part of Einstein's 1916 theory of radiation that enabled physicists to use a version of classical electrodynamics to describe mathematically the electron's unvisualizable quantum jumps.

In its first 10 years Bohr's atomic theory achieved impressive successes. For example, it could account for most of the characteristics of spectral lines in the hydrogen atom, in addition to accounting reasonably well for the place of elements in the periodic table. By 1923, however, the image of the atom as a miniature Copernican system was beginning to lose its utility, owing to its inability to accommodate to certain theoretical problems such as the three-body problem (e.g., to calculate the characteristics of the helium atom).

There also was the light quantum. In 1905 Einstein had made the heuristic proposal that for certain physical processes it was helpful to consider that light possessed particulate properties, in other words, as if it were propagated through space like a hail of shot or light quanta. The notion of light quanta was resisted by such major physicists as H.A. Lorentz and Max Planck, and then by Bohr as well. Their principal reason was, as Bohr wrote in 1921, the "insurmountable difficulties from the point of view of optical interference"—that is, how particles of light can interfere with one another to produce such well-known phenomena of wave theory as diffraction. Yet the usefulness of the light quantum for describing the photoelectric effect was undeniable.

In 1923, in the face of the dissolution of the image of a Copernican atom, Bohr concluded that the problem of the interaction between light and matter (i.e., dispersion) was the key issue. For treating the problem of dispersion Bohr suggested a method to avoid the notions of orbiting electrons and light quanta; namely, that an atom responded to incident radiation as if it were composed of as many harmonic oscillators as there are atomic transitions; and, since there is a denumerable infinity of transitions, there would also be a denumerable infinity of oscillators. Which oscillators would be set into motion would depend upon the frequencies in the incidence radiation. The character of the scattered radiation would be calculated through the correspondence principle. As was his style, Bohr presented this method with few details. Soon afterwards the mathematics was formulated by Rudolf Ladenburg and Fritz Reiche, who referred to the oscillators as "*Ersatzoszillatoren.*" Needless to say, though one may not achieve the image of a denumerable infinity of oscillators, if desired one can imagine many oscillators—perhaps even a swarm of oscillators reaching to the horizon.[8] Whereas the pre-1923 image of the scheme of the Bohr atomic theory had been imposed on it, the post-1923 atom was no longer visualizable in any conventional sense of this term because the orbital electron was no longer localized. In the post-1923 theory the image is no longer of an object but of the scheme. In order to avoid including the light quantum in the atomic theory, Bohr took the drastic step of renouncing images abstracted from the world of perceptions. With the light quantum were associated pictures from classical mechanics—for example, the picture of billiard balls colliding in explanation of how the light quantum collides with an electron to produce the photoelectric effect. Since the laws of conservation of energy and momentum were also linked to billiard ball collisions, Bohr had noted, these laws might not be exactly valid for individual processes. The preconceptual thinking of Bohr and most other physicists was centered on light as a wave with its undulatory image. Consequently, assimilation was incomplete since data were assimilated to light as a wave (with the scheme of bound electrons as an ensemble of oscillators). Accommodation was incomplete owing to its limitation only to waves.

According to Piaget, the advent of decentering is signaled by the formation of a structure that is an image not of an object, but of a scheme; this portion of egocentric representation is "intuitive thinking."

In 1923 Arthur Holly Compton interpreted his data from the scattering of X-rays off metal foils as further evidence for the reality of light quanta. Bohr's 1924 assimilation of Compton's data into his atomic theory resulted in partial accommodation to a level more abstract than previously; specifically, according to Bohr, Kramers and Slater, the ensemble of harmonic oscillators that represented a bound electron, which they referred to as "virtual oscillators," emitted waves that carried only probability and not energy and momentum. For example, when excited by incident real radiation, a virtual oscillator emitted a probability wave that could induce an upward transition in a neighboring atom without the source atom's undergoing the corresponding downward transition. Consequently, energy and momentum were not conserved in the individual processes but only on the average.

Although by mid-1925 the Bohr, Kramers, and Slater theory of radiation was empirically disconfirmed, the principal physicists involved in quantum theory had yet to accept the light quantum's reality. For example, although Kramers and Heisenberg mentioned the light quantum in their widely read paper on dispersion, in which virtual oscillators emitted real waves of radiation and not waves of probability, the authors treated light like a wave. This conception illustrates semireversible, or intuitive, thinking. They ascribed some reality to the light quantum, but not equal to that of waves. The scheme of mechanics, that is, the Bohr atomic theory recast in virtual oscillators, assimilates data to one aspect of light—waves. Hence, there is no reversibility between waves and particles. The disequilibrium in mathematical language (i.e., notation) serves further to place the Kramers-Heisenberg paper, and thus quantum mechanics, too, in the intuitive-thinking part of the period of egocentric representative activity. For example, Kramers and Heisenberg used ν_e and ν_a for the frequency emitted or absorbed during an atomic transition, instead of notation such as ν_{ik} for a transition between atomic states i and k. The improper notation prevented them from realizing what it meant to represent the atom by an ensemble of harmonic oscillators. As Heisenberg recalled, we "practically [had] the matrix multiplication" (*AHQP*: 13 February 1963).

In late 1925 Bohr acknowledged that recent empirical data had forced on quantum physics a "corpuscular transmission of light"; however, he maintained that these data did not distinguish decisively between the wave and particle modes of light. Furthermore, continued Bohr, if one entertained seriously the notion of light quanta, then imagery such as that of billiard balls in collision were useless since light quanta fluctuate in volumes of the order of the electron's volume, a phenomenon that classical electrodynamics could not even describe. Bohr mentioned recent results of Louis de Broglie and Einstein that had served even further to disequilibrate the contemporary view of atomic physics.

In 1923 de Broglie had proposed that matter also possessed a wave-particle duality; and Einstein's 1924 theory of the ideal quantum gas had implied that free particles were indistinguishable. Although empirical confirmation of the wave-particle duality of matter was not published until 1927, by mid-1925 there already existed provisional data supporting this hypothesis. For Bohr the scheme of quantum mechanics was in flux, and he knew not how to rectify it; Werner Heisenberg did. Heisenberg realized the advantage of renouncing visualizability in the classical sense of the term—that is, via pictures or images from the world of perceptions. Although the fecundity of the virtual oscillators had been appreciated by most of the small group of physicists in the forefront of research in quantum theory, they had been unwilling to jettison visualizability completely.

Attempting to assimilate the scheme of quantum mechanics to recent puzzling data on dispersion, Heisenberg realized the necessity to discard the notion of orbital stationary states. He interpreted the virtual oscillator methods as the proper mathematical model to broaden and sharpen the correspondence principle, thereby also focusing on the wave mode of light. The fundamental importance of this interpretation was further demonstrated to Heisenberg through his subsequent unsuccessful attempts at resolving the long-standing puzzle of the anomalous Zeeman effect, by assimilating the problem to mechanical models. What occurred next in Heisenberg's research on quantum theory follows Piaget's definition of creativity—namely, "creative imagination, which is the assimilating activity in a state of spontaneity." In Heisenberg's thinking, the clue came from attempting to assimilate recent data on dispersion to the scheme of quantum mechanics. In mid-1925 Heisenberg formulated the matrix or quantum mechanics in which the virtual oscillator metaphor of the bound electron was incorporated into a mathematical description of atomic transitions with infinite dimensional matrices. Imagery associated with the world of perceptions was abandoned. A new structure was created which was explicated further in papers by Born, Pascual Jordan, and Heisenberg. In particular, the entire mathematical apparatus of classical mechanics, suitably reinterpreted, could be taken over into the new quantum mechanics; consequently, energy and momentum were conserved. Born, Jordan, and Heisenberg wrote that the theory labored "under the disadvantage of not being directly amenable to a geometrically intuitive interpretation." Thus the new structure of quantum mechanics exhibited intuitive thinking (i.e., semireversible) while explicitly avoiding imagery of any sort.

Erwin Schrödinger's response to Heisenberg's quantum mechanics was to formulate a wave mechanics; for, Schrödinger wrote, he was "repelled . . . by the lack of visualizability [*Anschaulichkeit*]" on which quantum mechanics was based. Schrödinger's wave mechanics assimilated atomic phenomena to wave properties of matter only. His accommodation of a wave image was that the orbital electron is a charged wave surrounding the nucleus; the image turned out to be incorrect. The mathematics of wave mechanics was the more familiar wave equation with its attendant classical notion of continuity in contrast to what Bohr, Born, and Heisenberg took to be the essential discontinuities of atomic physics. There were no quantum jumps in Schrödinger's wave mechanics: rather, atoms made transitions from one state to another, the way a struck membrane passes continuously between vibratory states. Schrödinger went on to prove the equivalence of the wave and quantum mechanics. Heisenberg demurred, claiming the equivalence to be only mathematical since Schrödinger's physical interpretation of the wave mechanics was incorrect.

In late fall 1926, Bohr and Heisenberg began their intense struggle toward a physical interpretation of the quantum mechanics. In November 1926 Heisenberg described what he saw as the principal problems: Although he accepted the physical reality of the light quantum, he did not grant to it the same degree of physical reality as "objects of the everyday world," owing to its inability to explain interference phenomena. Moreover, the electron had lost its individuality in the atomic domain and was, in addition, unvisualizable. Thus, nature at the submicroscopic level contradicted our "customary intuition" abstracted from the world of perceptions. He concluded by setting the goal of a "contradiction-free intuitive interpretation of experiments which . . . themselves are contradiction free."

Illustrative of the preconceptual stage of the quantum theory are the paradoxes with which Bohr and Heisenberg struggled. How can a light quantum become polarized? In the diffraction of a low-intensity beam of electrons by a double-slit grating, which slit does an electron pass through—or does it pass through both slits? Assimilation is partial because it is centered on one or the other mode of matter (i.e., scheme) with no reversibility; accommodation is partial since an image of an individual is evoked.

Bohr's approach to these thought experiments differed fundamentally from Heisenberg's. For, as Heisenberg recalled, by late 1926 Bohr had accepted the wave-particle duality of light and matter. Consequently, Bohr could deal effectively with thought experiments using pictures. Heisenberg, on the other hand, persisted in focusing on the corpuscular aspect of matter, for which there was a "consistent mathematical scheme," and the wave nature of light. Bohr's transition from preconceptual to intuitive and then to concrete and formal operational activity is difficult to fold into genetic epistemology—that is, Bohr's acceptance of the complete reversibility between the wave and particle aspects of light and matter, which resulted in his stating the complementarity principle of 1927. In the latter part of November 1926, after having read Dirac's procedure for transforming between Schrödinger's and Heisenberg's mechanics, Heisenberg began to see how "things are related"; he could mathematize Bohr's thought experiments and begin to separate his own thought from images—that is, Heisenberg could deal with thought experiments using mathematical models. Yet, wrote Heisenberg in 1926, "what the words wave and particle mean we know not any more." Heisenberg recalled that these discussions with Bohr left them in a "state of complete despair."

The situation as of February 1927 was that Heisenberg had achieved reversibility concerning mathematical schemes, but not the physical schemes of wave and particle. It was in this intuitive stage that Heisenberg realized the uncertainty principles. Heisenberg's analysis was marred, however, by his focusing on the particle aspect of matter, even though he included the wave-particle duality of light. For example, Heisenberg attributed the quantum theory's intrinsic restrictions on the accuracy of simultaneously measuring an electron's momentum and position to essential discontinuities in the atomic domain. Striving to obtain "complete freedom from the image" offered by the world of perceptions, Heisenberg redefined "intuition" [*Anschauung*] on the basis of the mathematical scheme of the quantum mechanics, instead of the world of perceptions. Consequently, although we may choose to visualize the electron as a solid sphere moving either freely or in a miniature Copernican system, the scheme of quantum mechanics informs us otherwise. Should we attempt to "see" an electron in a stationary state, we would have to knock it out of the atom because "seeing" an electron involves at least one light quantum whose energy turns out to be greater than the atom's ionization energy. In summary, Heisenberg took his quantum mechanics to be a corpuscular-based theory whose corpuscles defied visualization. Lack of visualization and the uncertainty relations led Heisenberg to reject the causal law on the microscopic level.

Bohr disagreed vehemently with Heisenberg's conclusions, for Bohr had arrived independently at a more far-reaching resolution of the paradoxes with which they had wrestled. Bohr realized that these paradoxes were rooted in the wave-particle duality of light and matter. He proposed a resolution that took into account both horns of the dilemma (i.e., wave and particle) instead of resolving it in terms of one of them, or emphasizing one mode over the other, as Heisenberg had done.

In September of 1927 Bohr presented his notion of complementarity. According to complementarity the scheme of the quantum theory can be applied without images, and the use of images from the world of perceptions is restricted by the scheme. A submicroscopic entity assumes the wave or particle mode in response to the experimental apparatus, but can never behave as both wave and particle in a single experiment. The wave and particle aspects of the atomic entity, however, serve to imbue it with the totality of its properties (e.g., its mass, charge, wavelength, and momentum).

At first Heisenberg resisted Bohr's tack and a tense atmosphere prevailed at Copenhagen. For example, on 31 May 1927 Heisenberg wrote to Wolfgang Pauli that he preferred the particle mode for matter with its "discontinuities" and that he could deal with the wave mode only via the mathematical transformation theory of Dirac because it is "unintuitive."

A conceptual-physical point that was of concern to Heisenberg was how the discreteness of electric charge could be obtained from the wave description of an electron. This problem was solved in 1927–1928 in the framework of the quantization of wave fields formulated by Dirac, Jordan, Oskar Klein, and Eugene Wigner. The quantization of wave fields provides the mathematical prescription for transforming the wave and particle aspects of matter and light into one another while these two modes of existence remain mutually exclusive; it is the mathematical formulation of the symmetry inherent in Bohr's complementarity principle. The notion of conservation found in the quantization of wave fields transcends the item in Piaget's theory of mental development wherein, for example, the child realizes that a piece of clay maintains its amount of substance when it is rolled into longer pieces. In the atomic domain the electron maintains its charge, mass, momentum, and spin whether it is a particle or wave, and whether these properties are imageable. Thus, extant historical evidence indicates that quantum theory passed directly from the preconceptual and intuitive stages to the formal operational period without an intervening concrete operational period. This is not unexpected for Heisenberg because he linked reversibility between particles and waves with a mathematical scheme, which turned out to be the quantization of wave fields. In Bohr's case the transition is less clear. For both men, models of elementary particles abstracted from the world of perceptions had proved to be unfruitful.

There are parallels between decisive developments in the quantum theory and structures in genetic epistemology. We have identified establishment of the permanent object (the Bohr atom), and the egocentric representative stage of thinking that results in reversibility and conservation. Furthermore, the development of quantum theory exhibits the assimilation-accommodation mechanism in opposition leading to equilibrium. The formalization that occurs in the formal operational period is illustrated strikingly in the thinking of Heisenberg who continued to strive for total independence of images from the world of perceptions. Starting in 1929 Heisenberg's pioneering researches in quantum electrodynamics, nuclear physics and quantum field theory exhibited the theme that first appeared in the 1927 uncertainty principle paper—namely, to permit the mathematical formalism, that is, the scheme to redefine what is to be taken as intuitive. This complete freedom from the image was not achieved by Bohr, who did not participate in the development of quantum field theory. However, here, as was true of Einstein, and contrary to theory of genetic epistemology, imaginal thinking continued to be of importance into the operational stage.

GESTALT PSYCHOLOGICAL SCENARIOS
Einstein's Invention of the Relativity of Simultaneity
A Gestalt scenario that squares with recent historical scholarship is the following. Except for some changes in dating, Wertheimer's first

three acts can be left unchanged (see Chapter 5); in fact, Einstein agreed with these acts. (I shall proceed without breaking up the scenario into acts.) From 1900 through mid-1905 Einstein's researches on statistical mechanics convinced him that electrodynamics and mechanics were inadequate to discuss the constitution of matter, and that the basic problem confronting physical theory was the nature of light. The negative results of ether-drift experiments and the "intuition" of Einstein's thought experimenter led Einstein to conclude that there was no such thing as absolute motion. In particular, Einstein was drawn to Lorentz's local time coordinate, which permitted a systematic explanation for the first-order ether-drift experiments and Fizeau's 1851 measurement of the velocity of light in a moving medium.

It was axiomatic in the ether-based system for the velocity of light to be exactly c, relative to observers in the ether. For observers in inertial reference systems it turned out to be c to first order in (v/c), owing to such calculational hypotheses as the local time coordinate for problems in the optics of moving bodies. Lorentz explained the principal second-order result of Michelson and Morley with the ad hoc contraction hypothesis. Using the modified Galilean transformation, Einstein deduced a velocity addition law in structure I (classical physics) that explained why the velocity of light was c in S_r, at least to order (v/c). But Einstein knew that within structure I there lurked velocities unknown relative to the ether. Consequently, he did not understand the relation of the new velocity addition law to the Gestalt of structure I because imposing a Newtonian unity on the modified Galilean transformations—that is, treating the local time as real time—implied discarding Lorentz's ether.

It was the thought experiment concerning the current generated in a conducting loop in inertial motion relative to a magnet (electromagnetic induction) that focused Einstein's thought on a section of the field of structure I. The Gestalt of structure I was not a well-formed structure owing to the asymmetries in a phenomenon that resulted in an effect (induced current) that depended on only the relative motion between the magnet and conductor.

In the relativity paper Einstein interpreted this redundancy as an asymmetry that is "not inherent in the phenomena." In an unpublished 1919 essay Einstein described the situation in electromagnetic induction as "unbearable." In a similar subjective manner he wrote that he had found it "intolerable" and "unnatural" that ether-based electromagnetic theories distinguished between reference systems.

Thus Einstein's pondering over the problem of electromagnetic induction caused stresses and strains in the field of structure I that aligned themselves in a direction (i.e., produced vectors) which indicated where the gap lay. To Einstein the difference between the conductor in motion and the magnet resting, and vice versa, was only a "difference in the choice of reference point": electromagnetic induction included notions of both mechanics and electromagnetism. According to Newton's exact principle of relativity the two reference systems were equivalent, but according to Lorentz's approximate theorem of corresponding states the ether-based systems were preferable. Einstein concluded that the clash of principles in electromagnetic induction cut to the heart of the basic problem confronting physical theory—namely, lack of clarity in notions fundamental to kinematics. Electromagnetic induction "forced" Einstein to "postulate the (special relativity) principle" and enabled him to discuss the equivalence of views between observers who relate their measurements using Lorentz's modified transformations. Einstein had used these transformations to deduce the new velocity addition law, and had then realized that the velocity of light should be taken as a quantity that measured exactly c in the two reference systems S and S_r. The principle of relativity asserts the equivalence of S and S_r. This being the case, the local time is indeed the physical time, and time is a relative quantity. The new Gestalt, or structure II, is special relativity theory and is centered about the two relativity principles.

Asymmetries remain which Einstein set out in 1907 to resolve— for example, the preferred status in the special relativity theory of inertial reference systems. The tendency toward a good Gestalt was irresistible (*Prägnanz* principle). Here, as in Gestalt psychology, the notion of a good Gestalt can be somewhat subjective. After all, Poincaré, Lorentz and most physicists of circa 1905 considered the contemporary electromagnetic world-picture to be a good Gestalt. By 1911 most serious physicists judged special relativity theory to be the only good Gestalt.

Thus, a Gestalt-psychological description of Einstein's invention of the relativity of simultaneity can be devised that includes some of the gross features of contemporary historical events. However, Einstein's study of such philosophers as Hume, Kant, Mach, and Poincaré is not included and finds no natural place to enter this historical drama. Nor have I mentioned the notion of *Anschauung*, which played an important role in research in German-speaking countries.

Development of Quantum Theory From 1913–1927

Perhaps nowhere else in the history of science is the Gestalt-psychological theory of thinking, replete with the Gestalt switch, illustrated so well as in the development of quantum theory during the period 1913–1927.[9]

Structure I is the Bohr theory of the atom that was elaborated during the period 1913 through early 1925. Already in 1923, however, drastic alterations were imposed on the theory in response to the data from dispersion. At this time the field of structure I began to be disequilibrated by the theory's inability to account for the characteristics of three-body systems such as the helium atom. These were theoretical problems, as was the so-called crossed-field problem which concerned the splitting of atomic spectral lines by external orthogonal uniform electric and magnetic fields that were imposed on the radiating atoms. These problems resisted solution in Bohr's theory. In addition there was the splitting of spectral lines in weak externally imposed magnetic fields (the anomalous Zeeman effect). Accounting for these data required imposing on Bohr's theory hypotheses and mechanical models with little theoretical foundation. By late 1924 and early 1925 the leaders of the scientific community found themselves in a crisis caused by the Bohr theory's inability to deal with certain key problems. The 1924 dispersion data focused the attention of Bohr and particularly of Heisenberg on a certain portion of the field of structure I, namely, the portion concerned with virtual oscillators. Stresses and strains in his thinking led Heisenberg to formulate the matrix mechanics. After struggling with the peculiar and difficult mathematics that seemed necessary for reinterpreting the transitions between the quantum states of virtual oscillators, Heisenberg recalled that "it came to me like a flash, thus I saw it, the energy [of the harmonic oscillator] was constant in time."[10] For Heisenberg the new Gestalt, or structure II (matrix mechanics), was centered about this quantity that can be taken to be the invariant required by the Gestalt theory of thinking.

CONCLUSION

Owing to the overlap between the Gestalt theories of perception and thinking, the role of pictures and images in the Gestalt dynamics of creative thinking is unclear. Gestalt psychology would have us believe that the transition from structure I to structure II was sudden and dramatic. There was, to use terminology from recent philosophy of science, a Gestalt switch. For example, Kuhn (1962) attributes progress in science to a shift from one paradigm to another. The mechanism for this change is roughly as follows: an established paradigm (e.g., Bohr's theory of the atom) on which normal science is practiced gradually enters a period of crisis owing to its inability to solve certain problems that the leaders of the science community deem to be key problems (e.g., the 1924 data on dispersion and the vexatious anomalous Zeeman effect). The crisis, Kuhn (1962) writes, is "terminated not by deliberation and interpretation but by a sudden and unstructured event like a Gestalt switch" (e.g., Heisenberg's discovery of the matrix mechanics). Kuhn continues: "Scientists then often speak of the 'scales falling from the eyes' or of the 'lightning flash' that 'inundates' a previously obscure puzzle, enabling its components to be seen in a new way that for the first time permits its solution." Some of the pioneers of the quantum theory have fondly recalled their discoveries in this manner (see Heisenberg's recollection in the section "Development of Quantum Theory From 1913–1927," having blocked out of their memories the struggles that constitute some of the fine structure in the transition between the old and new quantum theories. Einstein, on the other hand, never made such statements regarding the special or general theories of relativity.

An allusion to the Gestalt switch was made earlier by Norwood Hanson (1958), who referred to theories as "conceptual Gestalts": "Physical theories provide patterns within which data appear intelligible. They constitute a 'conceptual Gestalt'." According to Hanson the relevant data could be at hand, but unorganized. After much study of these data the investigator makes an intuitive leap and postulates a "pattern-statement" which serves to arrange these data into a pattern (e.g., Kepler's assertion that the orbit of Mars is *really* an ellipse).[11] The pattern statement enables the scientist to perceive these data as "seeing that" rather than viewing an unstructured field of knowledge, which Hanson referred to as "seeing as." Clearly Hanson's notion of scientific discovery is unabashedly Gestalt-oriented, and it thereby suffers from at least the same pitfalls as the Gestalt theory of thinking—for example, the emphasis on empirical data.

Hanson motivated his arguments for a Gestalt switch with the reversible perspective figures often associated with the Gestalt theory of perception. But our conceptual view of physical theory does not exhibit the flip-flop phenomenon that occurs in the visual or perceptual view of such famous Gestalt pictures as Köhler's goblet and faces. For example, the reception of the special theory of relativity does not support the notion of Gestalt switch because from 1905 to 1911 most of the physics community interpreted it as a generalization of Lorentz's theory of the electron. Therefore, the notion of the relativity of time was not clarified as being central to the special theory of relativity until 1911, and ever since time has been at the vortex of propaganda against and misunderstandings of this theory. On resistance to the new notion of time and simultaneity Einstein stressed (1946) that the "notion . . . of absolute simultaneity was anchored in the unconscious."[12]

In the light of Einstein's distinction between discovery and invention, what Hanson refers to as pattern statements are discoveries made from close inspection of empirical data; consequently, Einstein's two principles of the special theory of relativity would be taken to be pattern statements discovered through "putting experimental results in order"—in this case the results of ether-drift experiments and the thought experiment of magnet and conductor.[13] But interpreting these principles as pattern statements ignores Einstein's quest to enlarge on the "known facts" of 1905 in order to build a theory of principle that would exclude asymmetries that are "not inherent in the phenomena," and ignores his view that there is an essential abyss between data and concepts.

Today, whereas every physicist is aware of the consequences of relativity theory, many work in an essentially Newtonian world, as do most engineers. Indeed, nineteenth-century celestial mechanics is sufficient to put satellites into orbit. But in quantum mechanics it was a different story. For there were, in fact, several Gestalt switches through quantum mechanics, and here the metaphor is quite close to the actual phenomena: first, from the Copernican atom to no image; then, for the proponents of wave mechanics, from no image to an image of matter as waves; and, finally, for most other physicists, owing to the complementarity principle, the wave-particle duality of matter and light with its attendant restrictions on imagery and reinterpretation of causality. The notion of Gestalt switch becomes less discontinuous in genetic epistemology through its emphasis on the *construction* of invariant or conserved quantities. This modification, in turn, leads to a reassessment of Kuhn's notion of scientific progress, since it adds the fine structure that historical data require.

Genetic epistemology seems to reflect best the gradual transformation of knowledge through the assimilation-accommodation mechanism, with the appearance of several invariant or conserved quantities as hallmarks. It eliminates, thereby, the requirement for imposing crucial experiments and discontinuities on the historical scenario. Furthermore, genetic epistemology permits deeper insights into the formation and changes in the field of knowledge.

A merit of the Gestalt-psychological theory of thinking is that it permits an extraneous notion to enter the field of thinking for the purpose of precipitating the creative act. The constructive nature of genetic epistemology lessens the degree of free invention, although the Kantian notion of reflecting abstraction permits some flexibility in the structure formation.

Consequently, an incorporation of Gestalt-psychological notions with those of genetic epistemology could offer new insights into the progress of science by both individuals and groups. Case studies of episodes in the history of science, particularly the richly documented twentieth-century sciences, offer a fertile arena for further testing this proposal and, even more generally, for a necessary interaction between psychology and history of science. I can think of no better way to conclude this chapter than with the following quotation from Piaget (1970a):

> A great deal of work remains to be done in order to clarify this funda-
> mental process of intellectual creation, which is found at all the levels
> of cognition, from those of earliest childhood to those culminating in
> the most remarkable of scientific inventions.

NOTES

1. Holton (1973) quite rightly surmised that the "origin of themata will be best approached through studies concerned with the . . . psychological development of concepts in young children." The ellipses replace the phrase "nature of perception," which I do not consider as basic enough.

2. Meyerson (1962) concluded his analysis of conservation laws of mechanics thus: "These principles or laws are, as we see, among the most vast and the most important generalizations to which the human mind has attained to this day." Although Piaget dis-agreed fundamentally with Meyerson's conclusion (regarding conservation laws) that there are facets of reasoning that "will remain incomprehensible, inaccessible to reason, irreducible to purely rational elements" (Myerson), he agreed with Meyerson on the importance of conservation laws.

3. For other discussions of Piaget's work in relation to history of science see Kuhn's "Concept of Cause in the Development of Physics" (Kuhn briefly surveys the notions of cause and explanation through the history of science and finds no parallelisms with those in Piaget's psychological studies); and Kuhn's "Structure for Thought Experiments" (with the aid of notions that are theory laden from his view of the growth of science, Kuhn compares the development of the concept of speed from Piaget's experiments with how Aristotle and Galileo treat this concept)— both of these essays are in Kuhn (1977). See also Bohm (1965) for an analysis of "Physics and Perception" in which Bohm combines genetic epistemology with notions of perception to argue that "our actual mode of perception of the world [is closer to that] suggested by relativistic physics than it is to what is suggested by prerelativistic physics." Bohm, however, overemphasizes notions of perceptions not posed by Piaget (in genetic epistemology perception is a derivative notion), and the usefulness of genetic epistemology is to construct notions of permanence or of invariance. The upshot is a rather positivistic view of scientific research that, nevertheless, is intrinsically interesting for its foray into cognitive psychology.

4. The survey that follows of Piaget is based on the Piaget works listed in the Bibliography.

5. According to inversion, or negation, a displacement to the left can be compensated for by an equal displacement to the right. Or, in group-theoretical terms, if L is the quantity signifying motion to the left, then L^{-1} is its negation and $LL^{-1} = 1$—in other words, the system is restored to its original configuration. Reciprocity pertains to order relationships such as less than and greater than, and if $A = B$, then $B = A$; it has nothing to do with negation. Piaget associates inversion and reciprocity with two of the parent structures of the Bourbaki circle—algebraic and order structures, respectively.

6. It is apropos here to outline Gruber's (1974) application of Piaget's genetic epistemology to the case study of Darwin. Gruber uses Darwin's *B, C, D, E, M,* and *N* notebooks from 1837–1839 to analyze the transformation of Darwin's ideas on natural selection. A gradual transformation of Darwin's thinking can be discerned that exhibits the assimilation/accommodation mechanism with concommitant imagery both representative (e.g., Darwin's "tree of life") and metaphorical (e.g., wedging and

artificial selection). In the course of the transformations, according to Gruber, there are sets of ideas that "remain more or less intact even though the larger system of which they are a part changes appreciably; I therefore call them *invariant groups*" (italics in original; Gruber (1974). An invariant group in one of Darwin's early theories is conservation of the approximate number of extant species. Gruber refers to this sort of group as a "conservation schema" which covers also a "commitment to search for continuity in nature." A second invariant group is the "equilibration schema" which is comprised of adaptation, adaptive change and continuous series of forms. From this analysis Gruber has drawn useful insights on a theory as a "way of handling the personal flow of information."

7. It turned out that simultaneously and independently of Einstein, Poincaré discovered the Lorentz group and yet never attributed a relative nature to time. I have not found a plausible reply of genetic epistemology to this episode. Chapter 1 reveals that Poincaré ascribed to a sensation-based philosophy of science predicated on a notion of physical reality defined to be the relationship among perceptions; and that he practiced a constructive approach to physical theory. Consequently, Poincaré was satisfied with a mathematically symmetric theory of the electron, lacking symmetry in its physical interpretation. Einstein's view of electrodynamics was the one better suited to a consistent description of all physical theory because it emphasized an invariant quantity.

8. There are people who can visualize enormously large arrays of objects and symbols. In a classic psychological case study entitled *The Mind of a Mnemonist,* A.R. Luria described a mnemonist whose prodigious feats of memory were accomplished with the aid of visualization in the mind's eye. For example, the mnemonist S. memorized meaningless complex mathematical formulae by associating symbols with trees, houses, and rocks; he visualized large arrays of numbers, and could move blocks of them around; and he solved intricate arithmetic word problems by visualizing the concrete objects contained in the problem. S. said many times, "I can only understand what I can visualize." Hence, S. could not shift his thinking from concrete terms to abstractions; using notions from genetic epistemology, we can say that S. was unable to make the transition from the concrete to the formal operational levels. Luria writes that S. was "unable

to grasp an idea unless he could actually see it, and so he tried to visualize the idea of 'nothing,' to find an image with which to depict 'infinity.' And he persisted in these agonizing attempts all his life, forever coping with a basically adolescent conflict that made it impossible for him to cross that 'accursed' threshold to a higher level of thought." The results of Chapter 6, however, run somewhat counter to Luria's conclusion. Rather, we might say that S. could not achieve a higher level of mental imagery.

9. Not coincidentally, so do the gross features of the development of quantum theory during 1913–1927 exhibit the characteristics required by Kuhn's notion of the development of science.

10. Quoted from a letter of "many years later" from Heisenberg to Van der Waerden (VW).

11. Hanson also took certain laws to be conceptual Gestalts or pattern statements. For example, Newton's law of gravitation, writes Hanson, is a conceptual Gestalt because it "made the laws of Kepler cohere for Newton as they did not cohere for Kepler himself . . . " (Hanson, 1958).

12. Compare with B.L. Whorf (1956), who writes: "Newtonian space, time and matter are no intuitions. They are recepts from culture and language. That is where Newton got them."

13. Hanson, however, did not consider the pattern statement to result from induction, that is, by merely "enumerating and summarizing observables" (1958).

Concluding Remarks and Suggestions for Further Research

The genesis of mathematical invention is a problem that must inspire the psychologist with the keenest interest. For this is the process in which the human mind seems to borrow least from the exterior world, in which it acts, or appears to act, only by itself and on itself, so that by studying the process of geometric thought, we may hope to arrive at what is most essential in the human mind.

H. Poincaré (1908b)

IF ONLY THE ANALYSES in this book could provide definitive replies to the problems that I posed in the Introduction! The case studies analyzed here suggest that there is good reason to agree with Poincaré and Einstein, among others, that this shall not come to pass. But, like them, we are not deterred from exploring a phenomenon through which "we may hope to arrive at what is most essential in the human mind" (Poincaré, 1908b; see epigraph above). The methods of analysis in this book offer a step toward that goal because they indicate that when the problem of the "mystery" of creative thinking is properly defined, it can shed light on several other problems hitherto unrelated: the process of thinking itself, the changing views of physical reality required by progress in science, and the process by which theories supplant one another. It is mainly to these associations that I address the concluding remarks.

Cognitive psychological theories such as Gestalt psychology and genetic epistemology are useful for studying the dynamics of creative scientific thinking and the construction of scientific concepts, but they have not proved adequate for analyzing the role of mental images. The historical case studies and psychological analyses presented here show the necessity for combining logical reasoning with mental images that are also functional in thought processes. The mental images may be either constructed from objects that have actually been perceived or abstracted from the mathematical formalism of a physical theory.

The success with which the four extensions of genetic epistemology could be applied to historical studies leads me to suggest that the history of scientific thought can be viewed as a series of hierarchical levels constituted of theories that have evolved through stages which exhibit parallelisms with those of genetic epistemology, but differ in their inclusion of mental imagery in the formal operational stage. Kosslyn's postulate of a propositional and literal encoding of images may well be of importance here. In order to permit freer reign to the imagination, I propose to include a suitably modified version of Wertheimer's classical Gestalt psychological theory. One modification is to insist less on the crucial and critical discontinuities. Combining these three cognitive theories offers a means to formulate a closer approximation than has hitherto been available to the description of creative scientific thinking.

The psychological analyses in Chapters 6 and 7 lead me to conjecture that the creative act in science emphasizes invention over discovery, while the construction of knowledge emphasizes discovery over invention. Invention is similar to symbolic play in which information is assimilated to schemes—with their attendant imagery—to which they are only somewhat related; consider, for instance, Einstein's assimilation of the 1895 thought experiment to classical physics and Rutherford's assimilation of alpha particle data to classical physics. The construction of knowledge, or theory elaboration, is analogous to imitation in which information (here empirical evidence) is assimilated to particular signifiers or images; consider Schrödinger's construction of wave mechanics and Heisenberg's focus from late 1926 into 1927 on the particle mode of matter.

This study has found that each well-developed theory has images. Before the quantum theory, the images were determined by linguistic and cultural considerations. For example, although the quantities position and velocity have the same meaning in every language, this is not true of terms like "intuition" or the German, *Anschauung*. Whereas the guiding notion of *Anschauung* had in at least one instance, a negative effect on research, it played a key role in the development of the quantum theory.

The importance of mental imagery in scientific thinking, in conjunction with a hierarchical development of theories, throws further light on the relationship or commensurability between successive theories. There are correspondence rules for passing between theories. For example, in the limit that the velocity of light becomes infinite, the kinematical equations of special relativity take their mathematical form from classical mechanics. In the limit that Planck's constant goes to zero, the kinematical quantities of quantum theory become interpretable as classical quantities. Although these rules work in practice, they are somewhat cavalier because, as Hanson has put it so well (1958), "It need not follow that there is a logical staircase running from regions of 10^{-13} cm to 10^{-13} light-years." The knotty problems of exactly what happens in the limiting case, where physical quantities on the quantum level pass over into classical entities, takes on a new meaning when imagery is taken into account. In terms of imagery, quantum theory and classical theory are incommensurable, whereas special relativity and Newtonian mechanics are commensurable because their imagery is identical to that of the world of perceptions. In fact, the passage from equations of special relativity to those of classical physics can be accomplished not by tampering with special relativity's central quantity c, but by invoking cases in which the velocity of matter (v) is much less than that of light in vacuum (c), and by keeping in mind the relativity of time. After all, only in the domain $v \ll c$ is Newtonian mechanics adequate for most phenomena that concern macroscopic matter. Fortuitously the relativistic flavor of mechanics enters through the ratio of v to c. Although quantum theory admits of limiting procedures alternative to allowing Planck's constant h to go to zero, they run counter to our "customary intuition."[1] This characteristic of the quantum theoretical correspondence principle is probably tied to restrictions on imagery constructed from the world of perceptions.[2]

These results bear on the often-debated problem of whether there are scientific revolutions. The historical studies in this book have shown that in each of the seminal developments in early twentieth-century physics—special relativity (1905), general relativity (1915), and quantum mechanics (1925)—the stated desire by the principal scientists was initially to salvage notions based on intuitions constructed from the world of perceptions, and then gradually to transform them in such a way that the new ones were linked in a well-defined manner to the familiar linguistic-perceptual anchors to the world we live in. Subsequent extensions of the quantum theory to quantum field theory (1929), nuclear theory (1932), early elementary-particle theories (circa 1934), and then renormalization theory

(1948) required further transformation of concepts. But here, too, historical research shows that the transformations were gradual and, particularly for Heisenberg whose work was important for current fundamental theory, emphasis was on correspondence-limit rules. The notion of scientific revolutions describes only the gross structures of scientific change. In the fine structure, where change is gradual, resides the fascinating problem of the nature of creative scientific thinking.

In summary, when scientists hold a theory, they hold a particular mode of imagery as well. Examination of historical studies reveals that scientists' willingness to change their imagery is influenced by their research. Since mental imagery is a key ingredient in creative scientific thinking, it is reasonable to discuss an IMAGination. The limits of the IMAGination of Heisenberg's approach are as yet unclear, and when they become established, new vistas for mental imagery may be revealed. It is reasonable to suggest that through the research of Poincaré, Einstein, Bohr, and Heisenberg, the course of twentieth-century physics has merged with that of cognitive psychology. The progress of science is linked with transformations of perceptions and imagery—for is not the history of science also the history of theories of perception and imagery?

It is my hope that the interplay that has been demonstrated here between language and imagery, and the application of these data to current cognitive theories, will provide the stimulus for cognitive scientists and historians of science to investigate together a fascinating realm of thought for which no one of them has all the necessary tools. The results of such a collaboration can only be far-reaching— for is not the whole greater than the sum of its parts?

NOTES

1. For example, passing to the limit of an infinite number of photons in a given enclosure.
2. In the atomic domain there is the additional complication that the scale of the atom's energy levels is set by the fine structure constant e^2/hc, which renders even more artificial the limiting procedure of letting h pass to zero or c pass to infinity.

Edward M. Purcell informed me that Niels Bohr made a similar comment during a visit to the Physics Department at Harvard University in 1961. The place was Purcell's office where Purcell and others had taken Bohr for a few minutes of rest. They were in the midst of a general discussion when Bohr commented: "People say that classical mechanics is the limit of quantum mechanics when h goes to zero." Then, Purcell recalled, Bohr shook his finger and walked to the blackboard on which he wrote e^2/hc. As he made three strokes under h, Bohr turned around and said, "you see h is in the denominator." Later in the day Purcell and Norman Ramsey arranged for the blackboard to be photographed. The result is in the figure. (Courtesy of E.M. Purcell and N. Ramsey).

Bibliography

Bibliography

See the Author's Notes to the Reader for the code used in this listing. For brevity the following nonstandard abbreviations are used: *HSPS* = *Historical Studies in the Physical Sciences*, *VW* = *Van der Waerden* (1967), *ZsP* = *Zeitschrift für Physik*.

Abraham, Max, Dynamik des Electrons, *Göttinger Nachr.*, 20–41 (1902a).

——— Prinzipien der Dynamik des Elektrons, *Phys. Z.*, *4*, 57–63 (1902b).

——— Prinzipien der Dynamik des Elektrons, *Ann. Phys.*, *10*, 105–179 (1903).

——— Die Grundhypothesen der Elektronentheorie, *Phys. Z.*, *5*, 576–579 (1904a).

——— Zur Theorie der Strahlung und des Strahlungsdruckes, *Ann. Phys.*, *14*, 236–287 (1904b).

——— *Theorie der Elektrizität: Einführung in die Maxwellsche Theorie der Elektrizität* (Leipzig: Teubner, 1904c; second ed., 1907). Revision of Föppl (1894).

Anderson, J.R., Arguments Concerning Representations for Mental Imagery, *Psychological Review, 85*, 249–276 (1978).

——— Further Arguments Concerning Representations for Mental Imagery: A Response to Hayes-Roth and Pylyshyn, *Psychological Review, 86*, 395–406 (1979).

Arieti, A., *Creativity: The Magic Synthesis* (New York: Basic Books, 1976).

Arnheim, Rudolf, *Visual Thinking* (Berkeley: University of California Press, 1971).

Bethe, Hans, *Mesons and Fields, Volume II* (Evanston: Row, Peterson and Co. 1955) (with F. de Hoffmann).

——— *Quantum Mechanics of One- and Two-Electron Systems* (Berlin: Springer-Verlag, 1957) (with E. Salpeter). This is Volume 35 of S. Flügge (ed.), *Encyclopedia of Physics*.

Beveridge, W.I., *The Art of Scientific Investigation,* third ed. (New York: Vintage Books, 1957).

Beyer, R. (ed.), *Foundations of Nuclear Physics* (New York: Dover, 1949).

Beyerchen, Alan, D., *Scientists Under Hitler: Politics and the Physics Community in the Third Reich* (New Haven: Yale University, 1977).

Blackmore, John, *Ernst Mach: His Life, Work and Influence* (Berkeley: University of California Press, 1972).

Block, N. (ed.), *Imagery* (Cambridge, Massachusetts: MIT Press, 1981), with an introduction by N. Block, What is the Issue?

Bohm, David, *The Special Theory of Relativity* (New York: Benjamin, 1965).

Bohr, Niels, On the Constitution of Atoms and Molecules, *Phil. Mag.*, *26*, 1–25, 476–502, 857–875 (1913a). Reprinted with an introduction by Léon Rosenfeld in *On the Constitution of Atoms and Molecules* (New York: Benjamin, 1963).

———— On the Spectrum of Hydrogen. Lecture delivered to the Physical Society of Copenhagen on 20 December 1913. Reprinted in Bohr (1924a), pp. 1–19. [1913b]

———— On the Quantum Theory of Line-Spectra. *Kgl. Danske Vid. Selsk. Skr. nat.-mat. Afd.*, Series 8, *IV*, 1–118 (1918–1922). This paper appeared in 1918 and is reprinted in *VW* (1967), pp. 95–136. [1918]

———— On the Series Spectra of the Elements. Lecture delivered to the Physical Society in Berlin on 27 April 1920. Reprinted in Bohr's (1924a), pp. 20–60.

———— L'application de la théorie des quanta aux problèmes atomiques, in *Atomes et Electrons* (Paris: Gauthier-Villars, 1923), pp. 228–247. This is the proceedings of the 1921 Solvay conference. [1921a]

———— Zur Frage der Polarisation der Strahlung in der Quantentheorie, *ZsP*, *6*, 1–9 (1921b).

———— On the Application of the Quantum Theory to Atomic Structure: Part I. Postulates of the theory, *Proc. Camb. Phil. Soc.* (*Supplement*) (1924). Published in *ZsP*, *13*, 117–165 (1923).

———— *The Theory of Spectra and Atomic Constitution* (Cambridge, England: Cambridge University Press, 1924a).

———— The Quantum Theory of Radiation, *Phil. Mag.*, *47*, 785–802 (1924); an *almost* identical version is in *ZsP*, *24*, 69–87 (1924b). Reprinted from the *Phil. Mag.* in *VW*, pp. 159–176 (with H. Kramers and J. C. Slater).

———— Zur Polarisation des Fluorescenzlichtes, *Die Naturwissenschaften*, *49*, 115–117 (1924c).

———— Über die Wirkung von Atomen bei Stössen, *ZsP*, *34*, 142–157 (1925a).

———— Atomic Theory and Mechanics, *Nature* (*Supplement*) (5 December 1925). [1925b]

———— The Quantum Postulate and the Recent Development of Atomic Theory, *Nature* (*Supplement*) 580–590 (14 April 1928); reprinted with a different introduction and no footnotes in Bohr (1961), pp. 52–91. This is the published version of Bohr's lecture delivered on 16 September 1927 to the International Congress of Physics, Como, Italy.

———— *Atomic Theory and the Description of Nature* (Cambridge, England: Cambridge University Press, 1961).

Boltzmann, Ludwig [Most of Boltzmann's philosophical writings are in Boltzmann's (1905a) with English versions of selected essays in Boltzmann (1974). Some of Boltzmann's scientific papers are in Boltzmann (1909)].

————— On the Significance of Physical Theories, in Boltzmann (1905a), pp. 54–58 and Boltzmann (1974), pp. 33–36.

————— *Vorlesungen über Maxwells Theorie* (two vols.; *Vol. I*, Leipzig: Barth, 1891; *Vol. II*, 1893). [1891; 1893a]

————— Über einige die Maxwellsche Elektrizitätstheorie betreffende Fragen, *Ann. Phys.*, *48*, 100–107 (1893); reprinted in Boltzmann (1909), *Vol. 3*, pp. 398–405. [1893b]

————— Über die neueren Theorien der Elektrizität und des Magnetismus, in Boltzmann (1909), *Vol. 3*, pp. 502–503. [1893c]

————— *Vorlesungen über die Principe der Mechanik* (two vols.; *Vol. I*, Leipzig: Barth, 1897; *Vol. II*, 1904). The preface and sections 1–12 of *Vol. I* are translated in Boltzmann (1974), pp. 223–254; the preface and sections 35, 77, and 88 of *Vol. II* are translated in Boltzmann (1974), pp. 255–265. [1897a].

————— On the Question of the Objective Existence of Processes in Inanimate Nature, in Boltzmann (1905a), pp. 94–119 and Boltzmann (1974), pp. 57–76. [1897b]

————— Anfrage, die Hertzsche Mechanik betreffend, *Verh. d. 70 Vers. D. Naturf. u. Ärzte*, pp. 65–68, 74, Dusseldorf, 1898. Reprinted in Boltzmann (1909), *Vol. 3*, esp., p. 641.

————— On the Fundamental Principles and Equations of Mechanics, in Boltzmann (1905a), pp. 160–169 and Boltzmann (1974), pp. 101–128. [1899a]

————— On the Development of the Methods of Theoretical Physics in Recent Times, in Boltzmann (1905a), pp. 120–149 and Boltzmann (1974), pp. 77–100. [1899b]

————— On the Principles of Mechanics, in Boltzmann (1905a), pp. 120–191 and Boltzmann (1974), pp. 129–146. [1900]

————— On the Principles of Mechanics, in Boltzmann (1905a), pp. 191–198 and Boltzmann (1974), pp. 146–152. [1902]

————— An Inaugural Lecture on Natural Philosophy, in Boltzmann (1905a), pp. 199–205 and Boltzmann (1974), pp. 153–158. [1903]

————— On Statistical Mechanics, in Boltzmann (1905a), pp. 206–224 and Boltzmann (1974), pp. 159–172. [1904]

————— *Populäre Schriften* (*first ed.*: Leipzig: Barth, 1905; *second edition*: Braunschweig: Vieweg, 1979). The second edition contains an introduction by L. Broda. [1905a]

————— On a Thesis of Schopenhauer's, in Boltzmann (1905a), pp. 240–257 and Boltzmann (1974), pp. 185–198. [1905b]

————— *Wissenschaftliche Abhandlungen*, F. Hasenöhrl (ed.) (three vols.: Leipzig: Barth, 1909; reprinted with "editorial improvements and the correction of errata," New York: Chelsea Publishing Co., 1968). [1909]

————— *Ludwig Boltzmann: Theoretical Physics and Philosophical Problems*, B. McGuiness (ed.) (Boston: Reidel, 1974); translated by P. Foulkes.

Born, Max, Quantentheorie und Störungsrechnung, *Die Naturwissenschaf-*

ten, 27, 537–550 (1923). This is a special issue of *Die Naturwissenschaften* entitled "The first ten years of Niels Bohr's theory of the structure of the atom."

—— Zur Quantenmechanik, *ZsP, 34,* 858–888 (1925) (with P. Jordan). Translated in part in *VW,* pp. 227–306.

—— Zur Quantenmechanik, II, *ZsP, 35,* 557–615 (1926a) (with W. Heisenberg and P. Jordan). Translated in *VW,* pp. 321–385. The paper was received 16 November 1925.

—— Zur Quantenmechanik der Stossvorgänge, *ZsP, 37,* 863–867 (1926b).

—— Quantenmechanik der Stossvorgänge, *ZsP, 38,* 803–827 (1926c). Reprinted in part in Ludwig (1968), pp. 206–225.

—— Physical Aspects of Quantum Mechanics, *Nature, 119,* 354–357 (1927).

—— *The Born-Einstein Letters* (New York: Walker, 1971), translated by I. Born.

Bothe, W., Experimentells zur Theorie von Bohr, Kramers und Slater, *Die Naturwissenschaften, 13,* 440–441 (1925) (with H. Geiger).

Brace, Dewitt Bristol, On Double Refraction in Matter Moving through the Aether, *Phil. Mag., 7,* 317–329 (1904).

Braithwaite, R.B., *Scientific Explanation: A Study of the Function of Theory, Probability and Law in Science* (Cambridge, England: Cambridge University Press, 1968).

Breit, Gregory, The Quantum Theory of Dispersion, *Nature, 114,* 310 (1924).

Broda, Engelbert, *Ludwig Boltzmann: Mensch-Physiker-Philosoph* (Vienna: Deuticke, 1955).

de Broglie, Louis, Thèse (Paris, 1924) and Recherches sur la théorie des quanta. *Annales de Physique, 3,* 22–128 (1925).

Bromberg, Joan, The Impact of the Neutron, *HSPS, 3,* 307–341 (1971).

Brown, R., Icons and Images, in Block (1981) (with R. Herrnstein), pp. 19–49.

Brush, Stephen G., *The Kind of Motion We Call Heat* (two vols.; Amsterdam: North-Holland, 1976).

Bucherer, Alfred H., Messungen an Becquerelstrahlen. Die experimentelle Bestätigung der Lorentz-Einsteinschen Theorie, *Phys. Z., 9,* 755–762 (1908).

Calinon, A., *Etude Critique sur la Mécanique* (Nancy: Berger-Levrault, 1885).

—— *Etude sur les Diverses Grandeurs* (Paris: Gauthier-Villars, 1897).

Capek, M., *The Concepts of Space and Time: Their Structure and Their Development* (Dordrecht: Reidel, 1976). This is Volume *22* of *Boston Studies in the Philosophy of Science.*

Cassidy, David C., Cosmic Ray Showers; High Energy Physics, and

Quantum Field Theory; Programmatic Interactions in the 1930s, *HSPS, 12,* 1–40 (1981).

Cassirer, Ernst, *Determinism and Indeterminism in Modern Physics: Historical and Systemic Studies of the Problem of Causality* (1st ed. 1936: New Haven: Yale University Press, 1956), translated by O.T. Benfey with a Preface by H. Margenau.

Chadwick, James, The Existence of a Neutron, *Proceedings of the Royal Society of London (A), 136,* 692–708 (1932). Reprinted in Beyer (1949), pp. 5–21.

—— The Neutron and its properties, in *Nobel Lectures in Physics: 1922 to 1941* (New York: North-Holland, 1965), pp. 339–348. [1935]

Compton, Arthur H., A Quantum Theory of the Scattering of X-rays by Light Elements. *Phys. Rev., 21,* 483–502 (1923).

Darboux, Gaston, Eloge de Henri Poincaré, in *Oeuvres de Henri Poincaré* (eleven vols.; Paris: Gauthier-Villars, 1934–54), *Vol. 2,* pp. vii–lxxii. [1913]

Dirac, Paul A.M., The Physical Interpretation of the Quantum Mechanics, *Proc. Roy. Soc. (A), 113,* 621–641 (1926).

—— The Quantum Theory of the Emission and Absorption of Radiation, *Proc. Roy. Soc. (A), 114,* 243–265 (1927). Reprinted in Schwinger (1958), pp. 1–23.

—— *Principles of Quantum Mechanics* (first edition, Oxford, England: Oxford University Press, 1930).

—— The Lagrangian in Quantum Mechanics, *Physikalische Zeitschrift der Sowetunion, 3,* 64–72 (1932).

—— Discussion of the Infinite Distribution of Electrons in the Theory of the Positron, *Proc. Camb. Phil. Soc., 30,* 150–163 (1934).

Dugas, René, *La théorie physique au sens de Boltzmann* (Neuchâtel: Griffon, 1959).

Duhem, Pierre, *The Aim and Structure of Physical Theory* (1st. ed. 1906, 2nd ed. 1914: New York: Atheneum, 1962), translated by P.P. Wiener from the second French edition. [1962]

Dyson, F.J., The S-Matrix in Quantum Electrodynamics, *Phys. Rev., 75,* 1736–1755 (1949).

Einstein, Albert, Allgemeine molekulare Theorie der Wärme, *Ann. Phys., 14,* 354–362 (1904).

—— Eine neue Bestimmung der Moleküldimensionen, Inaugural-Dissertation, Zürich Universität. [1905a]

—— Über einen die Erzeugung und Verwandlung des Lichtes betreffenden heuristischen Gesichtspunkt, *Ann. Phys. 17,* 132–148 (1905b), translated by A.B. Arons and M.B. Peppard, *Am. J. Phys., 33,* 367–374 (1965).

—— Die von der molekularkinetischen Theorie der Wärme geforderte Bewegung von in ruhenden Flüssigkeiten suspendierten Teilchen,

Ann. Phys., *17*, 549–560 (1905c). Reprinted in A. Einstein, *Investigations on the Theory of the Brownian Movement* (New York: Dover, 1956), translated by A.D. Cowper, with notes by R. Furth.

———— Zur Elektrodynamik bewegter Körper, *Ann. Phys.*, *17*, 891–921 (1905d). Reprinted in H.A. Lorentz., A. Einstein, H. Minkowski, *Das Relativitätsprinzip, eine Sammlung von Abhandlungen* (Leipzig: Teubner; first ed., 1913; second and third enlarged eds., 1919, 1923), translated from the edition of 1923 by W. Perrett and G.B. Jeffrey as *The Principle of Relativity: A Collection of Original Memoirs on the Special and General Theories of Relativity* by H.A. Lorentz, A. Einstein, H. Minkowski, and H. Weyl (London: Methuen, 1923); the Methuen version was reprinted (New York: Dover, n.d.). Hereafter the Dover reprint volume is designated as *PRC*. Einstein's (1905d) is on pp. 37–65 of *PRC*.

For the purpose of serious historical analysis I had to retranslate Einstein's relativity paper from the *Annalen* version for my book, *Albert Einstein's Special Theory of Relativity: Emergence (1905) and Early Interpretation (1905–1911)*, where the retranslation appears in the Appendix (pp. 391–415). The hitherto most frequently quoted English translation of the relativity paper is in the Dover reprint volume. That version contains some substantive mistranslations, infelicities, and outdated Britishisms. For example, Einstein's second principle of the relativity theory is mistranslated as: "Any ray of light moves in the 'stationary' system of co-ordinates with the determined velocity, c, whether the ray be emitted by a stationary or by a moving body." The correct translation is: "Any ray of light moves in the 'resting' coordinate system with the definite velocity, c, *which is independent* of whether the ray was emitted by a resting or by a moving body" (italics added to indicate a key phrase that was omitted in the Dover translation).

The Dover translation was made from a retypeset version of Einstein's relativity paper that had appeared in a Teubner reprint volume, thereby adding to the misprints in the original *Annalen* version. In addition, the Dover translation does not distinguish between Einstein's footnotes and those added to the Teubner edition by Arnold Sommerfeld. This state of affairs is an example of the pitfalls inherent in using a translation that was not made from the original paper and of the importance of going back to the original papers.

Hereafter all citations to Einstein's relativity paper are to the translation in my book.

———— Ist die Trägheit eines Körpers von seinem Energieinhalt abhängig?, *Ann. Phys.*, *18*, 639–641 (1905e). Reprinted in *PRC*, pp. 69–71, where the volume number is stated incorrectly and the title misspelled.

———— Theorie der Lichterzeugung und Lichtabsorption, *Ann. Phys.*, *20*, 199–206 (1906).

———— Die vom Relativitätsprinzip geforderte Trägheit der Energie, *Ann. Phys.*, *23*, 371–384 (1907a).

——— Relativitätsprinzip und die aus demselben gezogenen Folgerungen, *Jahrb. Radioakt.*, *4*, 411–462 (1907b); *5*, 98–99 (Berichtigungen).

——— Zum gegenwärtigen Stande des Strahlungsproblems, *Phys. Z.*, *10*, 185–193 (1909a).

——— Entwicklung unserer Anschauungen über das Wesen und die Konstitution der Strahlung, *Phys. Z.*, *10*, 817–825 (1909b).

——— Zum Ehrenfestschen Paradoxon, *Phys. Z.*, *12*, 509–510 (1911).

——— Entwurf einer verallgemeinerten Relativitätstheorie und eine Theorie der Gravitation, I. Physikalischer Teil von A. Einstein. II. Mathematischer Teil von M. Grossmann (Leipzig: Teubner, 1913) and *Z. Math. und Phys.*, *62*, 225–261 (1913).

——— Zur allgemeinen Relativitätstheorie, *Verh. D. Phys. Ges.*, 778–786 and 799–801 (1915).

——— Grundlage der allgemeinen Relativitätstheorie, *Ann. Phys.*, *49*, 769–822 (1916a). Reprinted in part in *PRC*, pp. 111–173.

——— Ernst Mach, *Phys. Z.*, *17*, 101–104 (1916b).

——— Quantentheorie der Strahlung, *Physikalische Gesellschaft, Zürich, Mitteilungen*, *16*, pp. 47–62. [1916c]

——— *Relativity, The Special and the General Theory* (Braunschweig: Vieweg, 1917; New York: Holt, 1920), translated from the fifth German Edition by R.W. Lawson. [1917a]

——— Quantentheorie der Strahlung, *Phys. Z.*, *18*, 121–128 (1917b). Translated in *VW*, pp. 63–77.

——— Dialog über Einwände gegen die Relativitätstheorie, *Die Naturwissenschaften*, 697–702 (1918a).

——— Motiv des Forschens. Lecture delivered in honor of Max Planck's sixtieth birthday in 1918, and reprinted with the incorrect title, "Principles of Research" in A. Einstein, *Essays in Science* (New York: Philosophical Library, 1934), translated by A. Harris, pp. 1–5. Hereafter *Essays in Science* is designated as *ES*. [1918b]

——— What is the Theory of Relativity, written for the London *Times*, 28 November 1919. Versions appear in A. Einstein, *Ideas and Opinions* (New York: Bonanza Books, n.d.), pp. 227–232; A. Einstein, *Out of My Later Years* (Totowa, New Jersey: Littlefield Adams and Co., 1967), pp. 54–57. Hereafter *Ideas and Opinions* is designated as *IO*.

——— Relativity and the Ether. Lecture presented on 2 October 1920 at Leiden University, and reprinted in *ES*, pp. 98–111. [1920]

——— Geometry and Experience. Lecture presented on 27 January 1921 to the Prussian Academy of Sciences, and reprinted in *IO*, pp. 232–246. [1921a]

——— *The Meaning of Relativity: Four Lectures Delivered at Princeton University, May 1921* (fifth ed.; Princeton, New Jersey: Princeton University Press, 1970), translated by E.P. Adams. [1921b]

——— Fundamental Ideas and Problems of the Theory of Relativity. Lecture delivered on 11 July 1923 to the Nordic Assembly of Naturalists at Gothenburg, in acknowledgment of the Nobel Prize. Reprinted in *Nobel Lectures, Physics: 1901–1921* (New York: Elsevier, 1967), pp. 479–490. [1923]

—— Quantentheorie des einatomigen idealen Gases, *Berliner Berichte*, 261–267 (1924); *Ibid.*, 3–14 (1925); *Ibid.*, 18–25 (1925).

—— The Mechanics of Newton and Their Influence on the Development of Theoretical Physics, *Die Naturwissenschaften*, *15*, 273–276 (1927); reprinted in *ES*, pp. 28–39.

—— Maxwell's Influence on the Development of the Conception of Physical Reality, in *James Clerk Maxwell: A Commemoration Volume* (Cambridge, England: Cambridge University Press, 1931), pp. 66–73. Reprinted in *ES*, pp. 40–45.

—— On the Method of Theoretical Physics. Lecture delivered on 10 June 1933 at Oxford University. Reprinted in *ES*, pp. 12–21.

—— The Problem of Space, Ether and the Field in Physics. Reprinted in *ES*, pp. 61–77. [1934a]

—— Notes on the Origin of the General Theory of Relativity. Reprinted in *ES*, pp. 78–84. [1934b]

—— Physics and Reality, *Journal of the Franklin Institute*, *221*, 313–347 (1936). Reprinted in *IO*, pp. 290–323.

—— Remarks on Bertrand Russell's Theory of Knowledge, in P.A. Schilpp (ed.), *The Philosophy of Bertrand Russell* (Evanston, Illinois: The Library of Living Philosophers, 1944), pp. 277–291. Reprinted in *IO*, pp. 18–24.

—— Autobiographical Notes, in P.A. Schilpp (ed.), *Albert Einstein: Philosopher-Scientist* (Evanston, Illinois: The Library of Living Philosophers, 1949), pp. 2–94. [1946]

—— Reply to Criticisms, in P.A. Schilpp (ed.), *Albert Einstein: Philosopher-Scientist* (Evanston, Illinois: The Library of Living Philosophers, 1949), pp. 665–688.

—— *Albert Einstein: Lettres à Maurice Solovine* (Paris: Gauthier-Villars, 1956).

—— *Albert Einstein–Michele Besso: Correspondance 1903–1955* (Paris: Hermann, 1972), translated into French by P. Speziali, who also supplied notes and an introduction.

Eisler, R., *Handwörterbuch der Philosophie* (second ed.; Berlin: Mittler, 1922).

—— *Wörterbuch der Philosophischen Begriffe* (Berlin: Mittler, 1927).

Euler, H., Über die Streuung von Licht an Licht nach der Diracschen Theorie, *Die Naturwissenschaften*, *23*, 246–247 (1935) (with B. Kockel).

Farah, Martha, Concept Development, *Advances in Child Development and Behavior*, *16*, 125–167 (1982) (with S.M. Kosslyn).

Ferguson, E.S., The Mind's Eye: Nonverbal Thought in Technology, *Science*, *197*, 827–836 (1977).

Fermi, Enrico, Versuch einer Theorie der β-Strahlen, *ZsP, 88,* 161–177 (1934). Reprinted in Beyer (1949), pp. 161–177.

Feynman, Richard P., The Theory of Positrons, *Phys. Rev., 76,* 749–759 (1949). Reprinted in Schwinger (1958), pp. 225–235.

—— The Development of the Space-Time View of Quantum Electrodynamics, in *Nobel Lectures in Physics: 1963–1970* (New York: North-Holland, 1972), pp. 155–178. [1965]

Fierz, M. and V.F. Weisskopf (eds.), *Theoretical Physics in the Twentieth Century* (New York: Interscience, 1960).

Finke, R.A., Levels of Equivalence in Imagery and Perception, *Psychological Review, 87,* 113–132 (1980).

Fizeau, Hippolyte, Sur les hypothèses relatives a l'éther lumineaux, et sur une expérience qui paraît demontrer que le mouvement des corps change la vitesse avec laquelle la lumière se propage dans leur intérieur, *C.R. Acad. Sci., 33,* 349–355 (1851).

Föppl, August, *Einführung in die Maxwell'sche Theorie der Elektricität* (Leipzig: Teubner, 1894).

Forman, Paul, The Doublet Riddle and Atomic Physics *circa* 1924, *ISIS, 59,* 156–174 (1968).

Frank, A.I., Fundamental Properties of the Neutron: Fifty Years of Research, *USPEKHI, 25,* 280–297 (1982).

Frank, Philipp, *Einstein: Sein Leben und seine Zeit* (New York: Knopf, 1947), translated by G. Rosen, and edited and revised by S. Kusaka (New York: Knopf, 1953).

Furry, W., On the Theory of the Electron and the Positive, *Phys. Rev., 45,* 245–262 (1934) (with J.R. Oppenheimer).

Galison, Peter L., The Discovery of the Muon and the Failed Revolution Against Quantum Electrodynamics, *Centaurus, 26,* 262–283 (1983).

Galton, Francis, *Inquiries into Human Faculty and Its Development* (London: MacMillan, 1883).

Gamow, George, *Structure of Atomic Nuclei and Nuclear Transformations* (Oxford, England: Oxford University Press, 1937).

Gardner, Howard, *Art, Mind and Brain: A Cognitive Approach to Creativity* (New York: Basic Books, 1982).

Geschwind, Norman, Neurological Knowledge and Complex Behaviors, in Norman (1981), pp. 27–35.

Ghiselin, B. (ed.), *The Creative Process* (New York: Mentor, 1952).

Goldberg, Stanley, Poincaré's Silence and Einstein's Relativity: The Role of Theory and Experiment in Poincaré's Physics, *The British Journal for the History of Science, 5,* 73–84 (1970).

Gombrich, Ernst H., *Art and Illusion: A Study in the Psychology of Pictorial Representation* (Princeton, New Jersey: Princeton University Press, 1956).

Green, J.A., *Life and Work of Pestalozzi* (London: Universal Tutorial Press Ltd., 1913).

Gruber, Howard E., *Darwin on Man: A Psychological Study of Scientific Creativity* (Together with Darwin's Early and Unpublished Notes Transcribed and Annotated by Paul H. Barrett) (New York: Dutton, 1974).

———— On the Relation Between 'Aha Experiences' and the Construction of Ideas, *History of Science, 19,* 41–59 (1981).

Grünbaum, Adolf, *Philosophical Problems of Space and Time* (first ed., New York: Knopf, 1963; second enlarged ed., Dordrecht: Reidel, 1973).

Guidymin, Jerzy, On the Origin and Significance of Poincaré's Conventionalism, *Studies in History and Philosophy of Science, 8,* 271–301 (1977).

Gutting, Gary, Einstein's Discovery of Special Relativity, *Philos. Sci., 39,* 51–67 (1972).

Hadamard, Jacques, *The Psychology of Invention in the Mathematical Field* (Princeton, New Jersey: Princeton University Press, 1945; reprinted from the 2nd enlarged edition of 1949, New York: Dover, 1954). [1954]

Hanson, Norwood R., *Patterns of Discovery* (Cambridge, England: Cambridge University Press, 1958).

Hardy, G.H., *A Mathematician's Apology* (Cambridge, England: Cambridge University Press, 1967: first edition, 1940).

Haugeland, J. (ed.), *Mind Design* (Cambridge, Massachusetts: MIT Press, 1981), with an introduction by J. Haugeland, Semantic Engines: An Introduction to Mind Design.

Heilbron, John L., The Genesis of the Bohr Atom, *HSPS, 1,* 211–290 (1969) (with T.S. Kuhn).

———— The Origins of the Exclusion Principle, *HSPS, 13,* 261–310 (1983).

Heisenberg, Werner, Über eine Anwendung des Korrespondenzprinzips auf die Frage nach der Polarisation des Fluoreszenzlichts, *ZsP, 31,* 617–622 (1925a).

———— Über quantentheoretische Umdeutung kinematischer und mechanischer Beziehungen, *ZsP, 33,* 879–893 (1925b). Translated in *VW,* pp. 261–276.

———— Mehrkörperproblem und Resonanz in der Quantenmechanik, *ZsP, 38,* 411–426 (1926a).

———— Quantenmechanik, *Die Naturwissenschaften, 14,* 899–994 (1926b).

———— Schwankungserscheinungen und Quantenmechanik, *ZsP, 40,* 501–506 (1926c).

———— Über den anschaulichen Inhalt der quantentheoretischen Kinematik und Mechanik, *ZsP, 43,* 172–198 (1927).

———— Zur Theorie des Ferromagnetismus, *ZsP, 49,* 619–636 (1928).

———— Die Entwicklung der Quantentheorie, 1918–1928, *Die Naturwissenschaften, 26,* 490–496 (1929).

———— *Die physikalischen Prinzipien der Quantentheorie* (Leipzig: Herzl, 1930), translated by C. Eckart and F.C. Hoyt as *The Physical Principles of the Quantum Theory* (New York: Dover, 1963). This text is from a set of lectures delivered by Heisenberg at the University of Chicago in Spring, 1929.

—— Über den Bau der Atomkerne. I, *ZsP*, *77*, 1–11 (1932a).

—— Über den Bau der Atomkerne. II, *ZsP*, *78*, 156–164 (1932b).

—— Über den Bau der Atomkerne, III, *ZsP*, *80*, 587–596 (1933a).

—— Considérations théoriques générales sur la structure du noyau, in *Structure et propriétés des noyaux atomiques* (Paris: Gauthier-villars, 1934), pp. 281–385. This is the Solvay Conference Proceedings of 1933. [1933b]

—— Bemerkungen zur Diracschen Theorie des Positrons, *ZsP*, *90*, 209–231 (1934).

—— Bemerkungen zur Theorie des Atomkerns, in *Pieter Zeeman* (The Hague: Martinus Nijhoff, 1935), pp. 108–116.

—— Folgerungen aus der Diracschen Theorie des Positrons, *ZsP*, *98*, 714–732 (1936) (with H. Euler).

—— Uber die in der Theorie der Elementarteilchen auftretende universelle Länge, *Ann. Phys.*, *32*, 20–33 (1938).

—— Die 'beobachtbaren Grössen' in der Theorie der Elementarteilchen, *ZsP*, *120*, 513–538 (1943a).

—— Die beobachtbaren Grössen in der Theorie der Elementarteilchen, II, *ZsP*, *120*, 673–702 (1943b).

—— Der mathematische Rahmen der Quantentheorie der Wellenfelder, *Zeitschrift für Naturforschung*, *1*, 608–622 (1946).

—— Zur Quantentheorie der Elementarteilchen, *Zeitschrift für Naturforschung*, *5*, 251–259 (1950).

—— Erinnerungen an die Zeit der Entwicklung der Quantenmechanik, in Fierz and Weisskopf (1960), pp. 40–47. [1960]

—— Quantum Theory and Its Interpretation, in Rozental (1967).

—— *Der Teil und das Ganze: Gespräche im Umkreis der Atomphysik* (München: Piper, 1969), translated by A.J. Pomerans as *Physics and Beyond: Encounters and Conversations* (New York: Harper, 1971). [1969]

—— Abstraction in Modern Art and Science, in W. Heisenberg, *Across the Frontiers* (New York: Harper Torchbooks, 1974), translated by P. Heath.

—— Discussion with Professor Werner Heisenberg, in O. Gingerich (ed.), *The Nature of Scientific Discovery: A Symposium Commemorating the 500th Anniversary of the Birth of Nicolaus Copernicus* (Washington: Smithsonian Institution Press, 1975), pp. 556–573.

—— Was ist ein Elementarteilchen, *Die Naturwissenschaften*, *63*, 1–7 (1976).

Heitler, Walter, Wechselwirkung neutraler Atome und homöopolare Bindung nach der Quantenmechanik, *ZsP*, *44*, 455–472 (1927) (with F. London).

—— Störungsenergie und austausch beim Mehrkörperproblem, *ZsP*, *46*, 47–72 (1928).

von Helmholtz, Hermann, The Origin and the Meaning of Geometric Axioms (I), *Mind*, 301–321 (1876). Translated in Helmholtz (1971).

—— An Autobiographical Sketch. Lecture delivered in 1891 on the occasion of his Jubilee in Berlin. Translated in Helmholtz (1971), pp. 466–478.

—— The Origin and Correct Interpretation of Our Sense Impressions, *Z. Psychologie und Physiologie der Sinnesorgane, VII*, 81–96 (1894).

—— R. Kahl (ed.), *Selected Writings of Hermann von Helmholtz*, (Wesleyan, Connecticut: Wesleyan University Press, 1971).

Hertz, Heinrich, *Electric Waves* (Leipzig: Teubner, 1892; London: MacMillan, 1893; New York: Dover, 1962), translated by D.E. Jones.

—— *The Principles of Mechanics* (Leipzig: Teubner, 1894; London: MacMillan, 1899; New York: Dover, 1956), translated by D.E. Jones and J.T. Walley, with a preface by Hermann von Helmholtz. The Dover edition contains an introduction by R.S. Cohen.

Hiebert, Erwin, Boltzmann's Conception of Theory Construction: the Promotion of Pluralism, Provisionalism, and Pragmatic Realism, in J. Hintikka, D. Gruender, and E. Agazzi (eds.), *Pisa Conference Proceedings, Vol. II* (Boston: Reidel, 1980), pp. 175–198.

Hildesheimer, Wolfgang, *Mozart* (New York: Vintage Books, 1983), translated by M. Faber.

Hindle, B., *Emulation and Invention* (New York: New York University Press, 1981).

Höffding, Harald, *A History of Modern Philosophy* (two vols.; 1900: New York: Dover, 1955).

Hoffmann, Banesh and Helen Dukas (eds.), *Albert Einstein: The Human Side* (Princeton, New Jersey: Princeton University Press, 1979).

Holton, Gerald, *Thematic Origins of Scientific Thought: Kepler to Einstein* (Cambridge, Massachusetts: Harvard University Press, 1973). The essays' original dates of publication are given in parentheses.
  (a) On the Origins of the Special Theory of Relativity, pp. 165–183 (1960).
  (b) Poincaré and Relativity, pp. 185–195 (1964).
  (c) Influences on Einstein's Early Work, pp. 197–217 (1967).
  (d) Mach, Einstein, and the Search for Reality, pp. 219–259 (1968).
  (e) Einstein, Michelson, and the 'Crucial' Experiment, pp. 261–352 (1969).
  (f) The Roots of Complementarity, pp. 115–161 (1970).
  (g) On Trying to Understand Scientific Genius, pp. 353–380 (1971).

—— Dyonesians, Apollonians, and the Scientific Imagination, in G. Holton, *The Scientific Imagination: Case Studies* (Cambridge, England: Cambridge University Press, 1978), pp. 84–110.

———— Einstein's Model for Constructing a Scientific Theory, in P. Aichelburg and R.U. Sexl (eds.), *Albert Einstein: His Influence on Physics, Philosophy and Politics* (Braunschweig: Vieweg, 1979), pp. 109–136.

———— Einstein's Scientific Program: The Formative Years, in H. Woolf (ed.), *Some Strangeness in the Proportion: A Centennial Symposium to Celebrate the Achievements of Albert Einstein*, (Reading, Massachusetts: Addison-Wesley, Advanced Book Program, 1980), pp. 49–65.

Hopf, Ludwig, *Die Relativitätstheorie* (Berlin: Springer-Verlag, 1931).

Iwanenko, D., The Neutron Hypothesis, *Nature, 129*, 798 (1932).

———— Interaction of Neutrons and Protons, *Nature, 133*, 981–982 (1934).

Jacobi, J., *Complex/Archetype/Symbol in the Psychology of C.G. Jung* (Princeton, New Jersey: Princeton University Press, 1971), translated by R. Mannheim.

Jammer, Max, *Concepts of Force: A Study in the Foundations of Dynamics* (Cambridge, Massachusetts: Harvard University Press, 1957).

———— *The Conceptual Development of Quantum Mechanics* (New York: McGraw-Hill, 1966).

Jensen, Hans J. D., Glimpses at the History of Nuclear Structure Theory, in *Nobel Lectures in Physics* (New York: North-Holland, 1972), pp. 40–50. [1963]

Johnson-Laird, P., Mental Models in Cognitive Science, in Norman (1981), pp. 147–191.

Jones, E., *The Life and Work of Sigmund Freud*, edited and abridged by L. Trilling and S. Marcus, with an introduction by L. Trilling (Garden City, New York: Doubleday, 1963).

Jordan, Pascuale, Über quantenmechanische Darstellung von Quantensprungen, *ZsP, 40*, 661–666 (1927a).

———— Über eine neue Begründung der Quantenmechanik, *ZsP, 40*, 809–838 (1927b).

———— Zum Mehrkörperproblem der Quantentheorie, *ZsP, 45*, 751–765 (1927c) (with O. Klein).

———— Über das Paulische Äquivalenzverbot, *ZsP, 47*, 631–651 (1928) (with E. Wigner). Reprinted in Schwinger (1958), pp. 41–61.

———— *Anschauliche Quantentheorie: eine Einführung in die moderne Auffassung der Quantenerscheinungen* (Berlin: Springer-Verlag, 1936).

Kant, Immanuel, *Kritik der reinen Vernunft* (1781; Kant's *Sämmtliche Werke*, part 2; Leipzig: Voss, 1838); Kant, *Critique of Pure Reason* (New York: St. Martin's Press, 1929), translated by N.K. Smith.

Kaufmann, Walter, Die Entwicklung des Elektronenbegriffs, *Phys. Z., 3*, 9–15 (1901a). Translated in *Electrician, 48*, 94–97 (1901).

———— Die magnetische und electrische Ablenkbarkeit der Becquerelstrahlen und die scheinbare Masse der Elektronen. *Göttinger Nachr.*, 143–155 (1901b).

———— Über die elektromagnetische Masse des Elektrons, *Göttinger Nachr.*, 291–303 (1902a).

——— Die elektromagnetische Masse des Elektrons, *Phys. Z.*, *4*, 54–57 (1902b).

——— Über die 'Elektromagnetische Masse' der Elektronen, *Göttinger Nachr.*, 90–103, 148-Berichtigung (1903).

——— Über die Konstitution des Elektrons, *Berl. Ber.*, *45*, 949–956 (1905).

——— Über die Konstitution des Elektrons, *Ann. Phys.*, *19*, 487–553 (1906); Nachtrag zu der Abhandlung: 'Über die Konstitution des Elektrons', *Ann. Phys.*, 20, 639–640 (1906).

Kirchhoff, Gustav, *Vorlesungen über mathematische Physik: Mechanik* (Leipzig: Teubner, 1876).

Klein, Martin J., Max Planck and the Beginnings of the Quantum Theory, *Arch. Hist. Exact Scis.*, *1*, 459–479 (1962).

——— Einstein and the Wave-Particle Duality, *The Natural Philosopher, 3*, 3–49 (1964).

——— Thermodynamics in Einstein's Thought, *Science, 157*, 509–516 (1967).

——— *Paul Ehrenfest: Vol. 1, The Making of a Theoretical Physicist* (Amsterdam: North-Holland, 1970a).

——— The First Stage of the Bohr-Einstein Dialogue, *HSPS, 2*, 1–39 (1970b).

——— Mechanical Explanation at the End of the Nineteenth Century, *Centaurus, 17*, 58–82 (1972).

Koffka, Kurt, *Principles of Gestalt Psychology* (New York: Harbinger, 1935).

Köhler, Wolfgang, *Die physischen Gestalten in Ruhe und im stationären Zustand* (Braunschweig: Vieweg, 1920).

——— *The Task of Gestalt Psychology* (Princeton, New Jersey: Princeton University Press, 1969).

Kosslyn, Stephen M., Imagery, Propositions, and the Form of Internal Representations, *Cognitive Psychology, 9*, 52–76 (1977) (with J.R. Pomerantz).

——— On the Demystification of Mental Imagery, *The Behavioral and Brain Sciences, 2*, 535–581 (1979) (with S. Pinker, G.E. Smith, and S.P. Schwartz).

——— *Image and Mind* (Cambridge, Massachusetts: Harvard University Press, 1981).

Kramers, Hendrik, *The Atom and the Bohr Theory of Its Structure* (London: Gyldendal, 1923), translated from the first Danish edition of 1922 by R.B. and R.T. Lindsay (with H. Holst).

——— Über die Streuung von Strahlung durch Atome, *ZsP, 31*, 681–707 (1925). Translated in *VW*, pp. 223–257 (with W. Heisenberg).

Kuhn, Thomas S., *Structure of Scientific Revolutions* (Chicago: University of Chicago Press, 1962; enlarged edition, 1970).

——— *The Essential Tension* (Chicago: University of Chicago Press, 1977).

——— *Black-Body Theory and the Quantum Discontinuity, 1894–1912* (Oxford, England: Oxford University Press, 1978).

Lachman, R., *Cognitive Psychology and Information Processing* (Hillsdale, New Jersey: Lawrence Erlbaum Associates, 1979) (with J.L. Lachman and E.C. Butterfield).

Ladenburg, Rudolf, Die quantentheoretische Deutung der Zahl der Dispersionselektronen, *ZsP, 4,* 451–468 (1921). Translated in *VW,* pp. 139–157.

——— Absorption, Zerstreuung und Dispersion in der Bohrschen Atomtheorie, *Die Naturwissenschaften, 11,* 584–598 (1923) (with F. Reiche).

Langevin, Paul, L'évolution de l'espace et du temps. Lecture delivered on 10 April 1911 at the Philosophy Congress at Bologna, and published in *Scientia, 10,* 31–54 (1911).

Lebon, Ernest, *Savants du Jour: Henri Poincaré, Biographie, Bibliographie analytique des écrits* (second ed., Paris: Gauthier-Villars, 1912).

Lorentz, Hendrik Antoon [Most of Lorentz's published papers are in H.A. Lorentz, *Collected Papers* (9 vols.; The Hague: Nijhoff, 1935–1939).]

——— La théorie électromagnetique de Maxwell et son application aux corps mouvants, *Arch. Néerl., 25,* 363 (1892a). Reprinted in *Collected Papers, Vol. 2,* pp. 164–343.

——— The Relative Motion of the Earth and the Ether, *Versl. Kon. Akad. Wetensch. Amsterdam, 1,* 74 (1892b). Reprinted in *Collected Papers, Vol. 4,* pp. 219–223.

——— Versuch einer Theorie der elektrischen und optischen Erscheinungen in bewegten Körpern (Leiden: Brill, 1895). Reprinted in *Collected Papers, Vol. 5,* pp. 1–137.

——— Weiterbildung der Maxwellschen Theorie. Elektronentheorie, in *Encykl. Math. Wiss., 14,* 145–288 (1904a).

——— Electromagnetic Phenomena in a System Moving with any Velocity Less than that of Light, *Proc. R. Acad. Amsterdam, 6,* 809 (1904b). Reprinted in *Collected Papers, Vol. 5,* pp. 172–197, and in part in PRC, pp. 11–34.

——— Ludwig Boltzmann, *Verh. D. Phys. Ges., 9,* 206 (1907). Reprinted in *Collected Papers, Vol. 9,* pp. 359–390.

——— Alte und neue Fragen der Physik., *Phys. Z., 11,* 1234–1257 (1910).

Ludwig, G., *Wave Mechanics* (New York: Pergamon, 1968).

Luria, A.R., *The Mind of a Mnemonist* (Chicago: Regnery, 1968), translated by L. Solotaroff.

Mach, Ernst, *Die Mechanik in ihrer Entwickelung historisch-kritisch dargestellt* (Leipzig: Brockhaus, 1883, 1889, 1897, 1901, 1904, 1908, 1912, 1921, 1933), translated by T. McCormack in 1893 from the second German edition and revised in 1942 to include additions and alterations up to the ninth German edition as E. Mach., *Science of Mechanics: A Critical and Historical Account of Its Development* (La Salle, Illinois: Open Court, 1960).

—————— *Erkenntnis und Irrtum: Skizzen zur Psychologie der Forschung* (Leipzig: Barth, 1905). English translation is E. Mach, *Knowledge and Error: Sketches on the Psychology of Inquiry* (Dordrecht: Reidel, 1976). [1905a]

—————— Sensation, Intuition and Phantasy, in Mach (1905a), pp. 142–161, and in the English version, pp. 105–119. [1905b]

—————— Über Gedankenexperimente, in Mach (1905a), pp. 180–197, and in English version, pp. 130–147. [1905c]

Mahoney, M.J., *Scientist as Subject: The Psychological Imperative* (Cambridge, Massachusetts: Ballinger, 1976).

Majorana, E., Über die Kerntheorie, *ZsP, 82,* 137–145 (1933).

Mansfield, R.S., *The Psychology of Creativity and Discovery* (Chicago: Nelson Hall, 1981) (with T.V. Busse).

Manuel, Frank E., *Portrait of Isaac Newton* (Cambridge, Massachusetts: Harvard University Press, 1968).

Margenau, Henry, *The Nature of Physical Reality: A Philosophy of Modern Physics* (New York: McGraw-Hill, 1950).

Maxwell, James Clerk, On Physical Lines of Force, *Phil. Mag., 21,* 161–175, 281–291, 338–348 (1861); *ibid., 22,* 12–24, 85–95 (1862). Reprinted in W.D. Niven, (ed.), *The Scientific Papers of James Clerk Maxwell* (two vols; New York: Dover, 1965), *Vol. 1,* pp. 451–513. [1861–1862]

—————— A Dynamical Theory of the Electromagnetic Field, *Phil. Trans., 155,* 495–512 (1865). Reprinted in *The Scientific Papers of James Clerk Maxwell, Vol. 2,* pp. 526–597.

—————— Attraction, originally published in 1878 in *Encyclopaedia Britannica.* Reprinted in *The Scientific Papers of James Clerk Maxwell, Vol. 2,* pp. 485–491. [1878a]

—————— Ether, originally published in 1878 in *Encyclopaedia Britannica.* Reprinted in *The Scientific Papers of James Clerk Maxwell, Vol. 2,* pp. 763–775. [1878b]

McCormmach, Russell, Einstein, Lorentz and the Electromagnetic View of Nature, *HSPS, 2,* 41–87 (1970).

McDermott, D., Artificial Intelligence Meets Natural Stupidity, *SIGART Newsletter, 57,* (1976). Reprinted in Haugeland (1981), pp. 143–160.

McMullin, Ernan, The History and Philosophy of Science: A Taxonomy, in R. Stuewer (ed.), *Historical and Philosophical Perspectives of Science: Vol. V, Minnesota Studies in the Philosophy of Science* (Minneapolis: University of Minnesota Press, 1970), pp. 12–67.

Medawar, Peter B., *Induction and Intuition in Scientific Thought* (Philadelphia: American Philosophical Society, 1969).

Merz, John Theodore, *A History of European Scientific Thought in the Nineteenth Century* (four vols., 1904–1912: New York: Dover, 1965); Vols. 3 and 4 of this set are entitled, *A History of European Thought in the Nineteenth Century.*

Meyer-Abich, K., *Korrespondenz, Individualität und Komplementarität* (Wiesbaden: Steiner, 1965).

Meyerson, Emil, *Identity and Reality* (New York: Dover, 1962), translated by K. Loewenberg.

Michelson, Albert A., Influence of Motion of the Medium on the Velocity of Light, *Amer. J. Sci., 31,* 377–386 (1886) (with Edward W. Morley).

⸻ On the Relative Motion of the Earth and the Luminferous Ether, *Amer. J. Sci., 34,* 333–345 (1887) (with Edward W. Morley).

Miller, Arthur I., A Study of Henri Poincaré's 'Sur la dynamique de l'électron', *Arch. Hist. Exact Scis. 10,* 207–328 (1973).

⸻ On Lorentz's Methodology, *Brit. J. Phil. Sci., 25,* 29–45 (1974).

⸻ Visualization Lost and Regained: The Genesis of the Quantum Theory in the Period 1913–1927, in J. Wechsler (ed.), *On Aesthetics in Science* (Cambridge, Massachusetts: MIT Press, 1978), pp. 72–102.

⸻ On Some Other Approaches to Electrodynamics in 1905, in H. Woolf (ed.), *Some Strangeness in the Proportion: A Centennial Symposium to Celebrate the Achievements of Albert Einstein* (Reading, Massachusetts: Addison-Wesley, 1980), pp. 66–91.

⸻ Unipolar Induction: A Case Study of the Interaction Between Science and Technology, *Annals of Science, 38,* 155–189 (1981a).

⸻ *Albert Einstein's Special Theory of Relativity: Emergence (1905) and Early Interpretation (1905–1911)* (Reading, Massachusetts: Addison-Wesley, 1981b).

⸻ Introduction to the second edition of P.W. Bridgman, *A Sophisticate's Primer of Relativity* (first ed., 1962; second ed., Wesleyan, Connecticut: Wesleyan University Press, 1983).

⸻ *Frontiers of Physics, 1900–1911* (Boston: Birkhäuser Boston, Inc., 1985). Among my essays in this volume are (1973), (1974), (1980), (1981a) and (1983).

⸻ Symmetry and Imagery in the Physics of Bohr, Einstein, and Heisenberg. Scheduled for publication in the *Proceedings of the 1st International Conference on the History of Scientific Ideas: Symmetries in Physics (1600–1980),* 20–26 September 1983, Sant Filiu de Guixols, Spain. [1984b]

von Mises, Richard, *Positivism: A Study in Human Understanding* (Cambridge, Massachusetts: Harvard University Press, 1951; unabridged republication with minor corrections, Dover, New York, 1968).

Mooij, J.J.A., *La Philosophie des Mathématiques de Henri Poincaré* (Paris: Gauthier-Villars, 1966).

Neisser, U., *Cognitive Psychology* (New York: Appleton-Century-Crofts, 1967).

Norman, D.A. (ed.), *Perspectives on Cognitive Science* (Norwood, New Jersey: Ablex, 1981), with an introduction by D.A. Norman, What is Cognitive Science?

Nye, M.J., The Boutroux Circle and Poincaré's Conventionalism, *Journal of the History of Ideas, 40,* 107–120 (1970).

Pais, Abraham, *Subtle is the Lord . . . The Science and the Life of Albert Einstein* (Oxford: Oxford University Press, 1982).

Paivio, A., *Imagery and Verbal Processes* (New York: Holt, Rinehart and Winston, 1971).

Pauli, Wolfgang, Über das Modell des Wasserstoffmolekülions, *Ann. Phys. 68,* 177–240 (1922).

———— Über das Wasserstoffspektrum vom Standpunkt der neuen Quantenmechanik *ZsP, 36,* 336–363 (1926). Translated in *VW*, pp. 387–415.

———— Über Gasentartung und Paramagnetismus, *ZsP, 41,* 1–102 (1927).

———— Remarks on the Polarization Effects in the Positron Theory, *Phys. Rev. 49,* 462–465 (1936).

———— Exclusion Principle, Lorentz Group and Reflection of Space-Time and Charge, in W. Pauli (ed.), *Niels Bohr and the Development of Physics* (New York: Pergamon, 1955), pp. 30–51.

———— *Wissenschaftlicher Briefwechsel mit Bohr, Einstein, Heisenberg, U.A.,* A. Hermann, K.v. Meyenn, V.F. Weisskopf (eds.) (Berlin: Springer-Verlag, 1979).

Peierls, R., The Vacuum in Dirac's Theory of the Positive Electron, *Proc. Roy. Soc. (A), 146,* 420–441 (1934).

Pestalozzi, Johann Heinrich, *Wie Gertrud ihre Kinder lehrt,* in Johann Heinrich Pestalozzi, *Schriften: 1798–1804* (Zurich: Rotapfel, 1946). Translated as J.H. Pestalozzi, *How Gertrude Teaches Her Children* (New York: Gordon Press, 1973, first edition 1894), translated by L.E. Holland and F.C. Turner; edited with an introduction and notes by E. Cooke.

Piaget, Jean, *Play Dreams and Imitation in Childhood,* translated by C. Gattegno and F.M. Hodgson (New York: Norton, 1962) (originally published, 1946).

———— *The Origins of Intelligence in Children* (New York: Norton, 1963), translated by M. Cook (originally published, 1936).

———— *The Child's Conception of Number* (New York: Norton, 1965), translated by C. Gattegno and F.M. Hodgson (originally published, 1941) (with A. Szeminska).

———— *The Child's Conception of Space* (New York: Norton, 1967), translated by F.J. Langdon and J.L. Lunzer (originally published, 1948) (with B. Inhelder).

———— *The Psychology of the Child* (New York: Basic Books, 1969), translated by H. Weaver (originally published, 1966) (with B. Inhelder).

———— *Genetic Epistemology* (New York: Columbia University Press, 1970a), translated by E. Duckworth (originally published, 1970).

———— *Structuralism,* (New York: Basic Books, 1970b) translated and edited by C. Maschler (originally published, 1968).

———— *Biology and Knowledge,* (Chicago: University of Chicago Press, 1971) translated by B. Walsh (originally published, 1967).

Planck, Max, Über eine Verbesserung der Wienschen Spektralgleichung, *Verh. D. Phys. Ges.*, *2*, 202–204 (1900).

——— Über das Gesetz der Energieverteilung im Normalspektrum, *Ann. Phys.*, *4*, 553–563 (1901).

Plato *Timaeus* (circa 350 B.C.: New York: Bobbs-Merrill, 1959), translated by F.M. Cornford.

Poincaré, Henri [Most of Poincaré's published scientific papers are in *Oeuvres de Henri Poincaré* (eleven vols.; Paris: Gauthier-Villars, 1934–1953)].

——— Une note sur les principes de la mécanique dans Descartes et dans Leibnitz, in Leibnitz, *La Monodologie* (Paris: Delgrave, 1881), with notes by E. Boutroux.

——— Sur les hypothèses fondamentales de la géométrie, *Bull. Soc. Math. France*, *15*, 203–216 (1887).

——— *Théorie mathématique de la lumière* (two vols., Paris: Naud, Vol. 1, 1889; Vol. 2, 1892). Part of the preface to Volume 1 is in Chapter 12 of Poincaré's (1902a).

——— Les Géométries non-euclidiènnes, *Rev. Générale Sci. Pures Appl.*, *2*, 769–774 (1891). A version of this paper is in Chapter III of Poincaré's (1902a).

——— Sur la nature du raisonnement mathématique, *Rev. Mét. Mor.* 2, 371–384 (1894). Reprinted in Chapter I of Poincaré's (1902a).

——— Les idées de Hertz sur la mécanique, *Revue générale des sciences*, *8*, 734–743 (1897). Reprinted in *Oeuvres, Vol.* 7, pp. 231–250. Portions of this book review of Hertz's (1894) is in Chapter VI of Poincaré's (1902a).

——— On The Foundations of Geometry, translated by T.J. McCormack, *Monist*, *9*, 1–43 (1898a).

——— La mesure de temps, *Rev. Mét. Mor.*, *6*, 371–384 (1898b). Reprinted in Chapter II of Poincaré's (1904a).

——— Des fondements de la Géométrie: A propos d'un livre de M. Russell, *Rev. Mét. Mor.*, *7*, 251–279 (1899). This is a book review of Russell's (1897). A version is in Chapter V of Poincaré's (1902a).

——— Sur les rapports de la Physique expérimentale et de la Physique mathématique, *Rapports présentés au Congrès international de Physique réuni à Paris en 1900* (four vols.; Paris: Gauthier-Villars, 1900), *Vol. 1*, pp. 1–29. Reprinted in German as Über die Beziehungen zwischen der experimentellen und der mathematischen Physik, *Phys. Z.*, *2*, 166–171, 182–186, and 196–201 (1900–1901). This lecture forms the substance of Chaps. IX and X of Poincaré's (1902a). [1900a]

——— Intuition and Logic in Mathematics, in Poincaré (1905b), pp. 15–25. This essay is from an address entitled, Du rôle de l'intuition et de la logique en Mathématiques, and presented 11 August 1900 to the International Congress of Mathematicians, Paris. [1900b]

——— La théorie de Lorentz et le principe de réaction, in *Recueil de travaux offerts par les auteurs à H.A. Lorentz* (The Hague: Nijhoff, 1900), pp. 252–278. Reprinted in *Oeuvres*, pp. 464–488. [1900c]

—— Sur les principes de la Mécanique, *Bibliothèque du Congress International de Philosophie tenu à Paris du 1er au 5 août 1900* (Paris: Colin, 1901), pp. 457–494. [1900d]

—— *Electricité et Optique* (Paris: Gauthier-Villars, 1901). (Poincaré's lectures at the Sorbonne from 1888, 1890 and 1899.) [1901]

—— *Science and Hypothesis* (New York: Dover, 1952), translator unknown. This is a translation of Poincaré's, *La Science et l'Hypothèse* (Paris: Ernest Flammarion, 1902a).

—— Sur la valeur objective de la science, *Rev. Mét. Mor.*, 10, 263–293 (1902b). A version of this paper is in Part III of Poincaré's (1905b).

—— Poincaré's Review of Hilbert's 'Foundations of Geometry', *Bulletin of the American Mathematical Society*, 10, 1–23 (1903).

—— L'état actuel et l'avenir de la Physique mathématique. Lecture delivered on 24 September 1904 to the International Congress of Arts and Science, Saint Louis, Missouri, and published in *Bull. Sci. Mat.*, 28, 302–324 (1904). Reprinted in Chapters VII–IX of Poincaré's (1905b). [1904a]

—— Mathematical Definitions in Education, in Poincaré (1908a), pp. 117–142. This essay is a version of Poincaré's, Les définitions générales en mathématiques, *L'Enseignement mathématique*, 6, 257–283 (1904b).

—— Sur la dynamique de l'électron, *C.R. Acad. Sci.*, 140, 1504–1508 (1905a). Reprinted in *Oeuvres, Vol. 9*, pp. 489–493.

—— *The Value of Science* (New York: Dover, 1958), translated by G. Halsted. This is a translation of Poincaré's *La Valeur de la Science* (Paris: Ernest Flammarion, 1905b).

—— Sur la dynamique de l'électron, *Rend. Circ. Mat. Palermo*, 21, 129–175 (1906). Reprinted in *Oeuvres, Vol. 9*, pp. 494–550.

—— The Choice of Facts, written in 1907 as a preface for the first American edition of Poincaré's (1905b).

—— *Science and Method* (New York: Dover, n.d.) translated by F. Maitland. This is a translation of Poincaré's *Science et Méthode* (Paris: Flammarion, 1908a).

—— L'invention mathématique. Lecture delivered on 23 May 1908 at l'Institut Général Psychologique, Paris. Reprinted in Chapter III of Poincaré's (1908a). [1908b]

—— La logique de l'infini, *Rev. Mét. Mor.*, 17, 461–482 (1909). Reprinted in Chapter IV of Poincaré's (1913).

—— La dynamique de l'électron (Paris: Dumas, 1913). [1912a]

—— Les rapports de la Matière de l'éther, *J. Phys. Théor. Appl. 2*, 347 (1912b). Reprinted in Chapter VII of Poincaré's (1913).

—— L'éspace et le temps. Lecture delivered on 4 May 1912 at the University of London. Reprinted in Chapter II of Poincaré's (1913). [1912c]

———— L'hypothèse des quanta, *Revue scientifique, Revue rose,* 24 February 1912, 225–232. A version is in Chapter VI of Poincaré's (1913). [1912d]

———— *Mathematics and Science: Last Essays* (New York: Dover, 1963), translated by J.W. Bolduc. This is a translation of Poincaré's, *Dernières Pensées* (Paris, Ernest Flammarion, 1913).

Polanyi, M., *Personal Knowledge: Towards a Post-Critical Philosophy* (New York: Harper Torchbooks, 1962; a revised version of the first edition of 1958).

Putnam, H., Reductionism and the Nature of Psychology, *Cognition, 2,* 131–146 (1973); abridged version in Haugeland (1981), pp. 205–219.

Pylyshyn, Z.W., What the Mind's Eye Tells the Mind's Brain: A Critique of Mental Imagery, *Psychological Bulletin, 80,* 1–24, (1973).

———— Computation and Cognition: Issues in the Foundations of Cognitive Science, *The Behavioral and Brain Sciences, 3,* 11–169 (1980).

———— The Imagery Debate: Analog Media versus Tacit Knowledge, in Block (1981), pp. 151–206. [1981a]

———— Complexity and the Study of Artificial and Human Intelligence, in Haugeland (1981), pp. 67–94. [1981b]

Lord Rayleigh [J.W. Strutt], Does Motion through the Aether cause Double Refraction? *Phil. Mag., 4,* 678–683 (1902).

Richardson, A., *Mental Imagery* (London: Routledge and Kegan Paul, 1969).

Richardson, O.W., *The Electron Theory of Matter* (Cambridge, England: Cambridge University Press, 1916).

Robertson, H.P., Geometry as a Branch of Physics, in P.A. Schilpp (ed.), *Albert Einstein: Philosopher-Scientist* (La Salle, Illinois: Open Court, 1949), pp. 315–332.

Roe, A., *The Making of a Scientist* (New York: Dodd, Mead & Co., 1952).

Rosenfeld, Léon, Niels Bohr in the Thirties, in Rozental (1967), pp. 114–136.

Rozental, S. (ed.), *Niels Bohr: His Life and Work as Seen by his Friends and Colleagues* (New York: Wiley, 1967).

Russell, Bertrand, *An Essay on the Foundations of Geometry* (Cambridge, England: Cambridge University Press, 1897; New York: Dover, 1952).

Ryle, G., *The Concept of Mind* (New York: Harper and Row, 1949).

Sauter, Joseph, *Erinnerungen an Albert Einstein.* This pamphlet (unpaginated) was published in 1965 by the Patent Office in Bern, and contains documents pertaining to Einstein's years at that office as well as a note by Sauter.

Schaffner, Kenneth, Einstein versus Lorentz: Research Programmes and the Logic of Comparative Theory Evaluation, *Brit. J. Phil. Sci., 25,* 45–78 (1974).

Schmid, Anne-Françoise, *Une Philosophie de Savant: Henri Poincaré & la Logique Mathématique* (Paris: Maspero, 1978).

Schrödinger, Erwin, Über das Verhältnis der Heisenberg-Born-Jordanschen Quantenmechanik zu der meinen, *Ann. Phys. 70,* 734–756 (1926a). Translated in part in Ludwig (1968), pp. 127–150.

—— Der stetige Übergang von der Mikro-zur Makromechanik, *Die Naturwissenschaften, 14,* 664–666 (1926b).

—— *Schrödinger, Planck, Einstein, Lorentz: Briefe zur Wellenmechanik,* K. Prizbaum (ed.) (Vienna: Springer-Verlag, 1963), translated by M.J. Klein as *Letters on Wave Mechanics* (New York: Philosophical Library, 1967). [1963]

Schwinger, Julian (ed.), *Selected Papers on Quantum Electrodynamics* (New York: Dover, 1958).

Searle, J.R., Minds, Brains and Programs, in Haugeland (1981), pp. 282–306.

—— The Myth of the Computer an Exchange, *New York Review of Books,* 24 June 1982, pp. 56–57.

Seelig, Carl, *Albert Einstein: Eine dokumentarische Biographie* (Zürich: Europa, 1954).

Serwer, Daniel, *Unmechanischer Zwang:* Pauli, Heisenberg, and the rejection of the mechanical atom, 1923–1925, *HSPS, 8,* 189–256 (1977).

Shankland, Robert S., Michelson 1852–1931, Expérience de base la relativité, in *Les inventeurs Célèbres—Sciences Physiques et Applications* (Paris: Lucien Mazenod, 1950), pp. 254–255.

—— Conversations with Albert Einstein, *Am. J. Phys., 31,* 47–57 (1963).

—— Conversations with Albert Einstein, II, *Am. J. Phys.,* 44, 895–901 (1973).

Shepard, Roger N., Mental Rotation of Three-Dimensional Objects, *Science, 171,* 701–703 (1971) (with J. Metzler).

—— Externalization of Mental Images and the Act of Creation, in B.S. Randhawas, W.E. Coffman, (eds.), *Visual Learning, Thinking, and Communication* (New York: Academic Press, 1978a), pp. 133–189.

—— The Mental Image, *The American Psychologist, 33,* 125–137 (1978b).

Silber, K. *Pestalozzi: The Man and His Work* (London: Routledge and Kegan Paul, 1960).

Simon, Herbert A., Scientific Discovery and the Psychology of Problem Solving, in H.A. Simon (ed.) *Models of Discovery* (Boston/Dordrecht: Reidel, 1977), pp. 22–39; originally published in 1966 in *Mind and Cosmos: Essays in Contemporary Science and Philosophy, Volume III* (Pittsburgh: University of Pittsburgh Press, 1966). [1966]

—— Does Scientific Discovery Have a Logic?, *Philosophy of Science, 40,* 471–480 (1973).

—— Computer Science as Empirical Inquiry: Symbols and Search, in Haugeland (1981), pp. 35–66 (with Allan Newell).

Sklar, Lawrence, *Space, Time and Spacetime* (Berkeley: University of California Press, 1974).

Sommerfeld, Arnold, Das Plancksche Wirkungsquantum und seine allgemeine Bedeutung für die Molekularphysik, *Phys. Z.*, *12*, 1057–1069 (1911).

―――― *Atomic Structure and Spectral Lines*, translated by H.L. Brose from the third German edition of 1922 (New York: Dutton, 1922).

―――― Das Werk Boltzmanns, *Wiener Chem.-Ztg.*, *47*, 25 (1944).

―――― *Mechanics: Lectures on Theoretical Physics* (New York: Academic Press, 1952), translated by M.O. Stern.

Stachel, John, Einstein and the Rigidly Rotating Disk, in A. Held (ed.), *General Relativity and Gravitation* (two vols., New York: Plenum, 1980), *Vol. 1*, pp. 1–15.

Stuewer, Roger, *The Compton Effect: Turning Point in Physics* (New York: Science History Publications, 1975).

―――― The Nuclear Electron Hypothesis, in W. Shea (ed.), *Otto Hahn and the Rise of Nuclear Physics* (Boston/Dordrecht: Reidel, 1983).

Swann, W.F.G., The Trend of Thought in Physics, *Science*, *61*, 425–435 (1925).

Tamm, Igor, Exchange Forces between Neutrons and Protons, *Nature*, *133*, 981 (1934).

Toulmin, Stephen, *Human Understanding: Volume I* (Princeton, New Jersey: Princeton University Press, 1972).

Toulouse, E., *Henri Poincaré* (Paris: Flammarion, 1910).

Trouton, Frederick Thomas, The Mechanical Forces Acting on a Charged Electric Condensor Moving Through Space, *Philos. Trans. R. Soc.*, *202*, 165–181 (1903) (with H.R. Noble).

Tweney, R.D., M.E. Doherty and C.R. Mynatt (eds.), *On Scientific Thinking* (New York: Columbia University Press, 1981), with an introduction by the editors.

Uehling, E.A., Polarization Effects in the Positron Theory, *Phys. Rev.*, *48*, 55–63 (1935).

Uhlenbeck, George and S. Goudsmit, Ersetzung der Hypothese vom unmechanischen Zwang durch eine Forderung bezüglich des inneren Verhaltens jedes einzelnen Elektrons, *Die Naturwissenschaften*, *13*, 953–954 (1925).

―――― The Spinning Electron and the Theory of Spectra, *Nature*, *117*, 264–265 (1926).

Van Vleck, John H., The Absorption of Radiation by Multiply Periodic Orbits, and Its Relation to the Correspondence Principle and the Rayleigh-Jeans Law, *Phys. Rev.*, *24*, 330–365 (1924). Reprinted in part in *VW*, pp. 203–222.

van der Waerden, B.L. Exclusion Principle and Spin, in Fierz and Weisskopf (1960), pp. 199–244.

―――― *Sources of Quantum Mechanics* (New York: Dover, 1967).

Wallas, Graham, *The Art of Thought* (New York: Harcourt & Brace, 1926).

Weber, C.L., Über unipolare Induktion, *Elektrot. Z.*, *16*, 513–514 (1895).

Weinberg, Steven, The Search for Unity: Notes for a History of Quantum Theory, *Daedalus, 106,* 17–35 (1977).

Weisskopf, Victor and E. Wigner, Berechnung der natürlichen Linienbreite auf Grund der Diracschen Lichttheorie, *ZsP, 63,* 54–73 (1930).

Wentzel, Gregor, Zur Theorie der β-Umwandlung und der Kernkräfte. I, *ZsP, 104,* 34–47 (1936).

―――― *Einführung in die Quantentheorie der Wellenfelder* (Vienna: Deuticke, 1943), translated by C. Houtermans and J.M. Jauch as G. Wentzel, *Quantum Theory of Fields* (New York: Interscience, 1949).

―――― Quantum Theory of Fields (until 1927), in Fierz and Weisskopf (1960), pp. 380–403.

Wertheimer, Max, *Productive Thinking* (first ed., New York: Harper, 1945; enlarged edition, 1959).

Wertheimer, Michael, Relativity and Gestalt: A Note on Albert Einstein and Max Wertheimer, *Journal of the History of the Behavioral Sciences, i,* 86–87 (1965).

Whorf, Benjamin L., *Language, Thought and Reality* (Cambridge, Massachusetts: MIT Press, 1964).

Wiechert, Emil, Relativitätsprinzip und Aether. I, *Phys. Z., 12,* 689–707 (1911); *Ibid., 12,* 737–758 (1911).

Wien, Wilhelm, Über die Möglichkeit einer elektromagnetischen Begrundung der Mechanik, in *Recueil de travaux offerts par les auteurs à H.A. Lorentz* (The Hague: Nijhoff, 1900), pp. 96–107. Reprinted in *Ann. Phys., 5,* 501–513 (1901).

Wigner, Eugene, On the Mass Defect of Helium, *Phys. Rev., 43,* 252–257 (1933).

Wisdom, J.O., Four Contemporary Interpretations of the Nature of Space, *Foundations of Physics, 1,* 269–284 (1971).

Wittgenstein, Ludwig, *The Blue and Brown Books* (New York: Harper Torchbooks, 1958).

―――― *Tractatus Logico-Philosophicus,* translated from the first German edition by D.F. Pears and B.F. McGuinnes (London: Routledge and Kegan Paul, 1961).

Wood, R.W. and A. Ellett, On the influence of magnetic fields on the polarisation of resonance radiation, *Proc. Roy. Soc. (A), 103,* 396–403 (1923).

―――― Polarized Radiation in Weak Magnetic Fields, *Phys. Rev., 24,* 243–254 (1924).

Yukawa, Hideki, On the Interaction of Elementary Particles. I, *Proceedings of the Physico-Mathematical Society of Japan, 3, 17,* 48–57 (1935). Reprinted in Beyer (1949), pp. 139–148. Yukawa read this paper on 17 November 1934.

Zahn, C.T. and A.A. Spees, A Critical Analysis of the Classical Experiments on the Variation of Electron Mass, *Phys. Rev., 53,* 511–521 (1938).

Index

Index

Numbers set in *italics* designate pages on which literature citations are given.